CLYMER®

POLARIS

SNOWMOBILE SHOP MANUAL
1984-1989

The world's finest publisher of mechanical how-to manuals

PRIMEDIA
Business Magazines & Media

P.O. Box 12901, Overland Park, KS 66282-2901

Copyright © 1990 PRIMEDIA Business Magazines & Media Inc.

FIRST EDITION
First Printing December, 1990
Second Printing September, 1992
Third Printing December, 1993
Fourth Printing November, 1994
Fifth Printing November, 1996
Sixth Printing November, 2001

Printed in U.S.A.

CLYMER and colophon are registered trademarks of PRIMEDIA Business Magazines & Media Inc.

ISBN: 0-89287-537-2

Library of Congress: 90-56147

Technical photography by Ron Wright.

Technical illustrations by Steve Amos.

Snowmobile Code of Ethics provided by the International Snowmobile Industry Association, Fairfax, Virginia.

Technical assistance provided by:
Don Anderson, Instructor/Technical Supervisor
Detroit Lakes Technical College
Outdoor Power Equipment & Snowmobile Technology Department
Detroit Lakes, Minnesota 56501

Dale Fett, Technical Advisor
Fett Racing
Route 4, Box 316
Frazee, Minnesota 56544

Rancheria Garage & Marina
P.O. Box 157
62311 Huntington Lake Rd.
Lakeshore, California 93634

With special thanks to Don Anderson, Dale Fett and Detroit Lakes Technical College.

CLYMER PUBLICATIONS
PRIMEDIA Business Magazines & Media

The following product lines are published by PRIMEDIA Business Directories & Books.

More information available at *primediabusiness.com*

Contents

Quick Reference Data

RECOMMENDED LUBRICANTS AND FUEL

Engine oil	Polaris Injection Oil
Coolant	Glycol-based automotive type antifreeze compounded for aluminum engines
Chaincase	Polaris Chaincase Oil
Grease	Low-temperature grease
Fuel	88 minimum octane leaded*
Brake fluid	DOT 3
Throttle cable	Polaris Clutch and Cable Lubricant
Choke cable	Polaris Clutch and Cable Lubricant
Clutch and drive system	Polaris Clutch and Cable Lubricant

* Unleaded gasoline with a minimum 88 octane rating can be used in place of leaded gasoline. However, do not use gasoline containing ethanol or methanol. In addition, Polaris specifies that gasohol, white gas or gasoline containing additives must not be used.

APPROXIMATE REFILL CAPACITY

Chaincase housing	3 oz. (88.8 cc)
Oil tank	2.5 qts. (2.4 l)
Fuel tank	
1984-1987	7 gal. (26.5 l)
1988-1989	7.3 gal. (27.7 l)
Cooling system	
Twins	4 quarts
Triples	5 quarts

SPARK PLUGS

Year	NGK	Champion	Gap
1984-1986	BR9ES	RN-2C	0.5 mm (0.020 in.)
1987-on	BR9ES	RN-2C	0.6 mm (0.025 in.)

BATTERY

Battery capacity	12 volt, 18 amp hour

OPERATING RPM

Model	RPM
Trail	
1984-1985	6,800
1986-1988	7,000
1989	6,600
Sport	6,500
400	
1984	*
1985	7,800
1986-1987	7,700
1988-1989	8,000
500	7,800
600	7,800
650	7,800
* Not specified.	

PILOT AIR SCREW ADJUSTMENT

Model	Turns out
Trail	1.0
Sport	1.0
400	
1984-1986	1.0
1987-1989	1 1/2
500	1.0
600	
1984	1.0
1985-1986	3/4
1987	1.0
650	1.0

IDLE SPEED ADJUSTMENT

Model	RPM
Trail	
1984-1985	2,300
1986	2,000
1987-1989	1,900
Sport	2,100
400	
1984-1986	2,200
1987-1989	1,900
500	1,900
600	
1984-1986	2,000
1987	1,900
650	1,900

EXCITER COIL TESTING

Model/wire color	Resistance reading
Sport	
Brown/white-to-black/red	225 ohms
Trail	
Brown/white-to-white	164 ohms
400/500	
Brown/white-to-white	164 ohms

PULSER COIL TESTING

Model/wire color	Resistance reading
Sport	—
Trail	
1983-1985	
Brown/white-to-black/red	45 ohms
1986-on	
Brown/white-to-black/red	17 ohms
400/500	
Brown/white-to-black/red	45 ohms

600/650 ELECTRICAL SPECIFICATIONS

Coil/wire color	Resistance reading
Exciter coil	
Black-to-white	261 ohms
Pulser coil	
Red-to-white	20 ohms
Control coil	
Green-to-blue	29.4 ohms
Ignition coil	
Primary	
Orange-to-black	0.106 ohms
Secondary	
Black-to-high tension lead	2,016 ohms

DRIVE BELT SPECIFICATIONS

	mm	in.
Belt width limit	30.16	1 3/16
Belt tension	31.75	1 1/4

CLUTCH ALIGNMENT

	mm	in.
Offset	15.88	5/8
Center-to-center distance	304.8	12.0

CLYMER®

POLARIS

SNOWMOBILE SHOP MANUAL
1984-1989

Chapter One

General Information

This Clymer shop manual covers the service and repair of the Polaris Indy series snowmobiles from 1984 through 1989.

Troubleshooting, tune-up, maintenance and repair are not difficult, if you know what tools and equipment to use and what to do. Step-by-step instructions guide you through jobs ranging from simple maintenance to complete engine and suspension overhaul.

This manual can be used by anyone from a first time do-it-yourselfer to a professional mechanic. Detailed drawings and clear photographs give you all the information you need to do the work right.

Some of the procedures in this manual require the use of special tools. The resourceful mechanic can, in many cases, think of acceptable substitutes for special tools—there is always another way. This can be as simple as using a few pieces of threaded rod, washers and nuts to remove or install a bearing or fabricating a tool from scrap material. However, using a substitute for a special tool is not recommended as it can be dangerous to and may damage the part. If you

find that a tool can be designed and safely made, but will require some type of machine work, you may want to search out a local community college or high school that has a machine shop curriculum. Shop teachers sometimes welcome outside work that can be used as practical shop applications for advanced students.

Table 1 lists model coverage.

General specifications are listed in **Tables 2-7**.

Metric and U.S. standards are used throughout this manual. U.S. to metric conversion is given in **Table 8**.

Critical torque specifications are found in table form at the end of each chapter (as required). The general torque specifications found in **Table 9** (metric) and **Tables 10-12** (SAE) can be used when a torque specification is not listed for a specific component or assembly.

A list of general technical abbreviations are given in **Table 13**.

Tap drill sizes can be found in **Table 14** (metric) and **Table 15** (SAE).

Table 16 lists wind chill factors.

Tables 1-16 are found at the end of the chapter.

MANUAL ORGANIZATION

This chapter provides general information useful to snowmobile owners and mechanics. In addition, information in this chapter discusses the tools and techniques for preventive maintenance, troubleshooting and repair.

Chapter Two provides methods and suggestions for quick and accurate diagnosis and repair of problems. Troubleshooting procedures discuss typical symptoms and logical methods to pinpoint the trouble.

Chapter Three explains all periodic lubrication and routine maintenance necessary to keep your snowmobile operating well. Chapter Three also includes recommended tune-up procedures, eliminating the need to constantly consult other chapters on the various assemblies.

Subsequent chapters describe specific systems, providing disassembly, repair, assembly and adjustment procedures in simple step-by-step form. If a repair is impractical for a home mechanic, it is so indicated. It is usually faster and less expensive to take such repairs to a dealer or competent repair shop. Specifications concerning a specific system are included at the end of the appropriate chapter.

NOTES, CAUTIONS AND WARNINGS

The terms NOTE, CAUTION and WARNING have specific meanings in this manual. A NOTE provides additional information to make a step or procedure easier or clearer. Disregarding a NOTE could cause inconvenience, but would not cause damage or personal injury.

A CAUTION emphasizes areas where equipment damage could occur. Disregarding a CAUTION could cause permanent mechanical damage; however, personal injury is unlikely.

A WARNING emphasizes areas where personal injury or even death could result from negligence. Mechanical damage may also occur. WARNINGS *are to be taken seriously.* In some cases, serious injury and death has resulted from disregarding similar warnings.

SAFETY FIRST

Professional mechanics can work for years and never sustain a serious injury. If you observe a few rules of common sense and safety, you can enjoy many safe hours servicing your own machine. If you ignore these rules you can hurt yourself or damage the equipment.

1. Never use gasoline as a cleaning solvent.
2. Never smoke or use a torch in the vicinity of flammable liquids, such as cleaning solvent, in open containers.
3. If welding or brazing is required on the machine, remove the fuel tank to a safe distance, at least 50 feet away.
4. Use the proper sized wrenches to avoid damage to fasteners and injury to yourself.
5. When loosening a tight or stuck nut, be guided by what would happen if the wrench should slip. Be careful; protect yourself accordingly.
6. When replacing a fastener, make sure to use one with the same measurements and strength as the old one. Incorrect or mismatched fasteners can result in damage to the snowmobile and possible personal injury. Beware of fastener kits that are filled with cheap and poorly made nuts,

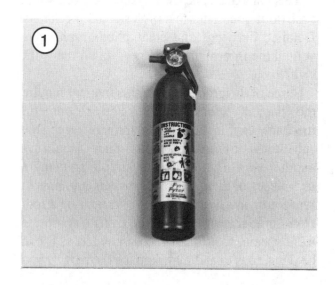

bolts, washers and cotter pins. Refer to *Fasteners* in this chapter for additional information.

7. Keep all hand and power tools in good condition. Wipe greasy and oily tools after using them. They are difficult to hold and can cause injury. Replace or repair worn or damaged tools.

8. Keep your work area clean and uncluttered.

9. Wear safety goggles during all operations involving drilling, grinding, the use of a cold chisel or *any* time you feel unsure about the safety of your eyes. Safety goggles should also be worn anytime solvent and compressed air is used to clean parts.

10. Keep an approved fire extinguisher (**Figure 1**) nearby. Be sure it is rated for gasoline (Class B) and electrical (Class C) fires.

11. When drying bearings or other rotating parts with compressed air, never allow the air jet to rotate the bearing or part. The air jet is capable of rotating them at speeds far in excess of those for which they were designed. The bearing or rotating part is very likely to disintegrate and cause serious injury and damage. To prevent bearing damage when using compressed air, hold the inner bearing race (**Figure 2**) by hand.

SERVICE HINTS

Most of the service procedures covered are straightforward and can be performed by anyone reasonably handy with tools. It is suggested, however, that you consider your own capabilities carefully before attempting any operation involving major disassembly.

1. "Front," as used in this manual, refers to the front of the snowmobile; the front of any component is the end closest to the front of the snowmobile. The "left-" and "right-hand" sides refer to the position of the parts as viewed by a rider sitting and facing forward. For example, the throttle control is on the right-hand side. These rules are simple, but confusion can cause a major inconvenience during service. See **Figure 3**.

2. When disassembling any engine or drive component, mark the parts for location and mark all parts which mate together. Small parts, such as bolts, can be identified by placing them in plastic sandwich bags (**Figure 4**). Seal the bags and label them with masking tape and a marking pen. When reassembly will take place immediately, an accepted practice is to place nuts and bolts in a cupcake tin or egg carton in the order of disassembly.

3. Finished surfaces should be protected from physical damage or corrosion. Keep gasoline off painted surfaces.

4. Use penetrating oil on frozen or tight bolts, then strike the bolt head a few times with a hammer and punch (use a screwdriver on screws). Avoid the use of heat where possible, as it can warp, melt or affect the temper of parts. Heat also ruins finishes, especially paint and plastics.

5. No parts removed or installed (other than bushings and bearings) in the procedures given in this manual should require unusual force during disassembly or assembly. If a part is difficult to remove or install, find out why before proceeding.

6. Cover all openings after removing parts or components to prevent dirt, small tools, etc. from falling in.

7. Read each procedure *completely* while looking at the actual parts before starting a job.

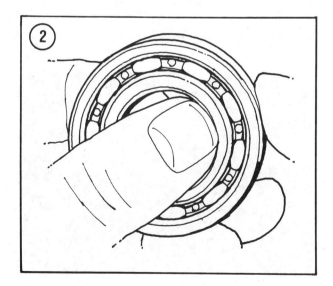

Make sure you *thoroughly* understand what is to be done and then carefully follow the procedure, step-by-step.

8. Recommendations are occasionally made to refer service or maintenance to a snowmobile dealer or a specialist in a particular field. In these cases, the work will be done more quickly and economically than if you performed the job yourself.

9. In procedural steps, the term "replace" means to discard a defective part and replace it with a new or exchange unit. "Overhaul" means to remove, disassemble, inspect, measure, repair or replace defective parts, reassemble and install major systems or parts.

10. Some operations require the use of a hydraulic press. It would be wiser to have these operations performed by a shop equipped for such work, rather than to try to do the job yourself with makeshift equipment that may damage your machine.

11. Repairs go much faster and easier if your machine is clean before you begin work. There are many special cleaners on the market, like Bel-Ray Degreaser, for washing the engine and related parts. Follow the manufacturer's directions on the container for the best results. Clean all oily or greasy parts with cleaning solvent as you remove them.

> *WARNING*
> *Never use gasoline as a cleaning agent.*
> *It presents an extreme fire hazard. Be*
> *sure to work in a well-ventilated area*
> *when using cleaning solvent. Keep a fire*
> *extinguisher, rated for gasoline fires,*
> *handy in any case.*

12. Much of the labor charges for repairs made by dealers are for the time involved in the removal, disassembly, assembly, and reinstallation of other parts in order to reach the defective part. It is frequently possible to perform the preliminary operations yourself and then take the defective unit to the dealer for repair at considerable savings.

13. If special tools are required, make arrangements to get them before you start. It is frustrating and time-consuming to get partly into a job and then be unable to complete it.

14. Make diagrams (or take a Polaroid picture) wherever similar-appearing parts are found. For instance, crankcase bolts are often not the same length. You may think you can remember where everything came from—but mistakes are costly. There is also the possibility that you may be sidetracked and not return to work for days or

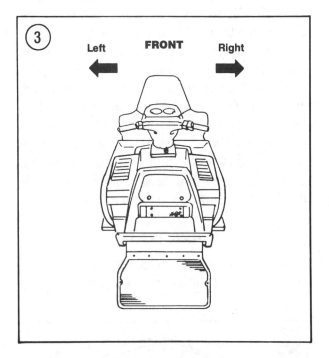

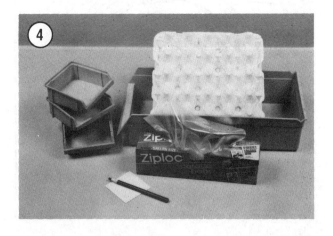

even weeks—in which time, carefully laid out parts may have become disturbed.

15. When assembling parts, be sure all shims and washers are replaced exactly as they came out.

16. Whenever a rotating part butts against a stationary part, look for a shim or washer. Use new gaskets if there is any doubt about the condition of the old ones. A thin coat of silicone sealant on non-pressure type gaskets may help them seal more effectively.

17. If it is necessary to make a cover gasket and you do not have a suitable old gasket to use as a guide, you can use the outline of the cover and gasket material to make a new gasket. Apply engine oil to the cover gasket surface. Then place the cover on the new gasket material and apply pressure with your hands. The oil will leave a very accurate outline on the gasket material that can be cut around.

CAUTION
When purchasing gasket material to make a gasket, measure the thickness of the old gasket (at an uncompressed point) and purchase gasket material with the same approximate thickness.

18. Heavy grease can be used to hold small parts in place if they tend to fall out during assembly. However, keep grease and oil away from electrical components.

19. A carburetor is best cleaned by disassembling it and cleaning the parts in hot soapy water. Never soak gaskets and rubber parts in commercial carburetor cleaners. Never use wire to clean out jets and air passages. They are easily damaged. Use compressed air to blow out the carburetor only if the float has been removed first.

20. Take your time and do the job right. Do not forget that a newly rebuilt engine must be broken in just like a new one.

ENGINE AND CHASSIS SERIAL NUMBERS

Polaris snowmobiles are identified by model and serial numbers. The model and serial numbers are found on a decal mounted on the right front side of the tunnel (**Figure 5**). The serial number is stamped into the tunnel. The model number is imprinted on the identification decal. See **Figure 6**. The engine is identified by a model number and a serial number. This

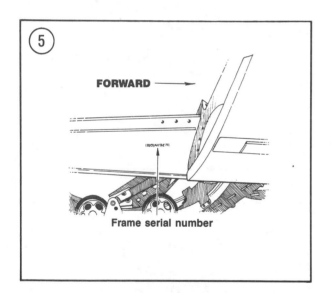

⑤

FORWARD →

Frame serial number

⑥

MODEL NO.   MADE IN U.S.A.   **POLARIS**

SERIAL NO.   PATENT NOTICE

THIS VEHICLE CONFORMS TO ALL APPLICABLE   Mfd. by Polaris Industries Inc.
U.S. FEDERAL AND STATE REQUIREMENTS AND   in Roseau, MN under one or more of the following patents
CANADIAN FEDERAL MOTOR VEHICLE STAND-   U.S. Patents   3,605,511   3,533,662   3,613,811   3,867,991
ARDS IN EFFECT ON THE DATE OF MANUFACTURE   3,580,647   3,613,810   3,605,510
                                                        3,483,766   3,545,821   3,525,412
MFD. DATE   Patented Canada   882,491/71   883,694/71   864,394/71
                                           Canadian Rd   34,573/71   34,572/71

information is listed on a decal found on the engine (**Figure 7**).

Write down all serial and model numbers applicable to your machine and carry the numbers with you when you order parts from a dealer. Always order by year and engine and machine numbers. If possible, compare the old parts with the new ones before purchasing them. If the parts are not alike, have the parts manager explain the reason for the difference and insist on assurance that the new parts will fit and are correct.

At the front of this book, write down the frame and engine identification numbers for your snowmobile.

ENGINE OPERATION

The following is a general discussion of a typical 2-stroke piston-ported engine. During this discussion, assume that the crankshaft is rotating counterclockwise in **Figure 8**. As the piston travels downward, a transfer port (A) between the crankcase and the cylinder is uncovered. The exhaust gases leave the cylinder through the exhaust port (B), which is also opened by the downward movement of the piston. A fresh air/fuel charge, which has previously been compressed slightly, travels from the crankcase (C) to the cylinder through the transfer port (A) as the port opens. Since the incoming charge is under pressure, it rushes into the cylinder quickly and helps to expel the exhaust gases from the previous cycle.

Figure 9 illustrates the next phase of the cycle. As the crankshaft continues to rotate, the piston moves upward, closing the exhaust and transfer ports. As the piston continues upward, the air/fuel mixture in the cylinder is compressed. Notice also that a vacuum is created in the crankcase at the same time. Further upward movement of the piston uncovers the intake port (D). A fresh air/fuel charge is then drawn into the crankcase through the intake port because of

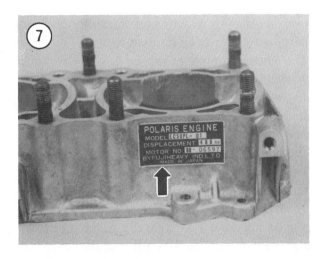

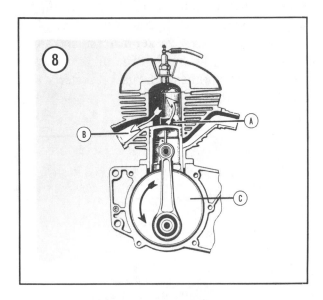

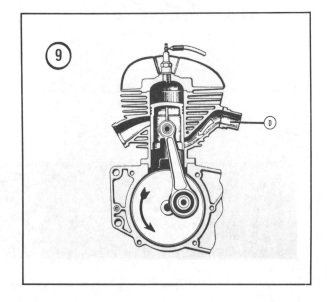

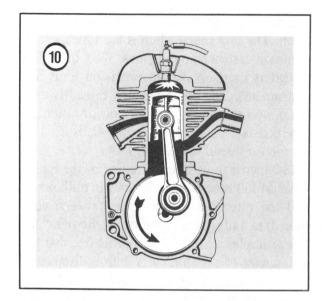

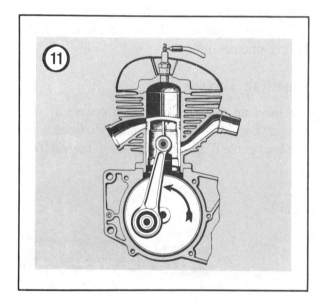

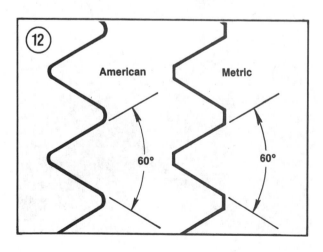

the vacuum created by the upward piston movement.

The third phase is shown in **Figure 10**. As the piston approaches top dead center, the spark plug fires, igniting the compressed mixture. The piston is then driven downward by the expanding gases.

When the top of the piston uncovers the exhaust port, the fourth phase begins, as shown in **Figure 11**. The exhaust gases leave the cylinder through the exhaust port. As the piston continues downward, the intake port is closed and the mixture in the crankcase is compressed in preparation for the next cycle.

It can be seen from this discussion that every downward stroke of the piston is a power stroke.

TORQUE SPECIFICATIONS

Torque specifications throughout this manual are given in Newton-meters (N·m) and foot-pounds (ft.-lb.). Newton meters are being adopted in place of meter-kilograms (mkg) in accordance with the International Modernized Metric system. Existing torque wrenches calibrated in meter-kilograms can be used by performing a simple conversion: move the decimal point one place to the right. For example, 4.7 mkg = 47 N·m. This conversion is accurate enough for mechanics' use even though the exact mathematical conversion is 3.5 mkg = 34.3 N·m.

FASTENERS

Polaris snowmobiles are manufactured with both ISO (International Organization for Standardization) metric fasteners and the American industrial standard called the Unified System or Unified thread form. Both threads are cut differently and are not interchangeable as shown in **Figure 12**.

All of the Polaris models covered in this manual are equipped with Fuji engines which use metric threads. All of the other components (steering, suspension, etc.) are assembled with American fasteners. When working on your Polaris snowmobile, keep track of fasteners removed from the different assemblies so that you do not try to interchange them. This caution is important when using air guns and similar tools because you can quickly strip or break the fastener before realizing you have matched 2 different thread types.

Threads

The materials and designs of the various fasteners used on your snowmobile are not arrived at by chance or accident. Fastener design determines the type of tool required to work the fastener. Fastener material is carefully selected to decrease the possibility of physical failure (**Figure 13**).

Nuts, bolts and screws are manufactured in a wide range of thread patterns. To join a nut and bolt, the diameter of the bolt and the diameter of the hole in the nut and the threads on both parts must be the same.

The best way to tell if the threads on 2 fasteners are matched is to turn the nut on the bolt (or the bolt into the threaded hole in a piece of equipment) with fingers only. Be sure both pieces are clean. If much force is required, check the thread condition on each fastener. If the thread condition is good but the fasteners jam, the threads are not compatible. A thread pitch gauge (**Figure 14**) can also be used to determine pitch. Most threads are cut so that the fastener must be turned clockwise to tighten it. These are called right-hand threads. Some fasteners have left-hand threads; they must be turned counterclockwise to be tightened. Left-hand threads are used in locations where normal rotation of the equipment would tend to loosen a right-hand threaded fastener.

ISO Metric Screw Threads

ISO (International Organization for Standardization) metric threads come in 3 standard thread sizes: coarse, fine and constant pitch. The ISO coarse pitch is used for most all common fastener applications. The fine pitch thread is used on certain precision tools and instruments. The constant pitch thread is used mainly on machine parts and not for fasteners. The constant pitch thread, however, is used on all metric thread spark plugs.

ISO metric threads are specified by the capital letter M followed by the diameter in millimeters and the pitch (or the distance between each thread) in millimeters separated by the sign "$\times$". For example, a M8 $\times$ 1.25 bolt is one that has a diameter of 8 millimeters with a distance of 1.25 millimeters between each thread. The measurement across 2 flats on the head of the bolt (**Figure 15**) indicates the proper wrench size to be used. **Figure 16** shows how to determine bolt diameter.

American Threads

American threads come in a coarse or fine thread. Because both coarse and fine threads are used for general use, it is important to match the

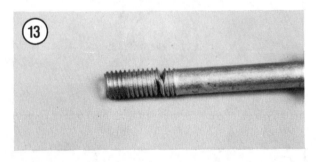

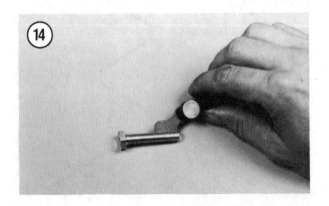

threads correctly so you do not strip the threads and damage one or both fasteners.

American fasteners are normally described by diameter, threads per inch and length. For example, 3/8-16 × 2 indicates a bolt 3/8 in. in diameter with 16 threads per inch, 2 inches long. The measurement across 2 flats on the head of the bolt or screw (**Figure 15**) indicates the proper wrench size to be used. **Figure 16** shows how to determine bolt diameter.

Markings found on American bolt heads indicate tensile strength. For example, a bolt with no head marking is usually made of mild steel,

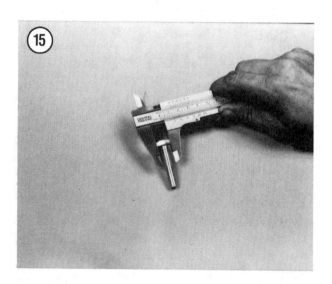

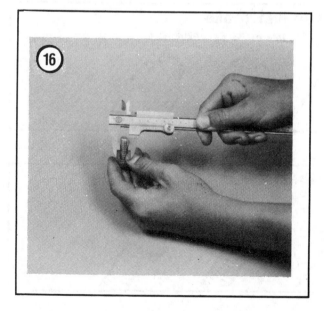

while a bolt with 2 or more markings indicates a higher grade material. **Figure 17** identifies the various head markings and SAE grade identification. When torquing SAE bolts, refer to the head marking (**Figure 17**) and then to the appropriate table (see **Tables 10-12**) for the torque specification.

Determine Bolt Length

When purchasing a bolt from a dealer or parts store, it is important to know how to specify bolt length. The correct way to measure bolt length is to measure the length starting from underneath the bolt head to the end of the bolt (**Figure 18**). Always measure bolt length in this manner to prevent from purchasing bolts that are too long.

Machine Screws

Machine screw refers to a numbering system used to identify screws smaller than 1/4 of an inch. Machine screws are identified by gage size (diameter) and threads per inch. For example, 12-28 indicates a 12 gage screw with 28 threads per inch.

There are many different types of machine screws. **Figure 19** shows a number of screw heads requiring different types of turning tools. Heads are also designed to protrude above the metal (round) or to be slightly recessed in the metal (flat). See **Figure 20**.

Bolts

Commonly called bolts, the technical name for these fasteners is cap screw. Refer to *ISO Metric Screw Threads* and *American Threads* in this section for additional information.

Nuts

Nuts are manufactured in a variety of types and sizes. Most are hexagonal (6-sided) and fit on bolts, screws and studs with the same diameter and pitch.

Figure 21 shows several types of nuts. The common nut is generally used with a lockwasher. Self-locking nuts have a nylon insert which prevents the nut from loosening; no lockwasher is required. Wing nuts are designed for fast removal by hand. Wing nuts are used for convenience in non-critical locations.

To indicate the size of a metric nut, manufacturers specify the diameter of the opening and the thread pitch. American nuts are identified by the diameter of the opening and the threads per inch. For both metric and American fasteners, this is similar to their respective bolt specifications, but without the length dimension. The measurement across 2 flats on the nut (**Figure 15**) indicates the proper wrench size to be used.

Prevailing Torque Fasteners

Several types of bolts, screws and nuts incorporate a system that develops an interference between the bolt, screw, nut or tapped hole threads. Interference is achieved in various ways: by distorting threads, coating threads with dry adhesive or nylon, distorting the top of an all-metal nut, using a nylon insert in the center or at the top of a nut, etc.

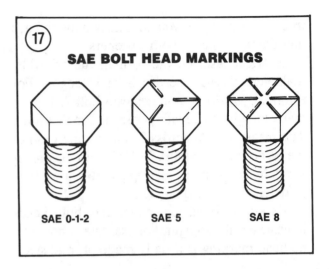

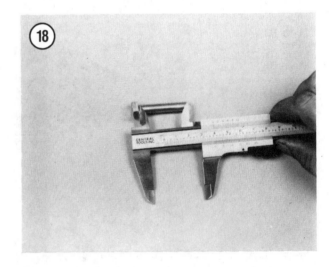

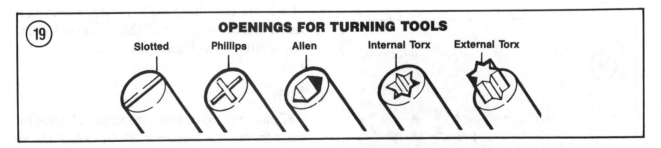

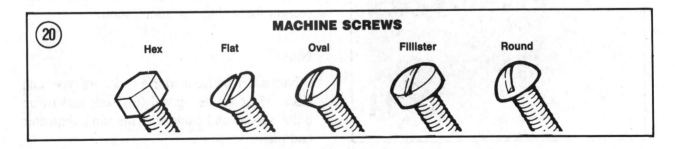

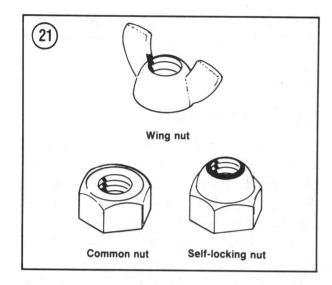

Wing nut

Common nut   Self-locking nut

LOCKWASHERS

Plain   Folding

Internal tooth   External tooth

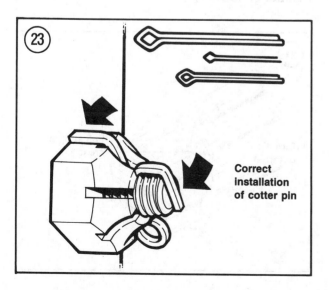

Correct installation of cotter pin

Prevailing torque fasteners offer greater holding strength and better vibration resistance. Some prevailing torque fasteners can be reused if in good condition. Others, like the nylon insert nut, form an initial locking condition when the nut is first installed; the nylon forms closely to the bolt thread pattern, thus reducing any tendency for the nut to loosen. When the nut is removed, the locking efficiency is greatly reduced. For greatest safety, it is recommended that you install new prevailing torque fasteners whenever they are removed.

Washers

There are 2 basic types of washers: flat washers and lockwashers. Flat washers are simple discs with a hole to fit a screw or bolt. Lockwashers are designed to prevent a fastener from working loose due to vibration, expansion and contraction. **Figure 22** shows several types of washers. Washers are also used in the following functions:

a. As spacers.
b. To prevent galling or damage of the equipment by the fastener.
c. To help distribute fastener load during torquing.
d. As fluid seals (copper, aluminum or laminated washers).

Note that flat washers are often used between a lockwasher and a fastener to provide a smooth bearing surface. This allows the fastener to be turned easily with a tool.

Cotter Pins

Cotter pins (**Figure 23**) are used to secure special kinds of fasteners. The threaded stud must have a hole in it; the nut or nut lock piece has castellations around which the cotter pin ends wrap. Cotter pins should not be reused after removal.

Circlips

Circlips can be internal or external design. They are used to retain items on shafts (external type) or within tubes (internal type). In some applications, circlips of varying thicknesses are used to control the end play of parts assemblies. These are often called selective circlips. Circlips should be replaced during installation, as removal weakens and deforms them.

Two basic styles of circlips are available: machined and stamped circlips. Machined circlips (**Figure 24**) can be installed in either direction (shaft or housing) because both faces are machined, thus creating two sharp edges. Stamped circlips (**Figure 25**) are manufactured with one sharp edge and one rounded edge. When installing stamped circlips in a thrust situation, the sharp edge must face away from the part producing the thrust. When installing circlips, observe the following:

   a. Circlips should be removed and installed with circlip pliers only. See *Circlip Pliers* in this chapter.

   b. Compress or expand circlips only enough to remove or install them.

   c. After the circlip is installed, make sure it is completely seated in its groove.

LUBRICANTS

Periodic lubrication assures long life for any type of equipment. The *type* of lubricant used is just as important as the lubrication service itself. The following paragraphs describe the types of lubricants most often used on snowmobiles. Be sure to follow the manufacturer's recommendations for lubricant types.

Generally, all liquid lubricants are called "oil." They may be mineral-based (including petroleum bases), natural-based (vegetable and animal bases), synthetic-based or emulsions (mixtures).

"Grease" is an oil to which a thickening base has been added so that the end product is semi-solid.

Grease is often classified by the type of thickener added; lithium soap is commonly used.

Engine Oil

Four-Stroke Engine Oil

Four-stroke oil for automotive engines is classified by the American Petroleum Institute (API) and the Society of Automotive Engineers (SAE) in several categories. Oil containers display these ratings on the top or label.

API oil grade is indicated by letters; oils for geasoline engines are identified by an "S".

Viscosity is an indication of the oil's thickness. The SAE uses numbers to indicate viscosity; thin oils have low numbers while thick oils have high numbers. A "W" after the number indicates that the viscosity testing was done at a low temperature to simulate cold-weather operation. Engine oils fall into the 5W-30 and 20W-50 range.

Multi-grade oils (for example 10W-40) are less viscous (thinner) at low temperatures and more viscous (thicker) at high temperatures. This allows the oil to perform efficiently across a wide range of engine operating conditions. The lower the number, the better the engine will start in cold climates. Higher numbers are usually recommended for engine running in hot weather conditions.

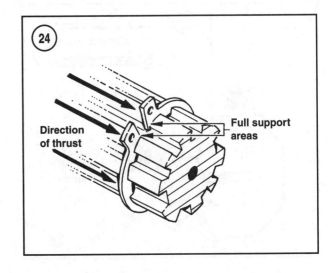

(24) Direction of thrust — Full support areas

CAUTION
Four-stroke oils are only discussed to provide a comparison. Polaris snowmobile engines are 2-stroke engines, thus only 2-stroke oil should be used.

Two-Stroke Engine Oil

Lubrication for a 2-stroke engine is provided by oil mixed with the incoming fuel-air mixture. Some of the oil mist settles out in the crankcase, lubricating the crankshaft and lower end of the connecting rods. The rest of the oil enters the combustion chamber to lubricate the piston rings and cylinder walls. This oil is burned during the combustion process.

Engine oil must have several special qualities to work well in a 2-stroke engine. It must mix easily and stay in suspension in gasoline. When burned, it can't leave behind excessive deposits. It must be appropriate for the high temperatures associated with 2-stroke engines.

All Polaris Indy models covered in this manual are equipped with an oil injection system. After engine break-in and during normal use, the oil injection system does not require pre-mixing of fuel and oil. However, during engine break-in, Polaris requires a 40:1 fuel/oil ratio to be used with the oil injection system.

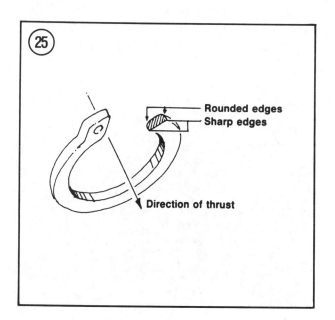

(25)

Rounded edges
Sharp edges

Direction of thrust

NOTE
*Refer to **Engine Lubrication** and **Break-In** in Chapter Three for additional information.*

Grease

Greases are graded by the National Lubricating Grease Institute (NLGI). Greases are graded by number according to the consistency of the grease; these range from No. 000 to No. 6, with No. 6 being the most solid. A typical multipurpose grease is NLGI No. 2. For specific applications, equipment manufacturers may require grease with an additive such as molybdenum disulfide (MOS2).

NOTE
A low-temperature grease should be used wheresoever grease is required on the snowmobile. Chapter Three lists the low-temperature greases recommended by Polaris.

RTV GASKET SEALANT

Room temperature vulcanizing (RTV) sealant is used on some pre-formed gaskets and to seal some components. RTV is a silicone gel supplied in tubes and can be purchased in a number of different colors. For most snowmobile use, the clear color is more preferable.

Moisture in the air causes RTV to cure. Always place the cap on the tube as soon as possible when using RTV. RTV has a shelf life of one year and will not cure properly when the shelf life has expired. Check the expiration date on RTV tubes before using and keep partially used tubes tightly sealed.

Applying RTV Sealant

Clean all gasket residue from mating surfaces. Surfaces should be clean and free of oil and dirt. Remove all RTV gasket material from blind attaching holes, as it can cause a "hydraulic" effect and affect bolt torque.

Apply RTV sealant in a continuous bead 2-3 mm (0.08-0.12 in.) thick. Circle all mounting holes unless otherwise specified. Torque mating parts within 10 minutes after application.

THREADLOCK

Because of the snowmobile's operating conditions, a threadlock (**Figure 26**) is required to help secure many of the fasteners. A threadlock will lock fasteners against vibration loosening and seal against leaks. Loctite 242 (blue) and 271 (red) are recommended for many threadlock requirements described in this manual.

Loctite 242 (blue) is a medium-strength threadlock for general purpose use. Component disassembly can be performed with normal hand tools. Loctite 271 (red) is a high-strength threadlock that is normally used on studs or critical fasteners. Heat or special tools, such as a press or puller, may be required for component disassembly.

Applying Threadlock

Surfaces should be clean and free of oil and dirt. If a threadlock was previously applied to the component, this residue should also be removed.

Shake the Loctite container thoroughly and apply to both parts. Assemble parts and/or tighten fasteners.

GASKET REMOVER

Stubborn gaskets can present a problem during engine service as they can take a long time to remove. Consequently, there is the added problem of secondary damage occurring to the gasket mating surfaces from the incorrect use of gasket scraping tools. To quickly and safely remove stubborn gaskets, use a spray gasket remover. Spray gasket remover can be purchased through snowmobile and automotive parts houses. Follow the manufacturer's directions for use.

BASIC HAND TOOLS

Many of the procedures in this manual can be carried out with simple hand tools and test equipment familiar to the average home mechanic. Keep your tools clean and in a tool box. Keep them organized with the sockets and related drives together, the open-end combination wrenches together, etc. After using a tool, wipe off dirt and grease with a clean cloth and return the tool to its correct place.

Top-quality tools are essential; they are also more economical in the long run. If you are now starting to build your tool collection, stay away from the "advertised specials" featured at some parts houses, discount stores and chain drug stores. These are usually a poor grade tool that can be sold cheaply and that is exactly what they are—*cheap*. They are usually made of inferior material, and are thick, heavy and clumsy. Their rough finish makes them difficult to clean and they usually don't last very long. If it is ever your misfortune to use such tools, you will probably find out that the wrenches do not fit the heads of bolts and nuts correctly and damage the fastener.

Quality tools are made of alloy steel and are heat treated for greater strength. They are lighter and better balanced than cheap ones. Their surface is smooth, making them a pleasure to work with and easy to clean. The initial cost of good-quality tools may be more but they are

cheaper in the long run. Don't try to buy everything in all sizes in the beginning; do it a little at a time until you have the necessary tools.

The following tools are required to perform virtually any repair job. Each tool is described and the recommended size given for starting a tool collection. Additional tools and some duplicates may be added as you become familiar with the vehicle.

Screwdrivers

The screwdriver is a very basic tool, but if used improperly it will do more damage than good. The slot on a screw has a definite dimension and shape. A screwdriver must be selected to conform with that shape. Use a small screwdriver for small screws and a large one for large screws or the screw head will be damaged.

Two basic types of screwdriver are required: common (flat-blade) screwdrivers (**Figure 27**) and Phillips screwdrivers (**Figure 28**).

Screwdrivers are available in sets which often include an assortment of common and Phillips blades. If you buy them individually, buy at least the following:

a. Common screwdriver—5/16 × 6 in. blade.
b. Common screwdriver—3/8 × 12 in. blade.
c. Phillips screwdriver—size 2 tip, 6 in. blade.
d. Phillips screwdriver—size 3 tip, 6 in. blade.

Use screwdrivers only for driving screws. Never use a screwdriver for prying or chiseling metal. Do not try to remove a Phillips or Allen head screw with a common screwdriver (unless the screw has a combination head that will accept either type); you can damage the head so that the proper tool will be unable to remove it.

Keep screwdrivers in the proper condition and they will last longer and perform better. Always keep the tip of a common screwdriver in good condition. **Figure 29** shows how to grind the tip to the proper shape if it becomes damaged. Note the symmetrical sides of the tip.

Pliers

Pliers come in a wide range of types and sizes. Pliers are useful for cutting, bending and crimping. They should never be used to cut hardened objects or to turn bolts or nuts. **Figure 30** shows several pliers useful in snowmobile repair.

Each type of pliers has a specialized function. Slip-joint pliers are used mainly for holding things and for bending. Needlenose pliers are used to hold or bend small objects. Waterpump pliers (commonly referred to as channel locks) can be adjusted to hold various sizes of objects; the jaws remain parallel to grip around objects such as pipe or tubing. There are many more types of pliers.

CAUTION
Pliers should not be used for loosening or tightening nuts or bolts. The pliers sharp teeth will grind off the nut or bolt corners and damage the fastener.

CAUTION
If slip-joint pliers are going to be used to hold an object with a finished surface that can be easily damaged, wrap the object with tape or cardboard for protection.

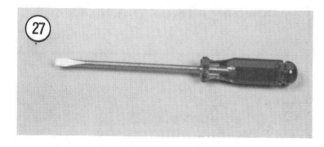

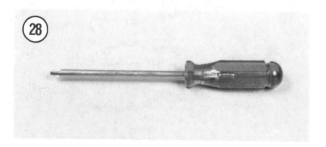

Vise-Grip Pliers

Vise-grip pliers (**Figure 31**) are used to hold objects very tight while another task is performed on the object. While Vise-grips work well, caution should be followed with their use. Because Vise-grip pliers exert more force than regular pliers, their sharp jaws will permanently scar the object. In addition, when Vise-grip pliers are locked in position, they can crush or deform thin wall material.

Vise-grips are available in many types for more specific tasks.

Circlip Pliers

Circlip pliers (**Figure 32**) are special in that they are only used to remove circlips from shafts or within engine or suspension housings. When purchasing circlip pliers, there are two kinds to distinguish from. External pliers (spreading) are used to remove circlips that fit on the outside of a shaft. Internal pliers (squeezing) are used to remove circlips which fit inside a housing.

Box-end, Open-end and Combination Wrenches

Box-end and open-end wrenches (**Figure 33**) are available in sets or separately in a variety of sizes. On open-end and box-end wrenches, the number stamped near the end refers to the distance between 2 parallel flats on the head of a nut or bolt. On combination wrenches, the number is stamped near the center.

Open-end wrenches are speedy and work best in areas with limited overhead access. Their wide jaws make them unsuitable for situations where the bolt or nut is sunken in a well or close to the edge of a casting. These wrenches only grip on two flats of a fastener so if either the fastener head or wrench jaws are worn, the wrench may slip off.

Box-end wrenches require clear overhead access to the fastener but can work well in situations where the fastener head is close to another part. They grip on all six edges of a

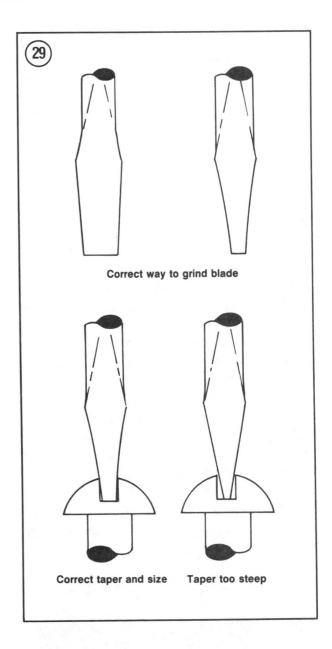

Correct way to grind blade

Correct taper and size Taper too steep

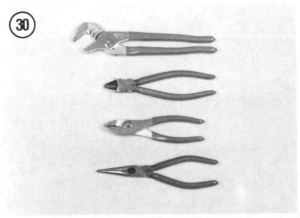

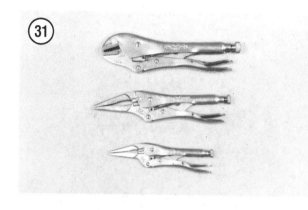

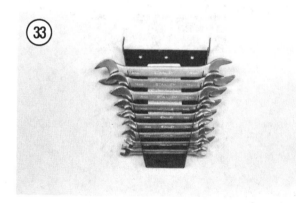

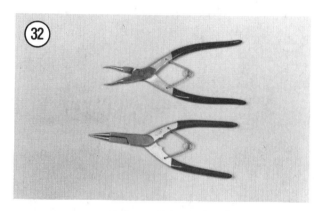

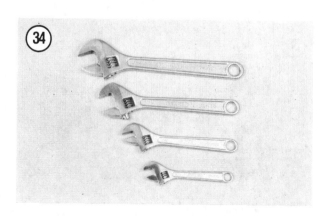

fastener for a very secure grip. They are available in either 6-point or 12-point. The 6-point gives superior holding power and durability but requires a greater swinging radius. The 12-point works better in situations with limited swinging radius.

Combination wrenches which are open on one side and boxed on the other are also available.

No matter what style of wrench you choose, proper use is important to prevent personal injury. When using a wrench, get in the habit of pulling the wrench toward you. This reduces the risk of injuring your hand if the wrench should slip. If you have to push the wrench away from you to loosen or tighten a fastener, open and push with the palm of your hand. This technique gets your fingers and knuckles out of the way should the wrench slip. Before using a wrench, always think ahead as to what could happen if the wrench should slip or if the bolt strips or breaks.

Adjustable Wrenches

An adjustable wrench (sometimes called a Crescent wrench) can be adjusted to fit nearly any nut or bolt head which has clear access around its entire perimeter. Adjustable wrenches are best used as a backup wrench to keep a large nut or bolt from turning while the other end is being loosened or tightened with a proper wrench. See **Figure 34**.

Adjustable wrenches have only two gripping surfaces which makes them more subject to slipping off the fastener and damaging the part and possibly your hand. See *Box-end, Open-end and Combination Wrenches* in this chapter.

These wrenches are directional; the solid jaw must be the one transmitting the force. If you use the adjustable jaw to transmit the force, it will loosen and possibly slip off.

Adjustable wrenches come in all sizes but something in the 6 to 8 inch range is recommended as an all-purpose wrench.

Socket Wrenches

This type is undoubtedly the fastest, safest and most convenient to use. Sockets which attach to a ratchet handle are available with 6-point or 12-point openings and 1/4, 3/8, 1/2 and 3/4 in. drives (**Figure 35**). The drive size indicates the size of the square hole which mates with the ratchet handle.

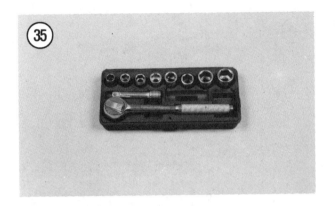

Torque Wrench

A torque wrench (**Figure 36**) is used with a socket to measure how tightly a nut or bolt is installed. They come in a wide price range and with either 3/8 or 1/2 in. square drive. The drive size indicates the size of the square drive which mates with the socket.

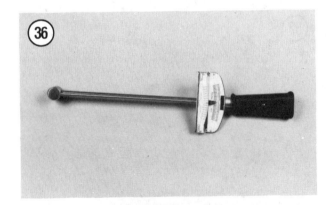

Impact Driver

This tool makes removal of tight fasteners easy and eliminates damage to bolts and screw slots. Impact drivers and interchangeable bits (**Figure 37**) are available at most large hardware, snowmobile and motorcycle dealers. Sockets can also be used with a hand impact driver. However, make sure the socket is designed for impact use. Do not use regular hand type sockets, as they may shatter.

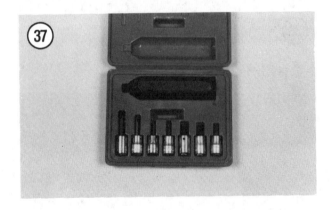

Hammers

The correct hammer (**Figure 38**) is necessary for repairs. Use only a hammer with a face (or head) of rubber or plastic or the soft-faced type that is filled with buckshot. These are sometimes necessary in engine teardowns. *Never* use a metal-faced hammer on engine or suspension parts, as severe damage will result in most cases. You can always produce the same amount of force with a soft-faced hammer. A metal-faced hammer, however, will be required when using a hand impact driver.

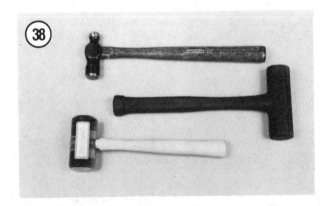

PRECISION MEASURING TOOLS

Measurement is an important part of snowmobile service. When performing many of the service procedures in this manual, you will be required to make a number of measurements. These include basic checks such as engine compression and spark plug gap. As you get deeper into engine disassembly and service, measurements will be required to determine the condition of the piston and cylinder bore,

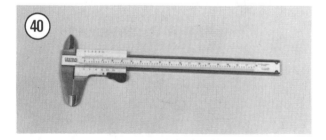

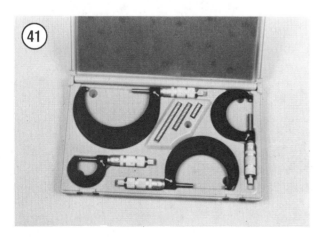

crankshaft runout and so on. When making these measurements, the degree of accuracy will dictate which tool is required. Precision measuring tools are expensive. If this is your first experience at engine service, it may be more worthwhile to have the checks made at a dealer. However, as your skills and enthusiasm increase for doing your own service work, you may want to begin purchasing some of these specialized tools. The following is a description of the measuring tools required in order to perform engine service described in this manual.

Feeler Gauge

The feeler gauge (**Figure 39**) is made of either a piece of a flat or round hardened steel of a specified thickness. Wire gauges are used to measure spark plug gap. Flat gauges are used for all other measurements.

Vernier Caliper

This tool is invaluable when reading inside, outside and depth measurements to within close precision. See **Figure 40**.

Outside Micrometers

One of the most reliable tools used for precision measurement is the outside micrometer. Outside micrometers will be required to measure piston diameter. Outside micrometers are also used with other tools to measure cylinder bore. Micrometers can be purchased individually or as a set (**Figure 41**).

Dial Indicator

Dial indicators (**Figure 42**) are precision tools used to check ignition timing and runout limits. For snowmobile repair, select a dial indicator with a continuous dial (**Figure 43**). This type of dial is required to accurately measure ignition timing and can be used for runout and height measurement checks.

Cylinder Bore Gauge

The cylinder bore gauge is a very specialized precision tool. The gauge set shown in **Figure 44** is comprised of a dial indicator, handle and a number of length adapters to adapt the gauge to different bore sizes. The bore gauge can be used to make cylinder bore measurements such as bore size, taper and out-of-round. An outside micrometer must be used together with the bore gauge to determine bore dimensions.

Small Hole Gauges

A set of small hole gauges (**Figure 45**) allow you to measure a hole, groove or slot ranging in size up to 13 mm (0.500 in.). An outside micrometer must be used together with the small hole gauge to determine bore dimensions.

Telescoping Gauges

Telescoping gauges (**Figure 46**) can be used to measure hole diameters from approximately 8 mm (5/16 in.) to 150 mm (6 in.). Like the small hole gauge, the telescoping gauge does not have a scale gauge for direct readings. Thus an outside micrometer is required to determine bore dimensions.

Compression Gauge

An engine with low compression cannot be properly tuned and will not develop full power. A compression gauge (**Figure 47**) measures engine compression. The one shown has a flexible stem with an extension that can allow

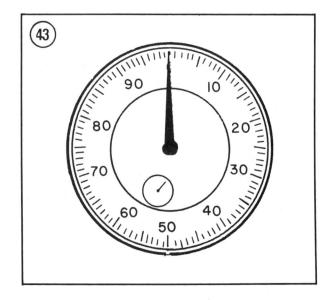

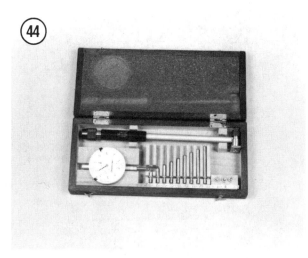

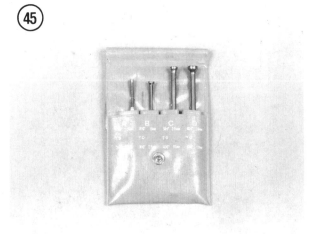

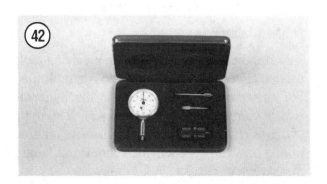

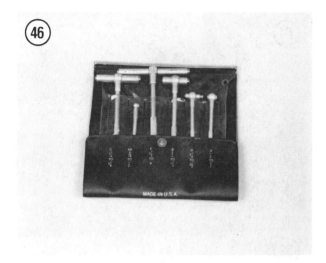

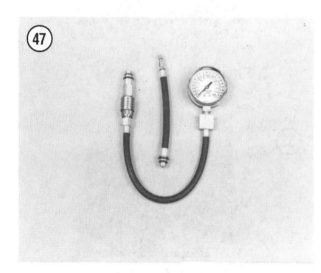

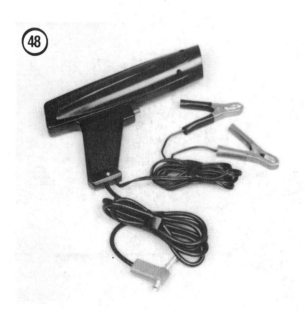

you to hold it while cycling the engine. Open the throttle all the way when checking engine compression. See Chapter Three.

Two-stroke Pressure Tester

Refer to *Two-Stroke Pressure Testing* in Chapter Two.

Strobe Timing Light

This instrument is useful for checking ignition timing. By flashing a light at the precise instant the spark plug fires, the position of the timing mark can be seen. The flashing light makes a moving mark appear to stand still opposite a stationary mark.

Suitable lights range from inexpensive neon bulb types to powerful xenon strobe lights. See **Figure 48**. A light with an inductive pickup is recommended to eliminate any possible damage to ignition wiring.

Multimeter or VOM

This instrument (**Figure 49**) is invaluable for electrical system troubleshooting.

Battery Hydrometer

A hydrometer is the best way to check a battery's state of charge. A hydrometer measures the weight or density of the sulfuric acid in the battery's electrolyte in specific gravity.

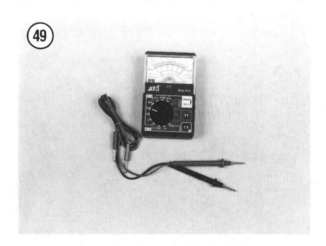

Screw Pitch Gauge

A screw pitch gauge (**Figure 50**) determines the thread pitch of bolts, screws, studs, etc. The gauge is made up of a number of thin plates. Each plate has a thread shape cut on one edge to match one thread pitch. When using a screw pitch gauge to determine a thread pitch size, try to fit different blade sizes onto the bolt thread until both threads match.

Magnetic Stand

A magnetic stand (**Figure 51**) is used to securely hold a dial indicator when checking the runout of a round object or when checking the end play of a shaft.

V-Blocks

V-blocks (**Figure 52**) are precision ground blocks used to hold a round object when checking its runout or condition.

Surface Plate

A surface plate is used to check the flatness of parts. While industrial quality surface plates are quite expensive, the home mechanic can improvise. A piece of thick metal can be put to use as a surface plate. The metal surface plate in **Figure 53** shows a piece of sandpaper glued to its surface that is used for cleaning and smoothing cylinder head and crankcase mating surfaces.

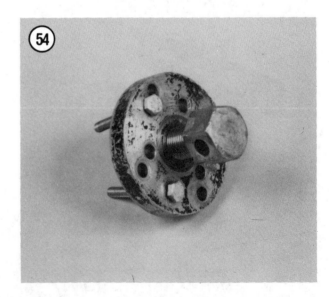

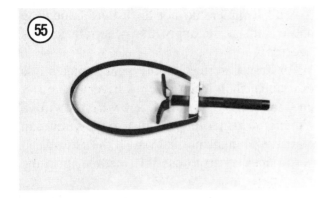

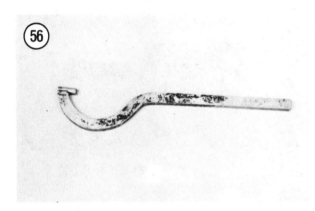

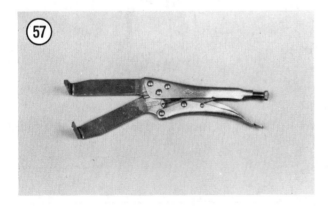

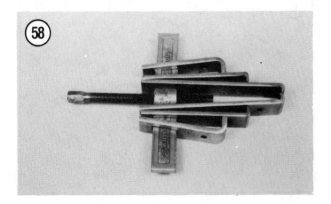

NOTE
Check with a local machine shop, fabricating shop or school offering a machine shop course for the availability of a metal plate that can be resurfaced.

SPECIAL TOOLS

This section describes special tools unique to snowmobile service and repair.

Flywheel Puller

A flywheel puller (**Figure 54**) will be required whenever it is necessary to remove the flywheel and service the stator plate assembly or when adjusting the ignition timing. In addition, when disassembling the engine, the flywheel must be removed before the crankcases can be split. There is no satisfactory substitute for this tool. Because the flywheel is a taper fit on the crankshaft, makeshift removal often results in crankshaft and flywheel damage. Don't think about removing the flywheel without this tool.

Strap Wrench

A strap wrench (**Figure 55**) can be used to hold the flywheel when loosening or tightening the flywheel bolt.

Starter Pulley Holder

A universal type holder (**Figure 56**) or the universal "Grabbit" (**Figure 57**) can be used to hold the recoil starter pulley during removal and installation.

Wheel Bearing Pullers

A puller set with long arms (**Figure 58**) will be required to remove suspension wheel bearings.

Track Clip Installer

A track clip installer (**Figure 59**) will be required to install track clips.

Spring Scale

A spring scale (**Figure 60**) will be required to check track tension.

Clutch Tools

A number of special tools will be required for clutch service. These are described in Chapter Twelve.

Expendable Supplies

Certain expendable supplies are also required. These include grease, oil, gasket cement, shop rags and cleaning solvent. Ask your dealer for the special locking compounds, silicone lubricants and lube products which make vehicle maintenance simpler and easier. Cleaning solvent is available at some service stations.

> *WARNING*
> *Having a stack of clean shop rags on hand is important when performing engine work. However, to prevent the possibility of fire damage from spontaneous combustion from a pile of solvent soaked rags, store them in a lid sealed metal container until they can be washed or properly discarded.*

> *NOTE*
> *To prevent absorbing solvent and other chemicals into your skin while cleaning parts, wear a pair of petroleum-resistant rubber gloves. These can be purchased through industrial supply houses or well-equipped hardware stores.*

MECHANIC'S TIPS

Removing Frozen Nuts and Screws

When a fastener rusts and cannot be removed, several methods may be used to loosen it. First, apply penetrating oil such as Liquid Wrench or WD-40 (available at hardware or auto supply stores). Apply it liberally and let it penetrate for 10-15 minutes. Rap the fastener several times with a small hammer; do not hit it hard enough to cause damage. Reapply the penetrating oil if necessary.

For frozen screws, apply penetrating oil as described, then insert a screwdriver in the slot and rap the top of the screwdriver with a hammer. This loosens the rust so the screw can be removed in the normal way. If the screw head is too chewed up to use this method, grip the

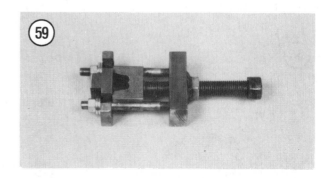

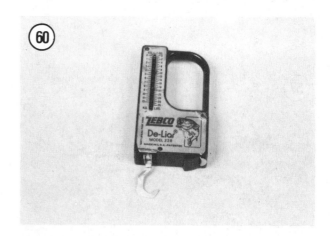

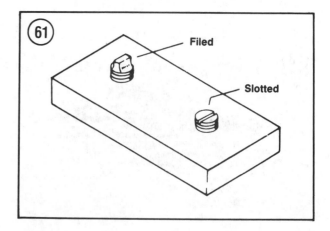

head with Vise-grip pliers and twist the screw out.

Avoid applying heat unless specifically instructed, as it may melt, warp or remove the temper from parts.

Removing Broken Screws or Bolts

When the head breaks off a screw or bolt, several methods are available for removing the remaining portion.

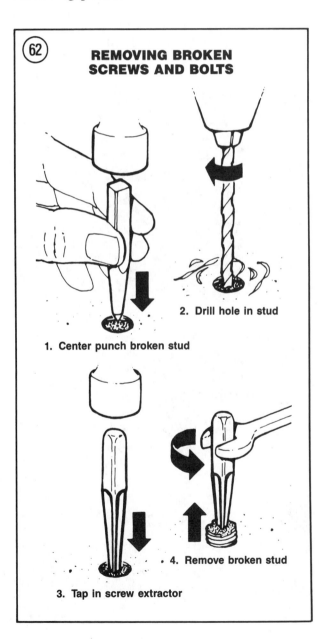

REMOVING BROKEN SCREWS AND BOLTS

1. Center punch broken stud

2. Drill hole in stud

3. Tap in screw extractor

4. Remove broken stud

If a large portion of the remainder projects out, try gripping it with Vise-grip pliers. If the projecting portion is too small, file it to fit a wrench or cut a slot in it to fit a screwdriver. See **Figure 61**.

If the head breaks off flush, use a screw extractor. To do this, centerpunch the exact center of the remaining portion of the screw or bolt. Drill a small hole in the screw and tap the extractor into the hole. Back the screw out with a wrench on the extractor. See **Figure 62**.

Remedying Stripped Threads

Occasionally, threads are stripped through carelessness or impact damage. Often the threads can be cleaned up by running a tap (for internal threads on nuts) or die (for external threads on bolts) through the threads. See **Figure 63**. To clean or repair spark plug threads, a spark plug tap can be used.

> *NOTE*
> *Tap and dies can be purchased individually or in a set as shown in* **Figure 64**.

If an internal thread is damaged, it may be necessary to install a Helicoil (**Figure 65**) or some other type of thread insert. Follow the manufacturer's instructions when installing their insert.

If it is necessary to drill and tap a hole, refer to **Table 14** (metric) or **Table 15** (American) for tap drill sizes.

Removing Broken or Damaged Studs

If a stud is broken or the threads severely damaged, perform the following. A tube of Loctite 271 (red), 2 nuts, 2 wrenches and a new stud will be required during this procedure (**Figure 66**).

1. Thread two nuts onto the damaged stud. Then tighten the 2 nuts against each other so that they are locked (**Figure 67**).

NOTE
*If the threads on the damaged stud do
not allow installation of the 2 nuts, you
will have to remove the stud with a pair
of Vise-grip pliers.*

2. Turn the bottom nut counterclockwise and unscrew the stud.

3. Clean the threads with solvent or electrical contact cleaner and allow to thoroughly dry.

4. Install 2 nuts on the top half of the new stud (**Figure 68**) as in Step 1. Make sure they are locked securely.

5. Coat the bottom half of a new stud with Loctite 271 (red).

6. Turn the top nut clockwise and thread the new stud securely.

7. Remove the nuts and repeat for each stud as required.

8. Follow Loctite's directions on cure time before assembling the component.

BALL BEARING REPLACEMENT

Ball bearings (**Figure 69**) are used throughout the snowmobile engine and chassis to reduce power loss, heat and noise resulting from friction. Because ball bearings are precision made parts, they must be maintained by proper lubrication and maintenance. When a bearing is found to be damaged, it should be replaced

immediately. However, when installing a new bearing, care should be taken to prevent damage to the new bearing. While bearing replacement is described in the individual chapters where applicable, the following should be used as a guideline.

NOTE
*Unless otherwise specified, install
bearings with the manufacturer's mark
or number facing outward.*

Bearing Removal

While bearings are normally removed only when damaged, there may be times when it is

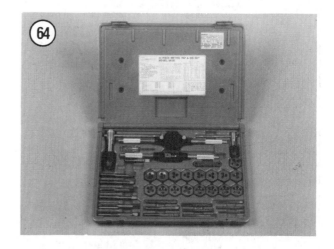

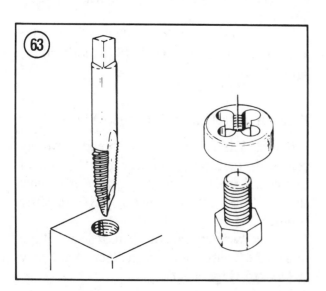

necessary to remove a bearing that is in good condition. However, improper bearing removal will damage the bearing and maybe the shaft or case half. Note the following when removing bearings.

1. When using a puller to remove a bearing on a shaft, care must be taken so that shaft damage does not occur. Always place a piece of metal

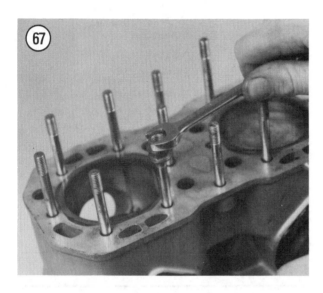

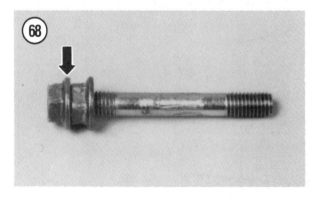

between the end of the shaft and the puller screw. In addition, place the puller arms next to the inner bearing race. See **Figure 70**.

2. When using a hammer to remove a bearing on a shaft, do not strike the hammer directly against the shaft. Instead, use a brass or aluminum rod between the hammer and shaft (**Figure 71**). In addition, make sure to support both bearing races with wood blocks as shown in **Figure 71**.

3. The most ideal method of bearing removal is with a hydraulic hand press. However, certain procedures must be followed or damage may occur to the bearing, shaft or case half. Note the following when using a press:

    a. Always support the inner and outer bearing races with a suitable size wood or aluminum ring (**Figure 72**). If only the outer race is supported, the balls and/or the inner race will be damaged.

    b. Always make sure the press ram (**Figure 72**) aligns with the center of the shaft. If the ram is not centered, it may damage the bearing and/or shaft.

    c. The moment the shaft is free of the bearing, it will drop to the floor. Secure or hold the shaft to prevent it from falling.

Bearing Installation

1. When installing a bearing into a housing, pressure must be applied to the *outer* bearing race (**Figure 73**). When installing a bearing onto a shaft, pressure must be applied to the *inner* bearing race (**Figure 74**).

2. When installing a bearing as described in Step 1, some type of driver will be required. Never strike the bearing directly with a hammer or the bearing will be damaged. When installing a bearing, a piece of pipe or a socket with an outer diameter that matches the bearing race will be required. **Figure 75** shows the correct way to use a socket and hammer when installing a bearing.

3. Step 1 describes how to install a bearing in a case half and over a shaft. However, when installing over a shaft and into a housing at the

same time, a snug fit will be required for both outer and inner bearing races. In this situation, a spacer must be installed underneath the driver tool so that pressure is applied evenly across *both* races. See **Figure 76**. If the outer race is not supported as shown in **Figure 76**, the balls will push against the outer bearing track and damage it.

Shrink Fit

1. *Installing a bearing over a shaft*: When a tight fit is required, the bearing inside diameter will be smaller than the shaft. In this case, driving the bearing on the shaft using normal methods may cause bearing damage. Instead, the bearing should be heated before installation. Note the following:

   a. Secure the shaft so that it can be ready for bearing installation.

   b. Clean the bearing surface on the shaft of all residue. Remove burrs with a file or sandpaper.

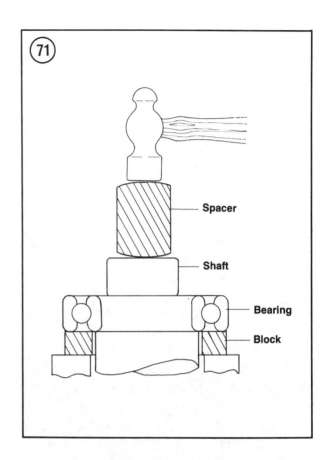

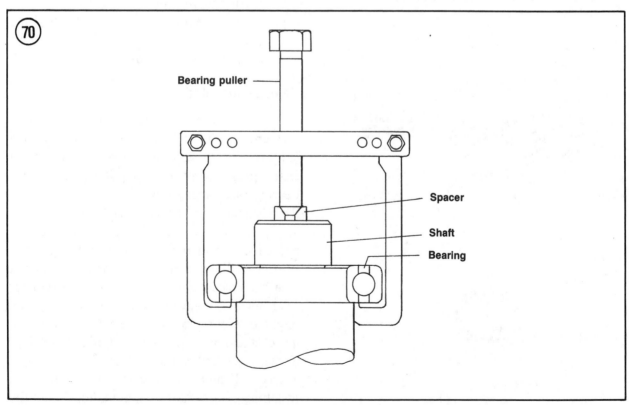

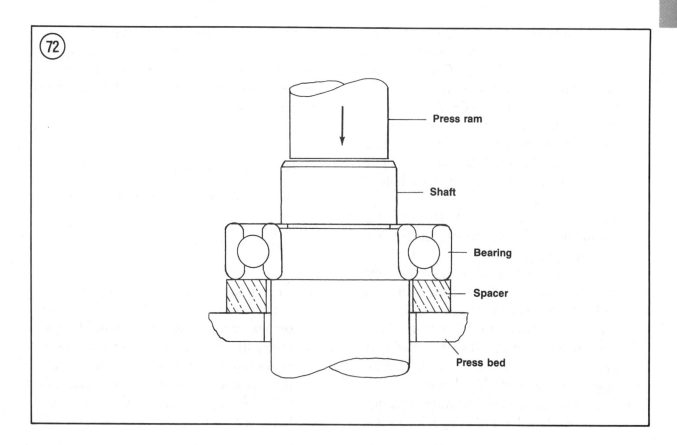

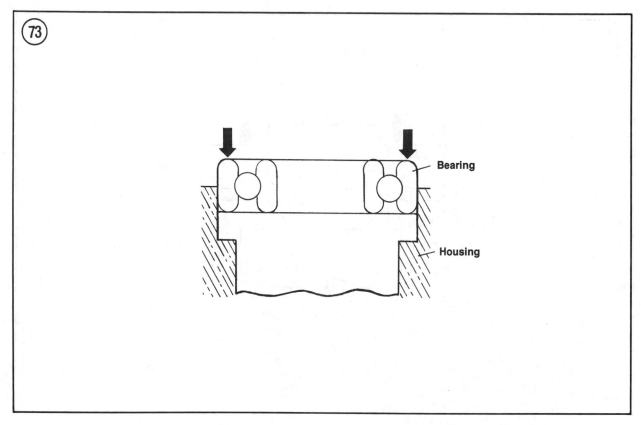

c. Fill a suitable pot or beaker with clean mineral oil. Place a thermometer (rated higher than 248° F [120° C]) in the oil. Support the thermometer so that it does not rest on the bottom or side of the pot.

d. Remove the bearing from its wrapper and secure it with a piece of heavy wire bent to hold it in the pot. Hang the bearing in the pot so that it does not touch the bottom or sides of the pot.

e. Turn the heat on and monitor the thermometer. When the oil temperature rises to approximately 248° F (120° C), remove the bearing from the pot and quickly install it. If necessary, place a socket on the inner bearing race and tap the bearing into place. As the bearing chills, it will tighten on the shaft so you must work quickly when installing it. Make sure the bearing is installed all the way.

2. *Installing a bearing in a housing*: Bearings are generally installed in a housing with a slight interference fit. Driving the bearing into the housing using normal methods may damage the housing or cause bearing damage. Instead, the housing should be heated before the bearing is installed. Note the following:

CAUTION
Before heating the crankcases in this procedure to remove the bearings, wash the cases thoroughly with detergent and water. Rinse and rewash the cases as required to remove all traces of oil and other chemical deposits.

a. The housing must be heated to a temperature of about 212° F (100° C) in an oven or on a hot plate. An easy way to check to see that it is at the proper temperature is to drop tiny drops of water on the case; if they sizzle and evaporate immediately, the temperature is correct. Heat only one housing at a time.

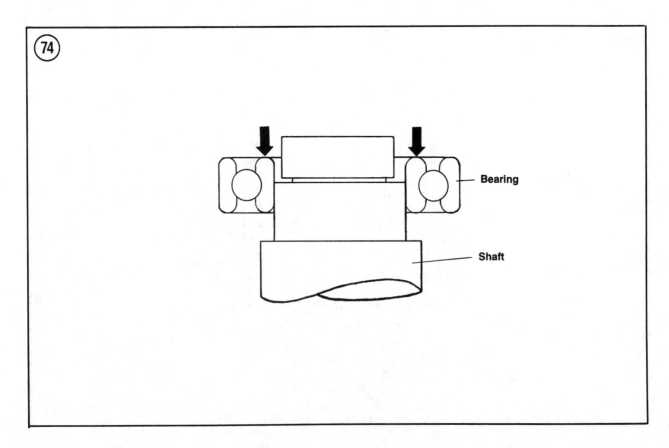

(74)

Bearing

Shaft

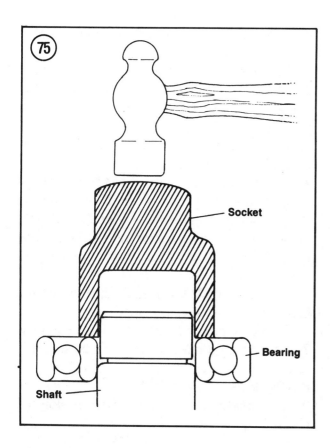

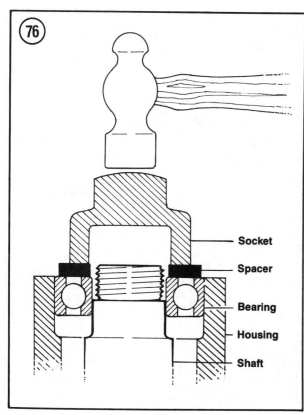

CAUTION
Do not heat the housing with a torch (propane or acetylene)—never bring a flame into contact with the bearing or housing. The direct heat will destroy the case hardening of the bearing and will likely warp the housing.

b. Remove the housing from the oven or hot plate and hold onto the housing with a kitchen pot holder, heavy gloves, or heavy shop cloths—*it is hot.*

NOTE
A suitable size socket and extension works well for removing and installing bearings.

c. Hold the housing with the bearing side down and tap the bearing out. Repeat for all bearings in the housing.

d. While heating up the housing halves, place the new bearings in a freezer if possible. Chilling them will slightly reduce their overall diameter while the hot housing assembly is slightly larger due to heat expansion. This will make installation much easier.

NOTE
Always install bearings with the manufacturer's mark or number facing outward.

e. While the housing is still hot, install the new bearing(s) into the housing. Install the bearings by hand, if possible. If necessary, lightly tap the bearing(s) into the housing with a socket placed on the outer bearing race. *Do not* install new bearings by driving on the inner bearing race. Install the bearing(s) until it seats completely.

OIL SEALS

Oil seals (**Figure 77**) are used to contain oil, water, grease or combustion gasses in a housing or shaft. Improper removal of a seal can damage the housing or shaft. Improper installation of the seal can damage the seal. Note the following:

a. Prying is generally the easiest and most effective method of removing a seal from a housing. However, always place a rag underneath the pry tool to prevent damage to the housing.

b. A low-temperature grease should be packed in the seal lips before the seal is installed.

c. Oil seals should always be installed so that the manufacturer's numbers or marks face out.

d. Oil seals should be installed with a socket placed on the outside of the seal as shown in **Figure 78**. Make sure the seal is driven squarely into the housing. Never install a seal by hitting against the top of the seal with a hammer.

SNOWMOBILE OPERATION

Snowmobiles are ideal machines for getting around during winter months. However, because snowmobiles are often operated in extreme weather conditions and over rough terrain, they should be checked before each ride and maintained on a periodic basics.

> *WARNING*
> *Never lean into a snowmobile's engine compartment while wearing a scarf or other loose clothing when the engine is running or when attempting to start the engine. If the scarf or clothing should catch in the drive belt or clutch, severe injury or death could result.*

Pre-start Inspection

A pre-start inspection should always be performed before heading out on your snowmobile. While the following list may look exhaustive, it can be performed rather quickly after a few times.

1. Familiarize yourself with your snowmobile.
2. Clean the windshield with a clean, damp cloth. Do not use gasoline, solvents or abrasive cleaners.
3. Check track tension (Chapter Three) and adjust if necessary.
4. Check the tether switch and the emergency stop switch for proper operation. If your machine is new or if you are using a friend's machine, practice using the tether or stop switch a few times so that its use will be automatic during an emergency.
5. Check the brake operation. Ensure that the brake system is correctly adjusted.
6. Check the fuel level and top it up if necessary.
7. Check the oil injection tank. Make sure it is full.
8. Check the coolant level.
9. *Air cooled models*: Check fan belt tension and adjust if necessary.

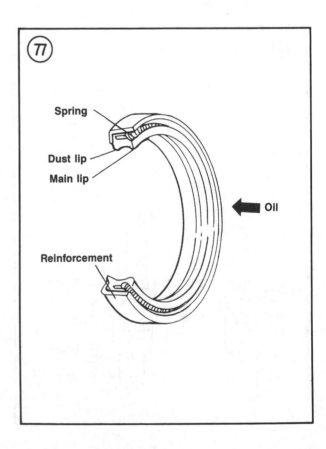

10. Operate the throttle lever. It should open and close smoothly.

11. Open the belt guard and visually inspect the drive belt. If the belt seems worn or damaged, replace it. Chapter Eleven lists drive belt wear limit specifications. Close the belt guard after inspecting the belt. Make sure the belt guard mounts are not loose or damaged.

12. While the engine shroud is open, visually inspect all hoses, fittings and parts for looseness or damage. Check the tightness of all bolts and nuts. Tighten as required.

13. Check the handlebar and steering components for looseness or damage. Do not ride the vehicle if any steering component is damaged. Tighten loose fasteners as required.

14. After closing the shroud, make sure the shroud latches are fastened securely.

15. Check the skis for proper alignment (Chapter Three). Check the ski pivot bolt for tightness or damage.

WARNING
When starting the engine, be sure that no bystanders are in front or behind the snowmobile. A sudden lurch of the machine could cause serious injury.

16. Make sure that all lights are working.

NOTE
If abnormal noises are detected after starting the engine, locate and repair the problem before starting out.

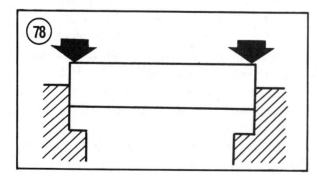

NOTE
Refer to the appropriate chapter for tightening torques and service procedures.

Tools and Spare Parts

Before leaving on a trip, make sure that you carry tools and spare parts in case of emergency. A tool kit should include the following:
   a. Flashlight.
   b. Rope.
   c. Tools.
   d. Tape.
A spare parts kit should include the following:
   a. Drive belt.
   b. Emergency starting strap.
   c. Light bulbs.
   d. Spark plugs.
   e. Main jets.
   f. Throttle cable.
   g. Brake cable.
   h. A good book…just in case.
If you are going out on a long trip, extra oil and fuel should be carried.

Emergency Starting

If your recoil starter rope should break, the engine can be started with the emergency starting strap stored in your snowmobile's tool kit.

1. Open the shroud.

WARNING
*The drive belt guard must be removed when starting the engine with the emergency starting strap. **Never** lean into the snowmobile's engine compartment while wearing a scarf or other loose clothing while the engine is running or when attempting to start the engine. If the scarf or clothing should catch in the drive belt or clutch, severe injury or death could result.*

2. Remove the belt guard pin and rotate the belt guard forward.

3. Remove the emergency starting strap from your tool kit.

4. Set all switches to ON.

WARNING
The emergency starting strap must be used as described in Step 5 only. Do not wrap the strap around the clutch tower or personal injury may occur when attempting to start the engine.

5. Engage the end of the emergency starting strap with one of the clutch towers on the primary sheave so that the end of the strap is on the outside of the clutch tower. Then wind the strap counterclockwise around the primary sheave.

6. Pull the strap upward and start the engine.

WARNING
Do not casually hold the emergency starting strap over the clutch assembly after it disengages from the primary sheave. If the end of the strap should fall into and engage with the rotating clutch or drive belt, personal injury to yourself or damage to the clutch or drive belt assembly may result.

7. Reinstall the drive belt guard after starting the engine.

8. Close and secure the shroud.

9. Store the emergency starting strap in your tool kit.

10. Repair the recoil starter assembly as soon as possible.

Clearing the Track

If the snowmobile has been operated in deep or slushy snow, it is necessary to clear the track after stopping to prevent the track from freezing. This condition would make starting and running difficult.

WARNING
Make sure no one is behind the machine when clearing the track. Ice and rocks thrown from the track can cause injury.

Tip the snowmobile on its side until the track clears the ground *completely*. Run the track at a moderate speed until all the ice and snow is thrown clear.

CAUTION
If the track does freeze, it must be broken loose manually with the engine turned OFF. Attempting to force a frozen track with the engine will burn and damage the drive belt.

SNOWMOBILE SAFETY

Proper Clothing

Warm and comfortable clothing are a must to provide protection from frostbite. Even mild temperatures can be very uncomfortable and dangerous when combined with a strong wind or when traveling at high speeds. See **Table 16** for wind chill factors. Always dress according to what the wind chill factor is, not the temperature. Check with an authorized dealer for suggested types of snowmobile clothing.

WARNING
To provide additional warmth as well as protection against head injury, always wear an approved helmet when snowmobiling.

Emergency Survival Techniques

1. Do not panic in the event of an emergency. Relax, think the situation over, then decide on a course of actions. You may be within a short distance of help. If possible, repair your snowmobile so you can drive to safety. Conserve your energy and stay warm.

2. Keep hands and feet active to promote circulation and avoid frostbite while servicing your machine.

3. Mentally retrace your route. Where was the last point where help could be located? Do not

attempt to walk long distances in deep snow. Make yourself comfortable until help arrives.

4. If you are properly equipped for your trip you can turn any undesirable area into a suitable campsite.

5. If necessary, build a small shelter with tree branches or evergreen boughs. Look for a sheltered area against a hill or cliff. Even burrowing in the snow offers protection from the cold and wind.

6. Prepare a signal fire using evergreen boughs and snowmobile oil. If you cannot build a fire, make an S-O-S in the snow.

7. Use a policeman's whistle or beat cooking utensils to attract attention.

8. When your camp is established, climb the nearest hill and determine your whereabouts. Observe landmarks on the way, so you can find your way back to your campsite. Do not rely on your footprints. They may be covered by blowing snow.

SNOWMOBILE CODE OF ETHICS

1. I will be a good sportsman and conservationist. I recognize that people judge all snowmobilers by my actions. I will use my influence with other snowmobile owners and operators to promote sportsmanlike conduct.

2. I will not litter any trails or areas, nor will I pollute streams or lakes. I will carry out what I carry in.

3. I will not damage living trees, shrubs or other natural features.

4. I will respect other people's properties and rights.

5. I will lend a helping hand when I see someone in need.

6. I will make myself and my vehicle available to assist in search and rescue operations.

7. I will not interfere with the activities of other winter sportsmen. I will respect their right to enjoy their recreational activity.

8. I will know and obey all federal, state or provincial and local rules regulating the operation of snowmobiles in areas where I use my vehicle.

9. I will not harass wildlife.

10. I will not snowmobile where prohibited.

Tables are on the following pages.

Table 1 MODEL LISTING

1984
Indy Trail
Indy 400
Indy 600
1985
Indy Trail
Indy 400
Indy 600
1986
Indy Trail
Indy 400
Indy 600
1987
Indy Sport
Indy Trail
Indy 400
Indy 600
1988
Indy Sport
Indy Trail
Indy 400
Indy 650
1989
Indy Sport
Indy Trail
Indy 400
Indy 500
Indy 650

Table 2 GENERAL SPECIFICATIONS—TRAIL

Item	Specification
Weight (dry)	
1984-1986	408 lbs. (185 kg)
1987	
Trail	420 lbs. (190 kg)
Trail ES	431 lbs. (195 kg)
1988-1989	
Trail & SP	420 lbs. (190 kg)
Trail ES	431 lbs. (195 kg)
Trail Deluxe & SKS	435 lbs. (197 kg)
Height	
1984	31.5 in. (80 cm)
1985-on	42 in. (106.7 cm)
Width	
1984-1987	*
1988-1989	
SP	43.12 in. (109.5 cm)
All other models	41.62 in. (105.7 cm)

(continued)

Table 2 GENERAL SPECIFICATIONS—TRAIL (continued)

Item	Specifications
Length	
1984-1986	106.25 in. (269.9 cm)
1987	109 in. (276.9 cm)
1988-1989	
SKS	113 in. (287 cm)
All other models	106.25 in. (269.9 cm)
Ski stance	
1984-1987	36.50 in. (92.7 cm)
1988-1989	
SP	38 in. (96.5 cm)
All other models	36.50 in. (92.7 cm)
Track width	15 in. (38.1 cm)
Track length	
1984-1987	120.96 in. (307.2 cm)
1988-1989	
SKS	133.56 in. (339.2 cm)
All other models	120.96 in. (307.2 cm)
* Not specified	

Table 3 GENERAL SPECIFICATIONS—SPORT

Item	Specification
Weight (dry)	
1987	408 lbs. (185 kg)
1988	390 lbs. (177 kg)
1989	
Sport	390 lbs. (177 kg)
GT	419 lbs. (190 kg)
Height	
1987-1988	42 in. (106.7 cm)
1989	
Sport	42 in. (106.7 cm)
GT	45 in. (114.3 cm)
Width	41.62 in. (105.7 cm)
Length	
1987	109 in. (276.9 cm)
1988	106.25 in. (269.9 cm)
1989	
Sport	106.25 in. (269.9 cm)
GT	118 in. (299.7 cm)
Ski stance	36.5 in. (92.7 cm)
Track width	15 in. (38.1 cm)
Track length	
1987-1988	120.96 in. (307.2 cm)
1989	
Sport	120.96 in. (307.2 cm)
GT	141 in. (358.1 cm)

Table 4 GENERAL SPECIFICATIONS—400

Item	Specification
Weight (dry)	
1984-1985	421 lbs. (191 kg)
1986-1987	440 lbs. (199 kg)
1988	
400	440 lbs. (199 kg)
Classic	451 lbs. (204 kg)
SKS	455 lbs. (206 kg)
1989	440 lbs. (199 kg)
Height	
1984	40 in. (101.6 cm)
1985-1986	31.5 in. (80 cm)
1987	42 in. (106.7 cm)
1988-1989	44 in. (111.8 cm)
Width	41.62 in. (105.6 cm)
Length	
1984-1986	106.25 in. (269.8 cm)
1987	109 in. (276.9 cm)
1988	
SKS	113 in. (287 cm)
All other models	106.25 in. (269.8 cm)
1989	106.25 in. (269.8 cm)
Ski stance	36.5 in. (92.7 cm)
Track width	15 in. (38.1 cm)
Track length	
1984-1987	120.96 in. (397.2 cm)
1988	
SKS	133.56 in. (339.2 cm)
All other models	120.96 in. (397.2 cm)

Table 5 GENERAL SPECIFICATIONS—500

Item	Specification
Weight (dry)	
500	442 lbs. (200 kg)
Classic	470 lbs. (213 kg)
SKS	447 lbs. (203 kg)
Height	
SP	43.12 in. (109.5 cm)
All other models	41.62 in. (105.7 cm)
Width	
SP	38 in. (96.5 cm)
All other models	36.5 in. (92.7 cm)
Length	
SKS	113 in. (287 cm)
All other models	106.25 in. (269.9 cm)
Ski stance	
SP	38 in. (96.5 cm)
All other models	36.5 in. (92.7 cm)
Track width	15 in. (38.1 cm)
Track length	
SKS	133.56 in. (339.2 cm)
All other models	120.96 in. (307.2 cm)

Table 6 GENERAL SPECIFICATIONS—600

Item	Specification
Weight (dry)	
1984-1985	463 lbs. (210 kg)
1986-1987	483 lbs. (219 kg)
Height	
1984-1986	31.5 in. (80 cm)
1987	42 in. (106.7 cm)
Width	41.62 in. (105.7 cm)
Length	
1984-1986	106.25 in. (270 cm)
1987	109 in. (276.9 cm)
Ski stance	36.50 in. (92.7 cm)
Track width	15 in. (38.1 cm)
Track length	120.96 in. (307.2 cm)

Table 7 GENERAL SPECIFICATIONS—650

Item	Specification
Weight (dry)	
650	483 lbs. (219 kg)
SKS	498 lbs. (226 kg)
Height	
1988	42 in. (106.7 cm)
1989	44 in. (111.8 cm)
Width	41.62 in. (105.7 cm)
Length	
650	106.25 in. (270 cm)
SKS	113 in. (287 cm)
Ski stance	36.5 in. (92.7 cm)
Track width	15 in. (38.1 cm)
Track length	
650	120.96 in. (307.2 cm)
SKS	133.56 in. (339.2 cm)

Table 8 U.S. STANDARDS AND METRIC EQUIVALENTS

Fractions	Decimal in.	Metric mm	Fractions	Decimal in.	Metric mm
1/64	0.015625	0.39688	3/16	0.1875	4.76250
1/32	0.03125	0.79375	13/64	0.203125	5.15937
3/64	0.046875	1.19062	7/32	0.21875	5.55625
1/16	0.0625	1.58750	15/64	0.234375	5.95312
5/64	0.078125	1.98437	1/4	0.250	6.35000
3/32	0.09375	2.38125	17/64	0.265625	6.74687
7/64	0.109375	2.77812	9/32	0.28125	7.14375
1/8	0.125	3.1750	19/64	0.296875	7.54062
9/64	0.140625	3.57187	5/16	0.3125	7.93750
5/32	0.15625	3.57187	21/64	0.328125	8.33437
11/64	0.171875	4.36562	11/32	0.34375	8.73125

(continued)

Table 8 U.S. STANDARDS AND METRIC EQUIVALENTS (continued)

Fractions	Decimal in.	Metric mm	Fractions	Decimal in.	Metric mm
23/64	0.359375	9.121812	11/16	0.6875	17.46250
3/8	0.375	9.52500	45/64	0.703125	17.85937
25/64	0.390625	9.92187	23/32	0.71875	18.25625
13/32	0.40625	10.31875	47/64	0.734375	18.65312
27/64	0.421875	10.71562	3/4	0.750	19.05000
7/16	0.4375	11.11250	49/64	0.765625	19.44687
29/64	0.453125	11.50937	25/32	0.78125	19.84375
15/32	0.46875	11.90625	51/64	0.796875	20.24062
31/64	0.484375	12.30312	13/16	0.8125	20.63750
1/2	0.500	12.70000	53/64	0.828125	21.03437
33/64	0.515625	13.09687	27/32	0.84375	21.82812
17/32	0.53125	13.49375	55/64	0.859375	21.82812
35/64	0.546875	13.89062	7/8	0.875	22.22500
9/16	0.5625	14.28750	57/64	0.890625	22.62187
37/64	0.578125	14.68437	29/32	0.90625	23.01875
19/32	0.59375	15.08125	59/64	0.921875	23.41562
39/64	0.609375	15.47812	15/16	0.9375	23.81250
5/8	0.625	15.87500	61/64	0.953125	24.20937
41/64	0.640625	16.27187	31/32	0.96875	24.60625
21/32	0.65625	16.66875	63/64	0.984375	25.00312
43/64	0.671875	17.06562	1	1.00	25.40000

Table 9 GENERAL TORQUE SPECIFICATIONS (METRIC)

Item	N·m	ft.-lb.
Bolt		
6 mm	6	4.3
8 mm	15	11
10 mm	30	22
12 mm	55	40
14 mm	85	61
16 mm	130	94
Nut		
6 mm	6	4.3
8 mm	15	11
10 mm	30	22
12 mm	55	40
14 mm	85	61
16 mm	130	94

Table 10 GENERAL TORQUE SPECIFICATIONS (SAE-GRADE 2)

Size	Dry ft.-lb.	Lubed ft.-lb.
1/4-20	66 in.-lb.	50 in.-lb.
1/4-28	76 in.-lb.	56 in.-lb.
5/16-18	11	8

(continued)

Table 10 GENERAL TORQUE SPECIFICATIONS (SAE-GRADE 2) (continued)

Size	Dry ft.-lbs.	Lubed ft.-lbs.
5/16-24	12	9
3/8-16	20	15
3/8-24	23	17
7/16-14	32	24
7/16-20	36	27
1/2-13	50	35
1/2-20	55	40

Table 11 GENERAL TORQUE SPECIFICATIONS (SAE-GRADE 5)

Size	Dry ft.-lb.	Lubed ft.-lb.
1/4-20	8	75 in.-lb.
1/4-28	10	86 in.-lb.
5/16-18	17	13
5/16-24	19	14
3/8-16	30	23
3/8-24	35	25
7/16-14	50	35
7/16-20	55	40
1/2-13	75	55
1/2-20	90	65

Table 12 GENERAL TORQUE SPECIFICATIONS (SAE-GRADE 8)

Size	Dry ft.-lb.	Lubed ft.-lb.
1/4-20	10	9
1/4-28	14	12
5/16-18	25	18
5/16-24	28	20
3/8-16	45	33
3/8-24	50	35
7/16-14	70	55
7/16-20	80	60
1/2-13	110	80
1/2-20	120	90

Table 13 TECHNICAL ABBREVIATIONS

ABDC	After bottom dead center
ATDC	After top dead center
BBDC	Before bottom dead center
BDC	Bottom dead center
BTDC	Before top dead center
C	Celsius (Centigrade)
cc	Cubic centimeters
CDI	Capacitor discharge ignition
cu. in.	Cubic inches
F	Fahrenheit
ft.-lb.	Foot-pounds
g	Gram
gal.	Gallons
hp	Horsepower
in.	Inches
kg	Kilogram
kg/cm^2	Kilograms per square centimeter
kgm	Kilogram meters
km	Kilometer
l	Liter
lbs.	Pounds
m	Meter
MAG	Magneto
mm	Millimeter
N·m	Newton-meters
oz.	Ounce
psi	Pounds per square inch
PTO	Power take off
pts.	Pints
qt.	Quarts
rpm	Revolutions per minute

Table 14 METRIC TAP DRILL SIZES

Metric (mm)	Drill size	Decimal equivalent	Nearest fraction
3×0.50	No. 39	0.0995	3/32
3×0.60	3/32	0.0937	3/32
4×0.70	No. 30	0.1285	1/8
4×0.75	1/8	0.125	1/8
5×0.80	No. 19	0.166	11/64
5×0.90	No. 20	0.161	5/32
6×1.00	No. 9	0.196	13/64
7×1.00	16/64	0.234	15/64
8×1.00	J	0.277	9/32
8×1.25	17/64	0.265	17/64
9×1.00	5/16	0.3125	5/16
9×1.25	5/16	0.3125	5/16
10×1.25	11/32	0.3437	11/32
10×1.50	R	0.339	11/32
11×1.50	3/8	0.375	3/8
12×1.50	13/32	0.406	13/32
12×1.75	13/32	0.406	13/32

Table 15 AMERICAN TAP DRILL SIZES

Tap thread	Drill size	Tap thread	Drill size
#0-80	3/64	1/4-28	No. 3
#1-64	No. 53	5/16-18	F
#1-72	No. 53	5/16-24	I
#2-56	No. 51	3/8-16	5/16
#2-64	No. 50	3/8-24	Q
#3-48	5/64	7/16-14	U
#3-56	No. 46	7/16-20	W
#4-40	No. 43	1/2-13	27/64
#4-48	No. 42	1/2-20	29/64
#5-40	No. 39	9/16-12	31/64
#5-44	No. 37	9/16-18	33/64
#6-32	No. 36	5/8-11	17/32
#6-40	No. 33	5/18-18	37/64
#8-32	No. 29	3/4-10	21/32
#8-36	No. 29	3/4-16	11/16
#10-24	No. 25	7/8-9	49/64
#10-32	No. 21	7/8-14	13/16
#12-24	No. 17	1-8	7/8
#12-28	No. 15	1-14	15/16
1/4-20	No. 8		

Table 16 WIND CHILL FACTOR

Estimated Wind Speed in MPH	Actual Thermometer Reading (°F)											
	50	40	30	20	10	0	−10	−20	−30	−40	−50	−60
	Equivalent Temperature (°F)											
Calm	50	40	30	20	10	0	−10	−20	−30	−40	−50	−60
5	48	37	27	16	6	−5	−15	−26	−36	−47	−57	−68
10	40	28	16	4	−9	−21	−33	−46	−58	−70	−83	−95
15	36	22	9	−5	−18	−36	−45	−58	−72	−85	−99	−112
20	32	18	4	−10	−25	−39	−53	−67	−82	−96	−110	−124
25	30	16	0	−15	−29	−44	−59	−74	−88	−104	−118	−133
30	28	13	−2	−18	−33	−48	−63	−79	−94	−109	−125	−140
35	27	11	−4	−20	−35	−49	−67	−82	−98	−113	−129	−145
40	26	10	−6	−21	−37	−53	−69	−85	−100	−116	−132	−148

*

Little Danger (for properly clothed person) | **Increasing Danger** | **Great Danger**

*Danger from freezing of exposed flesh.

*Wind speeds greater than 40 mph have little additional effect.

Chapter Two

Troubleshooting

Diagnosing mechanical problems is relatively simple if you use orderly procedures and keep a few basic principles in mind. The first step in any troubleshooting procedure is to define the symptoms as closely as possible and then localize the problem. Subsequent steps involve testing and analyzing those areas which could cause the symptoms. A haphazard approach may eventually solve the problem, but it can be very costly in terms of wasted time and unnecessary parts replacement.

Proper lubrication, maintenance and periodic tune-ups as described in Chapter Three will reduce the necessity for troubleshooting. Even with the best of care, however, all snowmobiles are prone to problems which will require troubleshooting.

Never assume anything. Do not overlook the obvious. If the engine won't start, check the position of the kill switch. Is the engine flooded with fuel from using the choke too much.

If the engine suddenly quits, check the easiest, most accessible problem first. Is there gasoline in the tank? Has a spark plug wire broken or fallen off?

If nothing obvious turns up in a quick check, look a little further. Learning to recognize and describe symptoms will make repairs easier for you or a mechanic at the shop. Describe problems accurately and fully.

Gather as many symptoms as possible to aid in diagnosis. Note whether the engine lost power gradually or all at once, what color smoke came from the exhaust and so on. Remember that the more complicated a machine is, the easier it is to troubleshoot because symptoms point to specific problems.

After the symptoms are defined, areas which could cause problems are tested and analyzed. Guessing at the cause of a problem may provide the solution, but it can easily lead to frustration, wasted time and a series of expensive, unnecessary parts replacements.

You do not need fancy equipment or complicated test gear to determine whether repairs can be attempted at home. A few simple checks could save a large repair bill and lost time while your snowmobile sits in a dealer's service department. On the other hand, be realistic and do not attempt repairs beyond your abilities.

Service departments tend to charge heavily for putting together a disassembled engine that may have been abused. Some won't even take on such a job—so use common sense, don't get in over your head.

Electrical specifications are listed in **Tables 1-3** at the end of this chapter.

OPERATING REQUIREMENTS

An engine needs 3 basics to run properly: correct fuel/air mixture, compression and a spark at the right time (**Figure 1**). If one basic requirement is missing, the engine will not run.

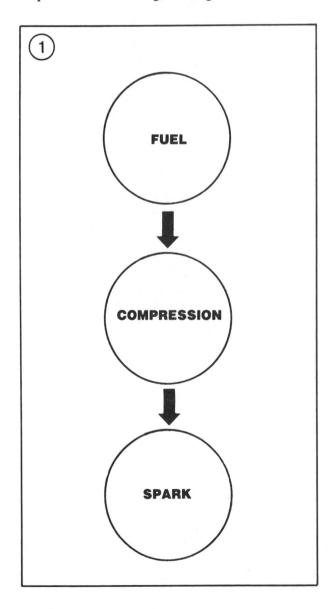

Two-stroke engine operating principles are described in Chapter One under *Engine Operation*. The ignition system is the weakest link of the 3 basics. More problems result from ignition breakdowns than from any other source. Keep that in mind before you begin tampering with carburetor adjustments and the like.

If the snowmobile has been sitting for any length of time and refuses to start, check and clean the spark plugs. Then check the condition of the battery (if so equipped) to make sure it has an adequate charge. If these are okay, then look to the gasoline delivery system. This includes the tank, fuel shutoff valve (**Figure 2**), pump and fuel line to the carburetor. Gasoline deposits may have gummed up carburetor jets and air passages. Gasoline tends to lose its potency after standing for long periods. Condensation may contaminate it with water. Drain the old gas and try starting with a fresh tankful.

TROUBLESHOOTING INSTRUMENTS

Chapter One lists the instruments needed and detailed instruction on their use.

TESTING ELECTRICAL COMPONENTS

Most dealers and parts houses will not accept returns on electrical parts purchased through their business. When testing electrical components, make sure that you perform the test procedures as described in this chapter and that your test equipment is working properly. If a test result shows that the component is defective but the reading is close to the service limit, have the component tested by a Polaris dealer to verify the test result before purchasing a new component.

ENGINE ELECTRICAL SYSTEM TROUBLESHOOTING

All models are equipped with a capacitor discharge ignition system. This section describes complete ignition system troubleshooting. See **Figure 3**, **Figure 4** or **Figure 5**.

This solid state system uses no contact breaker points or other moving parts. Because of the solid state design, problems with the capacitor discharge system are relatively few. However, when problems arise they stem from one of the following:

a. Weak spark.

b. No spark.

It is possible to check CDI systems that:

a. Do not spark.

b. Have broken or damaged wires.

c. Have a weak spark.

It is difficult to check CDI systems that malfunction due to:

a. Vibration problems.

b. Components that malfunction only when the engine is hot or under a load.

General troubleshooting procedures are provided in **Figure 6**.

Test Equipment

The Polaris Multitester (part No. 2870335) should be used for accurate testing of the ignition system. However, if you do not have access to the Polaris tester, a good-quality multitester can be substituted, though some of the readings may

**MAGNETO ASSEMBLY
(1984-1989 TRAIL; 1984-1989 400;
1989 500)**

1. Stator plate
2. Exciter coil
3. Pulser coil
4. Collar
5. Screw
6. Lighting coil
7. Clamp
8. Screw
9. Clamp
10. Screw
11. Grommet
12. Connector
13. Coupler
14. Flywheel
15. Fan
16. Bolt
17. Screw
18. Lockwasher
19. Washer
20. CDI control unit
21. Spark plug cap
22. Rubber nut
23. Screw

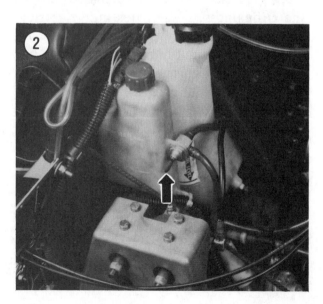

2

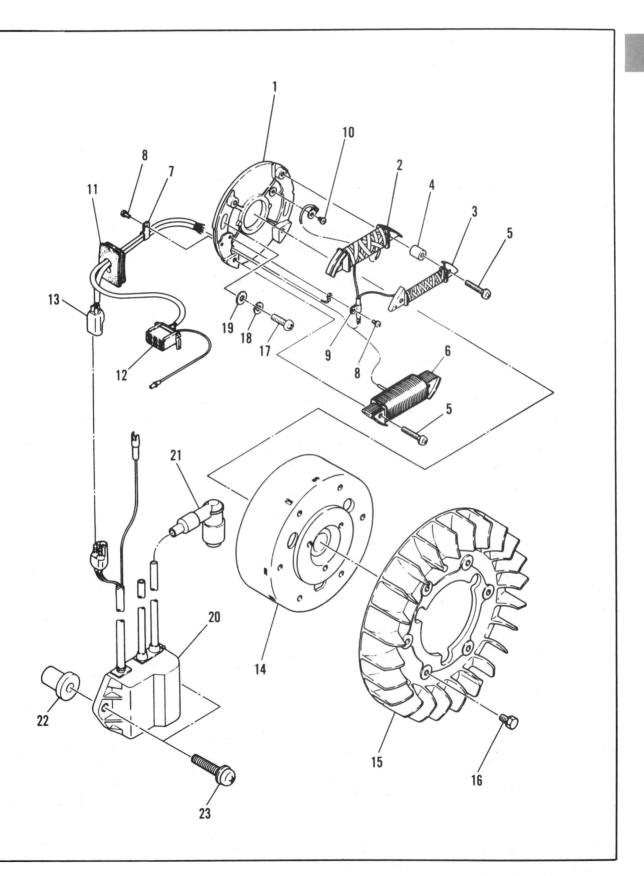

④

**MAGNETO ASSEMBLY
(1987-1989 SPORT)**

1. Flywheel
2. Fan
3. Bolt
4. Coil assembly
5. Exciter coil
6. Lighting coil
7. Grommet
8. Wire harness
9. Coupler
10. Coupler
11. Screw assembly
12. Screw assembly
13. Screw assembly
14. CDI control unit
15. Spark plug cap
16. Rubber nut
17. Screw assembly
18. Clamp

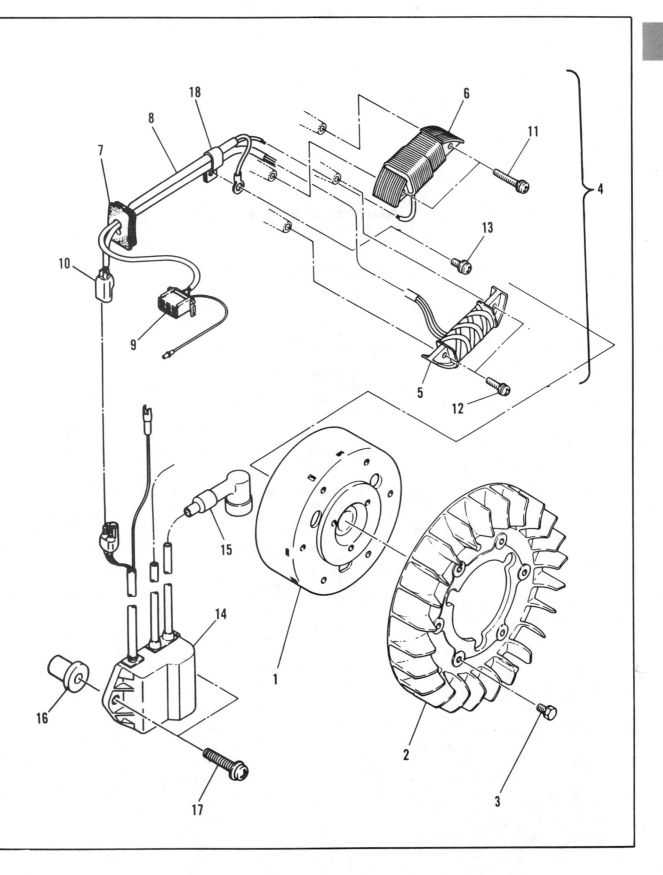

⑤

**MAGNETO ASSEMBLY
(1984-1989 600/650)**

1. Stator plate
2. Exciter coil
3. Pulser coil
4. Collar
5. Screw
6. Control coil
7. Lighting coil
8. Screw
9. Lighting coil
10. Collar
11. Screw
12. Clamp
13. Screw
14. Clamp
15. Grommet
16. Coupler
17. Flywheel
18. Screw assembly
19. Coupler
20. Spacers
21. CDI
22. Coupler
23. Ignition coils
24. Coupler
25. Screw assembly
26. Spark plug cap

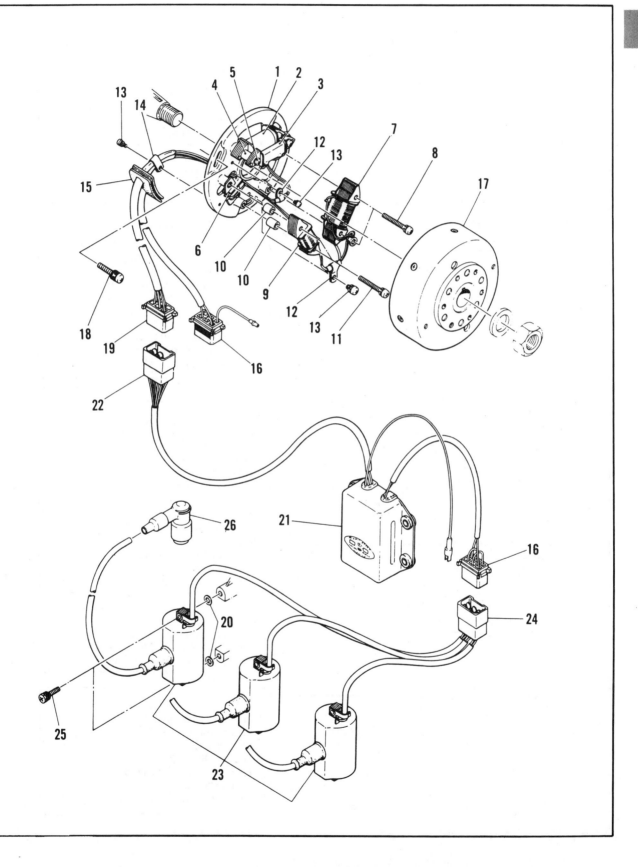

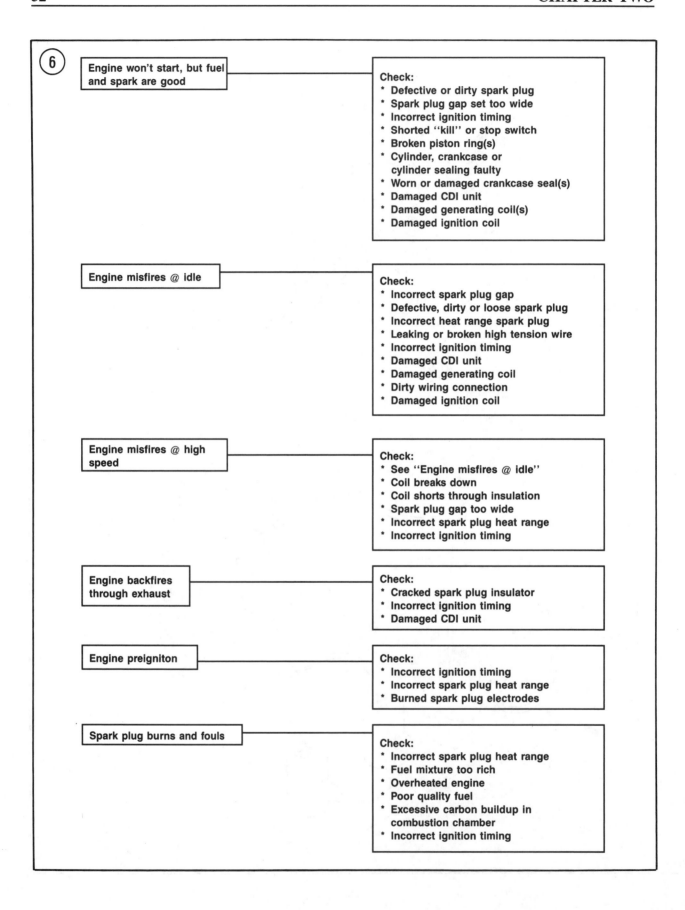

⑥ **Engine won't start, but fuel and spark are good**

Check:
* Defective or dirty spark plug
* Spark plug gap set too wide
* Incorrect ignition timing
* Shorted "kill" or stop switch
* Broken piston ring(s)
* Cylinder, crankcase or cylinder sealing faulty
* Worn or damaged crankcase seal(s)
* Damaged CDI unit
* Damaged generating coil(s)
* Damaged ignition coil

Engine misfires @ idle

Check:
* Incorrect spark plug gap
* Defective, dirty or loose spark plug
* Incorrect heat range spark plug
* Leaking or broken high tension wire
* Incorrect ignition timing
* Damaged CDI unit
* Damaged generating coil
* Dirty wiring connection
* Damaged ignition coil

Engine misfires @ high speed

Check:
* See "Engine misfires @ idle"
* Coil breaks down
* Coil shorts through insulation
* Spark plug gap too wide
* Incorrect spark plug heat range
* Incorrect ignition timing

Engine backfires through exhaust

Check:
* Cracked spark plug insulator
* Incorrect ignition timing
* Damaged CDI unit

Engine preigniton

Check:
* Incorrect ignition timing
* Incorrect spark plug heat range
* Burned spark plug electrodes

Spark plug burns and fouls

Check:
* Incorrect spark plug heat range
* Fuel mixture too rich
* Overheated engine
* Poor quality fuel
* Excessive carbon buildup in combustion chamber
* Incorrect ignition timing

differ slightly. Polaris allows a ±10% variance among all of the readings.

If you do not have access to a multitester, you can use visual inspection and an ohmmeter to pinpoint electrical problems caused by dirty or damaged connectors, faulty or damaged wiring or electrical components that may have cracked or broken. If basic checks fail to locate the problem, take your snowmobile to a Polaris dealer and have them troubleshoot the electrical system. Ohmmeter readings should be made when the engine is cold. Readings taken on a hot engine will show increased resistance caused by engine heat and result in unnecessary parts replacement without solving the basic problem. When switching between ohmmeter scales, always cross the test leads and zero the needle to assure a correct reading. Refer to your meter's instruction booklet or guide for additional information.

Precautions

Certain measures must be taken to protect the capacitor discharge system. Instantaneous damage to the semiconductors in the system will occur if the following is not observed.
1. Do not crank the engine if the CDI unit is not grounded to the engine.

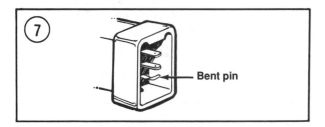
Bent pin

Loose connector

2. Do not touch or disconnect any ignition components when the engine is running or while the battery cables are connected (if so equipped).
3. Keep all connections between the various units clean and tight. Be sure that the wiring connectors are pushed together firmly.

Troubleshooting Preparation

Refer to the wiring diagram for your model at the end of this book when performing the following.

NOTE
To test the wiring harness for poor connections in Step 1, bend the molded rubber connector while checking each wire for resistance.

1. Check the wiring harness for visible signs of damage.
2. Check all of the connectors as follows:
   a. Disconnect each electrical connector in the ignition circuit. Check for bent or damaged pins in each male connector (**Figure 7**). A bent pin will not connect to its mating receptacle in the female end of the connector. This will cause an open circuit.
   b. Check each female connector end. Make sure the metal connector on the end of each wire (**Figure 8**) is pushed all the way into the plastic connector. If not, carefully push them in with a narrow-blade screwdriver. Make sure you do not pinch or cut the wire.
   c. Check all electrical wires where they enter the individual metal connector in both the male and female plastic connectors.
   d. Make sure all electrical connectors within the connector are clean and free of corrosion. If necessary, clean connectors with a spray electrical contact cleaner.
   e. After all is checked out, push the connectors together until they "click" and make sure they are fully engaged and locked together (**Figure 9**).

f. Never pull on the electrical wires when disconnecting an electrical connector—pull only on the connector plastic housing. See **Figure 10**.

3. Check all electrical components that are grounded to the engine for a good ground.

4. Check all wiring for disconnected wires or short or open circuits.

5. Make sure there is an adequate supply of fuel available to the engine. Make sure the oil tank is properly filled.

6. Check spark plug cable routing. Make sure the cables are properly connected to their respective spark plugs.

Spark Test

CAUTION
Before removing spark plugs, blow away any dirt that has accumulated next to the spark plug base. The dirt could fall into the cylinder when the plug is removed, causing serious engine damage.

1. Remove the spark plugs, keeping them in order. Check the condition of each plug. See Chapter Three. Replace fouled or damaged spark plugs as required.

WARNING
During this test do not hold the spark plug, wire or connector with fingers or a serious electrical shock may result. If necessary, use a pair of insulated pliers to hold the spark plug wire.

2. Open the shroud.

3. Connect the spark plug wire and connector to all of the spark plugs and touch the base of each spark plug to a good ground like the engine cylinder head. Position the spark plugs so you can see the electrodes.

4. Turn the ignition switch ON and pull the kill switch button up.

5. Crank the engine over with the pull starter. A fat blue spark should be evident across the spark plug electrodes.

6. If there is no spark, check again that all of the switches are ON. Then squeeze thumb throttle to half throttle (to tension throttle cable) and crank engine over again. If you now have spark, adjust the throttle cable to remove all slack from cable as described in Chapter Three.

NOTE
All models are equipped with a throttle safety switch; excessive throttle cable free play will deactivate the switch. Refer to Chapter Eight for additional information.

7. If there is still no spark, disconnect the throttle safety switch connector and check for spark. If a spark is now evident, check the throttle safety switch and kill switch as described in Chapter Eight. Replace the safety switch assembly if necessary.

8. If the throttle safety switch assembly is okay, test the ignition switch as described in Chapter Eight.

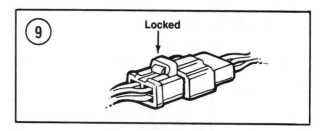

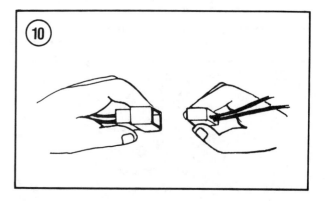

9. If you still do not have a spark, perform the following test procedures.

Ignition Component Resistance Test

An accurate ohmmeter will be required to perform the following tests. Refer to *Test Equipment* in this chapter. When switching between ohmmeter scales, always cross the test leads and zero the needle to assure a correct reading.

Twin-cylinder models

The following procedures describes resistance test procedures for the exciter and pulser coils. The exciter and pulser coils are mounted on the stator plate behind the flywheel. See **Figure 11**.
1. Open the shroud.
2. Disconnect the stator plate electrical connectors. See **Figure 3** or **Figure 4**.
3. Check the exciter coil resistance as follows:
   a. Switch an ohmmeter to the R × 1 scale.
   b. Measure resistance between the exciter coil wires indicated in **Table 1**. Compare actual readings to those given in **Table 1**.
   c. Disconnect the meter leads.
4. Check pulser coil resistance as follows:
   a. Switch an ohmmeter to the R × 1 scale.

b. Measure resistance between the pulser coil wires indicated in **Table 2**. Compare actual readings to those given in **Table 2**.
   c. Disconnect the meter leads.
5. If the exciter or pulser coil(s) did not provide the readings as specified in **Table 1**, replace the coil(s) as described in Chapter Eight.
6. Reconnect the stator plate electrical connectors.
7. Close and secure the shroud.

Three-cylinder models

The exciter, pulser and control coils are mounted on the stator plate behind the flywheel. See **Figure 12**.
1. Open the shroud.
2. Disconnect the stator plate, CDI box and ignition coil electrical connectors. See **Figure 5**.
3. Check the exciter coil resistance as follows:
   a. Switch an ohmmeter to the R × 100 scale.
   b. Connect an ohmmeter between the exciter coil black and white wires.
   c. Replace the exciter coil if the reading is not within specifications (**Table 3**). See Chapter Eight.
4. Check the pulser coil as follows:
   a. Switch an ohmmeter to the R × 1 scale.
   b. Connect an ohmmeter between the pulser coil red and white wires.
   c. Replace the pulser coil if the reading is not within specifications (**Table 3**). See Chapter Eight.
5. Check the control coil as follows:
   a. Switch an ohmmeter to the R × 1 scale.
   b. Connect an ohmmeter between the control coil green and blue wires.
   c. Replace the control coil if the reading is not within specifications (**Table 3**). See Chapter Eight.

NOTE
When checking the ignition coils in Step 6 and Step 7, keep in mind that normal resistance in both the primary and secondary coil winding is not a

guarantee that the unit is working properly; only an operational spark test can tell if a coil is producing an adequate spark from the input voltage. A Polaris dealer may have the equipment to test the coil's output. If not, it may be necessary to substitute a known good coil assembly to see if the problem goes away.

6. Check the ignition coil primary as follows:
   a. Switch an ohmmeter to the R × 1 scale.
   b. Connect an ohmmeter between the ignition coil orange and black wires.
   c. Repeat for each coil.
   d. Replace the ignition coil assembly if the primary reading for any one coil is not within specifications (**Table 3**). See Chapter Eight.
7. Check the ignition coil secondary as follows:
   a. Switch an ohmmeter to the R × 100 scale.
   b. Connect an ohmmeter between the secondary lead (spark plug lead) and the black wire.
   c. Repeat for each coil.
   d. Replace the ignition coil assembly if the secondary reading for any one coil is not within specifications (**Table 3**). See Chapter Eight.
8. Reconnect the ignition assembly electrical connectors.

CDI Control Unit

There is no acceptable test procedure available for the CDI control unit. See **Figure 13** (twin cylinder) or **Figure 14** (three cylinder). If you suspect the CDI control unit, check ignition timing as described in Chapter Three. If the ignition timing shows abnormal timing, replace the CDI control unit with a new unit and recheck ignition timing.

> *NOTE*
> *Before replacing the CDI control unit, make all of the previous ignition system tests as well as checking all of the ignition system wiring connectors and*

wiring. Replacing the CDI control unit should be your last step in troubleshooting the ignition system.

ALTERNATOR OUTPUT TEST

The following test checks the functional operation of the lighting coil. The lighting coil is mounted on the stator plate behind the flywheel.

1. Position the snowmobile so that the ski tips are placed against a stationary object. Then raise the rear of the snowmobile so that the track is clear of the ground.

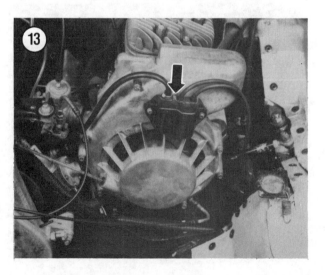

2. Open the shroud and secure it so that it cannot fall.

3. Disconnect the 6-prong alternator connector from the engine wiring harness. See **Figure 3**, **Figure 4** or **Figure 5**.

4. Set a voltmeter to the 100 ACV volt scale. Then connect the red voltmeter lead to the yellow connector and the black voltmeter lead to the either the engine or the brown wire at the connector block. See **Figure 15**.

NOTE
On twin-cylinder models, it will be necessary to connect the red voltmeter test leads to both yellow wires in the connector block when making the following test.

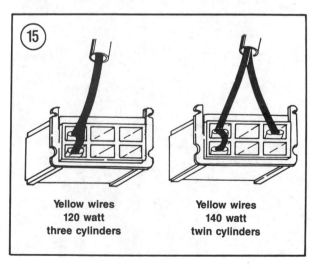

Yellow wires
120 watt
three cylinders

Yellow wires
140 watt
twin cylinders

WARNING
When performing the following steps, ensure that the track area is clear and that no one walks behind the track or serious injuries may result.

WARNING
***Never** lean into the snowmobile's engine compartment while wearing a scarf or other loose clothing when the engine is running or when attempting to start the engine. If the scarf or clothing should catch in the drive belt or clutch, severe injury or death could occur. Make sure the pulley guard is in place.*

5. Have an assistant start the engine.

6. Increase engine speed to 3,000 rpm; voltmeter should read approximately 20 volts. A voltmeter reading of 15-45 volts is considered normal.

7. Repeat for each yellow wire.

8. If the voltmeter reading was less than 15 volts, replace the lighting coil as described in Chapter Eight.

9. Turn the engine off and disconnect the voltmeter. Reconnect the alternator connector.

10. Close the shroud and lower the snowmobile track to the ground.

VOLTAGE REGULATOR

If you are experiencing blown bulbs or if all of the lights are dim (filaments barely light), test the voltage regulator (**Figure 16**) as follows. In addition, check the bulb filament; an overcharged condition will usually melt the filament rather then break it. If the lighting system fails completely, check for a defective lighting coil. See *Alternator Output Test* in this chapter.

NOTE
The headlight must be wired up and working properly when performing this procedure.

1. Position the snowmobile so that the ski tips are placed against a stationary object. Then raise

the rear of the snowmobile so that the track is clear of the ground.

2. Open the shroud and secure it so that it cannot fall.

3. Set a voltmeter to the AC 25 volt scale. Then insert one voltmeter lead through the voltage regulator plastic connector cover until it contacts the regulator terminal. Connect the opposite voltmeter lead to the heat sink placed between the console and voltage regulator (**Figure 17**).

WARNING
When performing the following steps, ensure that the track area is clear and that no one walks behind the track or serious injuries may result.

WARNING
Never lean into the snowmobile's engine compartment while wearing a scarf or other loose clothing when the engine is running or when attempting to start the engine. If the scarf or clothing should catch in the drive belt or clutch, severe injury or death could occur. Make sure the pulley guard is in place.

4. Have an assistant start the engine.

5. Turn on the lights.

6. Increase engine speed from idle to 3,000 rpm; voltmeter should read approximately 11-15 volts.

7. If voltmeter reads more than 15 volts, perform the following steps to isolate the problem.

8. Clean the voltage regulator terminals and retest. If the voltage reading is still excessive (16 volts or higher), perform Step 8.

CAUTION
When performing Step 8, run the engine only as long as it takes to check light operation. Excessive engine operation with the voltage regulator disconnected may damage the lights or light circuit.

9. Disconnect the voltage regulator connector from the voltage regulator and start the engine.

If the lights come on, the voltage regulator is faulty. Replace the voltage regulator as described in Chapter Eight.

10. Reconnect the voltage regulator connector.

11. Close the shroud and lower the snowmobile track to the ground.

CHARGING SYSTEM (MODELS WITH ELECTRIC STARTING)

Refer to **Figure 18** when troubleshooting the charging system.

Troubleshooting

Before testing the charging system, visually check the following:

1. Make sure the battery cables are properly connected. If polarity is reversed, check for a damaged rectifier.

2. Carefully inspect all wiring between the stator plate and battery for worn or cracked insulation and corroded or loose connections. Replace wiring or clean and tighten connections as required.

3. Check battery condition. Clean and recharge as required. See Chapter Eight.

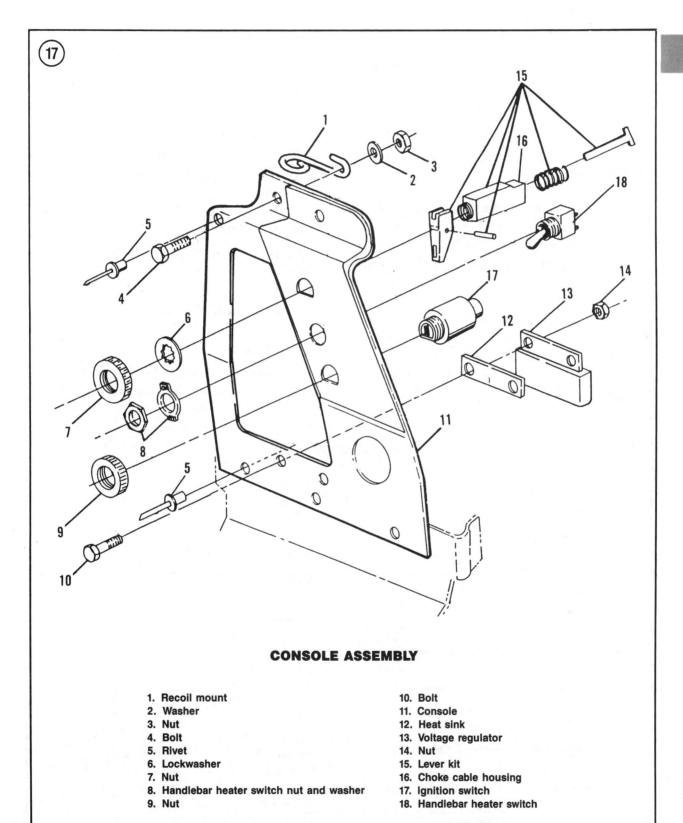

CONSOLE ASSEMBLY

1. Recoil mount
2. Washer
3. Nut
4. Bolt
5. Rivet
6. Lockwasher
7. Nut
8. Handlebar heater switch nut and washer
9. Nut
10. Bolt
11. Console
12. Heat sink
13. Voltage regulator
14. Nut
15. Lever kit
16. Choke cable housing
17. Ignition switch
18. Handlebar heater switch

Alternator Output Test

Test as described previously in this chapter.

Voltage Regulator Test

Test as described previously in this chapter.

Rectifier Test

1. Disconnect the 3 connectors from the rectifier.
2. Set an ohmmeter to the R × 1 scale. Set one ohmmeter lead onto the rectifier positive terminal (+) and the opposite lead onto one of the rectifier AC terminals. Note the reading and reverse the leads. The readings should be low one way and high when the leads are reversed.
3. Repeat Step 2 for the opposite rectifier AC terminal.
4. Replace the rectifier assembly if it failed to provide the test results in Step 2 or Step 3.

Circuit Breaker Test

The circuit breaker is mounted on the battery box.
1. Disconnect the electrical connectors from the circuit breaker.
2. Set an ohmmeter to the R × 1 scale. Connect the 2 ohmmeter leads across the 2 circuit breaker leads. If the reading is not 0 ohms, replace the circuit breaker.

Ignition Switch

Test the ignition switch as described in Chapter Eight.

ELECTRIC STARTING SYSTEM

Description

The electric starter motor is mounted horizontally to the front of the engine. When

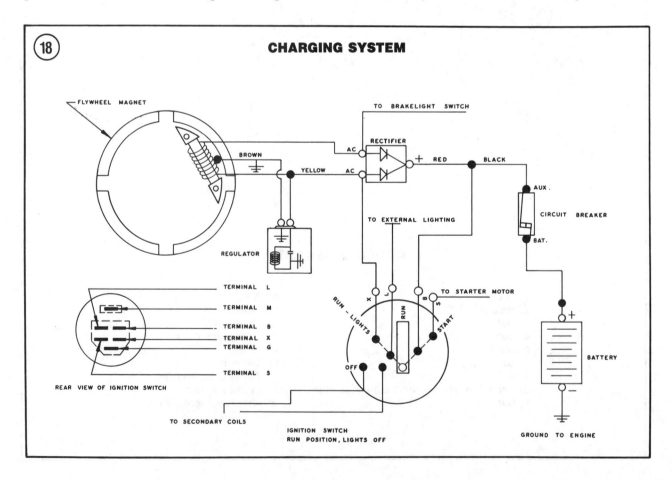

battery current is supplied to the starter motor, its pinion gear is thrust forward to engage the teeth on the engine flywheel. Once the engine starts, the pinion gear disengages from the flywheel.

The electric starting system requires a fully charged battery to provide the large amount of current required to operate the starter motor.

The starter relay carries the heavy electrical current to the motor. Depressing the starter switch allows current to flow through the relay coil. The relay contacts close and allow current to flow from the battery through the relay to the starter motor.

Refer to **Figure 19** when performing procedures in this section.

> *CAUTION*
> *Do not operate an electric starter motor continuously for more than 5 seconds. Allow the motor to cool for at least 15 seconds between attempts to start the engine.*

Troubleshooting

Before troubleshooting the starting circuit, make sure that:

a. The battery is fully charged.
b. Battery cables are the proper size and length. Replace cables that are undersize or damaged.
c. All electrical connections are clean and tight.
d. The wiring harness is in good condition, with no worn or frayed insulation or loose harness sockets.
e. The fuel system is filled with an adequate supply of fresh gasoline and that the oil tank is topped off.

Testing

Refer to **Figure 19** when performing the following tests:

Battery voltage test

1. Open the shroud.
2. Set a voltmeter to the 25 DC volt scale. Place the red voltmeter lead across the positive battery terminal and the black lead across the negative battery terminal (**Figure 20**). Battery voltage should be 11-13 volts. Interpret results as follows:
   a. If battery voltage is within 11-13 volts, proceed to Step 2.
   b. If battery voltage is low (less than 11 volts), service the battery as described in Chapter Eight.
2. With the voltmeter still attached to the battery as described in Step 2, crank the engine for 15 seconds. During cranking, battery voltage should not drop below 9 volts. After cranking, battery voltage should return to 11-13 volts. If battery voltage is low during or after cranking, check the battery as described in Chapter Eight. If cranking voltage dropped below 9 volts, also check the starter motor as described in Chapter Eight.
3. Disconnect the voltmeter leads and proceed with the following test procedures.

Solenoid does not engage starter

If the solenoid does not turn the starter over when the starter button is pressed, perform the following:

1. *Test 1:* Check the battery cables as follows:
   a. Set a voltmeter to the 25 DC volt scale. Place the red voltmeter lead to the circuit breaker on the battery side. Place the black lead to a good engine ground. The voltmeter should show battery voltage (11-13 volts).
   b. If the voltage reading is incorrect, check for loose or corroded connections or burned wires. Repair wiring as required.
2. *Test 2*: Check the circuit breaker as follows:
   a. Set a voltmeter to the 25 DC volt scale. Place the positive voltmeter lead across the

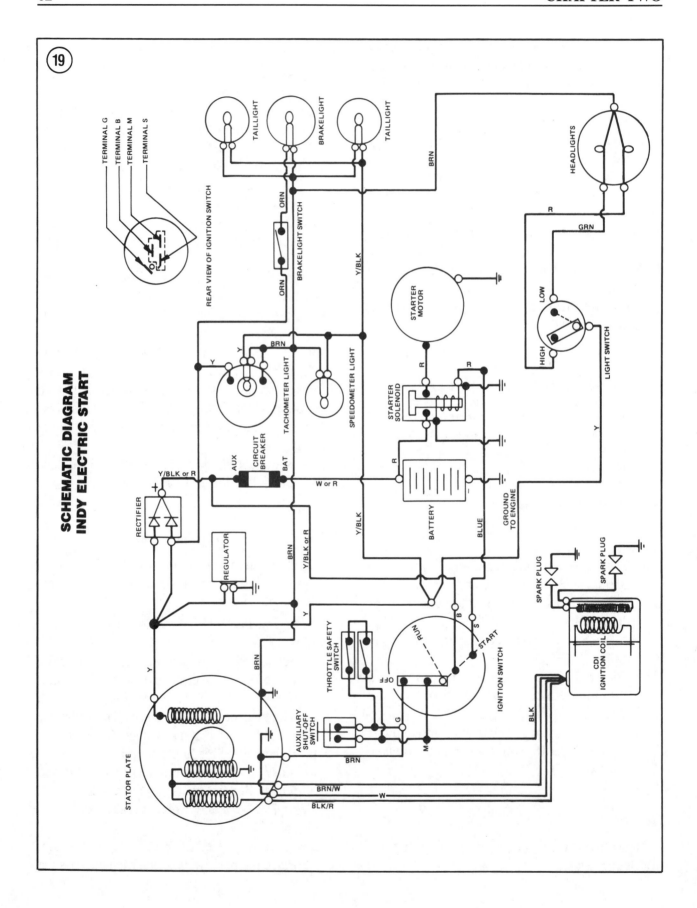

SCHEMATIC DIAGRAM
INDY ELECTRIC START

circuit breaker (auxiliary side) and the negative lead to a good engine ground. The voltmeter should show battery voltage (11-13 volts).

CAUTION
The circuit breaker must be removed from the battery circuit when performing sub-step b.

b. If the voltage reading is low, check the circuit breaker resistance as described under *Charging System* in this chapter. If the circuit breaker resistance reading is not 0 ohm, replace the circuit breaker.

3. *Test 3*: Check the ignition switch to circuit breaker wire as follows:
a. Set a voltmeter to the 25 DC volt scale. Place the positive voltmeter lead across ignition switch B terminal and the negative lead to a good engine ground. The voltmeter should show battery voltage (11-13 volts).

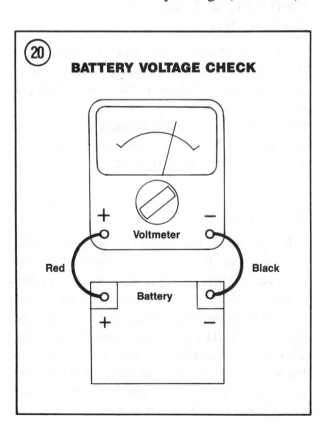

BATTERY VOLTAGE CHECK

b. If battery voltage is low, check the ignition switch to circuit breaker wire for loose, corroded or damaged connections. Check the wire for signs of burning or fraying.

4. *Test 4*: Check the ignition switch as follows:
a. Set a voltmeter to the 25 DC volt scale. Place the positive test lead to the ignition switch S terminal and the negative test lead to a good engine ground. Turn the ignition switch to its start position. Voltmeter should read battery voltage (11-13 volts).
b. If battery voltage is low, replace the ignition switch. See Chapter Eight.

5. *Test 5*: Check the ignition switch to solenoid wire as follows:
a. Set a voltmeter to the 25 DC volt scale. Place the positive voltmeter lead across the small solenoid terminal and the negative lead to a good engine ground. Turn the ignition switch to its start position. The voltmeter should read battery voltage (11-13 volts).
b. If the battery voltage is low, check for corroded, loose or damaged wiring.

6. If Tests 1-5 are correct, the problem is with the solenoid unit. Replace the solenoid as described in Chapter Eight.

Solenoid engages the starter motor but starter does not turn

1. Set a voltmeter to the 25 DC volt scale. Place the positive voltmeter lead to the large solenoid terminal (battery side) and the negative test lead to a good engine ground. Turn the ignition switch to its start position. Record the voltmeter reading.
2. With the voltmeter still set at 25 DC volts, place the positive lead to the large solenoid terminal on the starter motor side. Turn the ignition switch to its start position. Record the voltmeter reading.
3. If the voltage readings recorded in Step 1 and Step 2 are not equal, replace the solenoid. If the voltage readings are equal, refer to Chapter Eight and disassemble the starter assembly. Perform

the bench tests in Chapter Eight to isolate the problem.

4. Disconnect the voltmeter leads.

FUEL SYSTEM

Many snowmobile owners automatically assume that the carburetor is at fault when the engine does not run properly. While fuel system problems are not uncommon, carburetor adjustment is seldom the answer. In many cases, adjusting the carburetors only compounds the problem by making the engine run worse.

Fuel system troubleshooting should start at the gas tank and work through the system, reserving the carburetor as the final point. Most fuel system problems result from an empty fuel tank, a plugged fuel filter, malfunctioning fuel pump or sour fuel. **Figure 21** provides a series of symptoms and causes that can be useful in localizing fuel system problems.

Carburetor chokes can also present problems. A choke stuck open will show up as a hard starting problem; one that sticks closed will result in a flooding condition. Check choke operation and adjustment (Chapter Three).

Identifying Carburetor Conditions

The following list can be used as a guide when trying to determine rich and lean carburetor conditions.

When the engine is running rich, one or more of the following conditions may be present:
  a. The spark plug(s) will foul.
  b. The engine will miss and run rough when it is running under a load.
  c. As the throttle is increased, the exhaust smoke becomes more excessive.
  d. With the throttle open, the exhaust will sound choked or dull. Stopping the snowmobile and trying to clear the exhaust with the throttle held open will not clear up the sound.

When the engine is running lean, one or more of the following conditions may be present:

  a. The spark plug firing end will become very white or blistered in appearance.
  b. The engine overheats.
  c. Acceleration is slower.
  d. Flat spots are felt during operation that feel much like the engine is trying to run out of gas.
  e. Engine power is reduced.
  f. At full throttle, engine rpm will not hold steady.

ENGINE

Engine problems are generally symptoms of something wrong in another system, such as ignition, fuel or starting. If properly maintained and serviced, the engine should experience no problems other than those caused by age and wear.

Overheating and Lack of Lubrication

Overheating and lack of lubrication cause the majority of engine mechanical problems. Make sure the cooling system isn't damaged. Using a spark plug of the wrong heat range can burn a piston. Incorrect ignition timing, a faulty cooling system or an excessively lean fuel mixture can also cause the engine to overheat.

Preignition

Preignition is the premature burning of fuel and is caused by hot spots in the combustion chamber (**Figure 22**). The fuel actually ignites before it is supposed to. Glowing deposits in the combustion chamber, inadequate cooling or overheated spark plugs can all cause preignition. This is first noticed in the form of a power loss but will eventually result in extended damage to the internal parts of the engine because of higher combustion chamber temperatures.

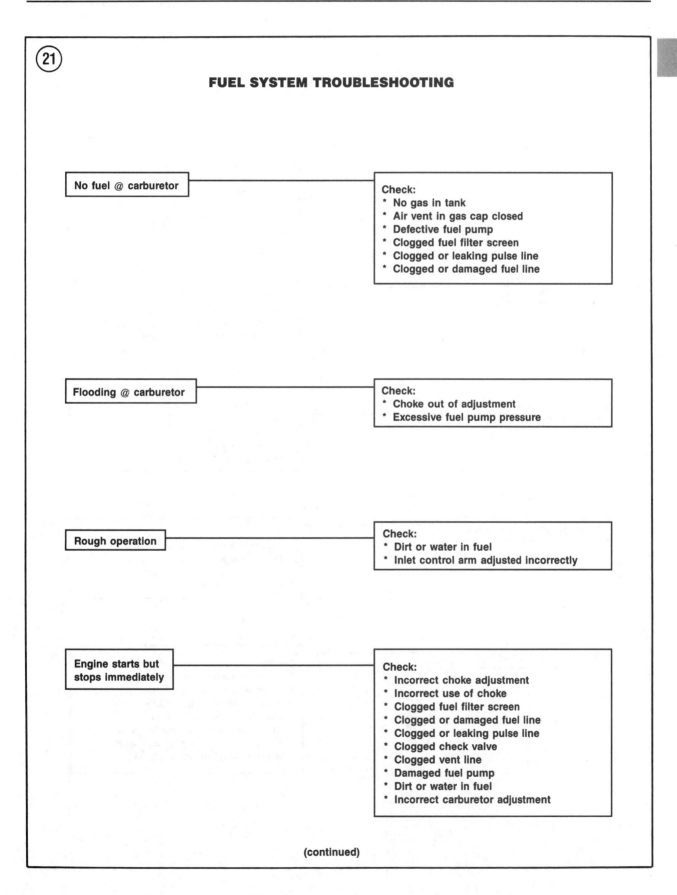

㉑

2

FUEL SYSTEM TROUBLESHOOTING

No fuel @ carburetor ——————

Check:
* No gas in tank
* Air vent in gas cap closed
* Defective fuel pump
* Clogged fuel filter screen
* Clogged or leaking pulse line
* Clogged or damaged fuel line

Flooding @ carburetor ——————

Check:
* Choke out of adjustment
* Excessive fuel pump pressure

Rough operation ——————

Check:
* Dirt or water in fuel
* Inlet control arm adjusted incorrectly

Engine starts but stops immediately ——————

Check:
* Incorrect choke adjustment
* Incorrect use of choke
* Clogged fuel filter screen
* Clogged or damaged fuel line
* Clogged or leaking pulse line
* Clogged check valve
* Clogged vent line
* Damaged fuel pump
* Dirt or water in fuel
* Incorrect carburetor adjustment

(continued)

(21) (continued)

Engine misfires ———————————

Check:
* Dirty carburetor
* Dirty or defective inlet seat
  or needle
* Choke out of adjustment
* Incorrect carburetor adjustment

Engine backfires ———————————

Check:
* Poor quality fuel
* Air/fuel mixture too rich or too lean
* Incorrect carburetor adjustment

Engine preignition ———————————

Check:
* Excessive oil in fuel
* Poor quality fuel
* Lean carburetor mixture

Spark plug burns and fouls ———————————

Check:
* Incorrect spark plug heat range
* Fuel mixture too rich
* Incorrect carburetor adjustment
* Poor quality fuel

High gas consumption ———————————

Check:
* Incorrect carburetor adjustment
* Clogged exhaust system
* Loose inlet seat and needle
* Defective inlet seat gasket
* Worn inlet seat and needle
* Foreign matter clogging inlet seat
* Leaks at fuel line connections

Detonation

Commonly called "spark knock" or "fuel knock," detonation is the violent explosion of fuel in the combustion chamber prior to the proper time of combustion (**Figure 23**). Severe damage can result. Use of low octane gasoline is a common cause of detonation.

Even when high octane gasoline is used, detonation can still occur if the engine is improperly timed. Other causes are over-advanced ignition timing, lean fuel mixture at or near full throttle, inadequate engine cooling, cross-firing of spark plugs, or the excessive accumulation of deposits on piston and combustion chamber.

Since the snowmobile engine is covered, engine knock or detonation is likely to go unnoticed, especially at high engine rpm when wind noise is also present. Such inaudible detonation, as it is called, is usually the cause when engine damage occurs for no apparent reason.

Poor Idling

A poor idle can be caused by improper carburetor adjustment, incorrect timing or ignition system malfunctions. Check the carburetor pulse and vent lines for an obstruction. Also check for loose carburetor hose clamps or a faulty carburetor flange gasket.

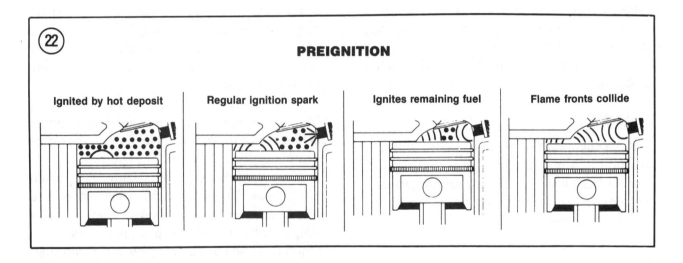

(22) **PREIGNITION**

Ignited by hot deposit   Regular ignition spark   Ignites remaining fuel   Flame fronts collide

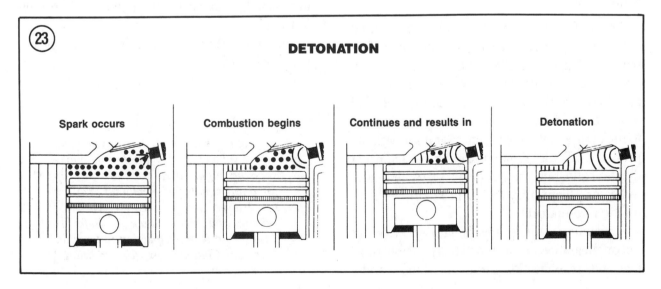

(23) **DETONATION**

Spark occurs   Combustion begins   Continues and results in   Detonation

Misfiring

Misfiring can result from a weak spark or a dirty spark plug. Check for fuel contamination. If misfiring occurs only under heavy load, as when accelerating, it is usually caused by a defective spark plug. Check for fuel contamination.

Water Leakage in Cylinder

The fastest and easiest way to check for water leakage in a cylinder is to check the spark plugs. Water will clean a spark plug. If one of the plugs is clean and the other is dirty, there is most likely a water leak in the cylinder with the clean plug. To check further, install a dirty plug in each cylinder. Run the engine for 5-10 minutes. Shut the engine of and remove the plugs. If one plug is clean and the other is dirty (or if all plugs are clean), a water leak in the cylinder is the problem.

Flat Spots

If the engine seems to die momentarily when the throttle is opened and then recovers, check for a dirty or contaminated carburetor, water in the fuel, plugged pilot jet(s) or an excessively lean or rich low speed mixture.

Power Loss

Several factors can cause a lack of power and speed. Look for air leaks in the fuel line or fuel pump, a clogged fuel filter or a throttle valve that does not operate properly. Dynamically (engine running) check the ignition timing at full advance. See Chapter Three. This will allow you to make sure that the ignition system is operating properly. If the ignition timing is incorrect dynamically but was properly set with a dial indicator, there may be a problem with an ignition component.

A piston or cylinder that is galling, incorrect piston clearance or worn or sticky piston rings may be responsible. Look for loose bolts, defective gaskets or leaking machined mating surfaces on the cylinder head, cylinder or crankcase. Also check the crankshaft seals. Refer to *Two-Stroke Pressure Testing* in this chapter.

Exhaust fumes leaking within the engine compartment can slow and even stop the engine.

Refer to **Figure 24** for a general listing of engine troubles.

Piston Seizure

Piston seizure or galling is the transfer of metal from the piston to the cylinder bore. Friction causes piston seizure. This is caused by one or more pistons with incorrect bore clearances, piston rings with an improper end gap, compression leak, incorrect type of oil, spark plug of the wrong heat range, incorrect ignition timing or an incorrectly operating oil injection pump. Overheating from any cause may result in piston seizure.

A noticeable reduction of speed may be your first sign of seizure while immediate stoppage indicates full lockup. A top end rattle should be considered as an early sign of seizure.

When diagnosing piston seizure, the pistons themselves can be used to troubleshoot and determine the failure cause. High cylinder temperatures normally cause seizure above the piston rings while seizure below the piston rings are usually caused by a lack of proper lubrication.

See **Figure 25** and **Figure 26** for examples of piston seizure.

> *NOTE*
> *After starting a water-cooled engine and before subjecting it to full throttle, make sure the engine is allowed to warm up for approximately 3-5 minutes. Failure to observe this step can cause cold seizure and piston damage. Refer to **Piston and Ring Inspection** in Chapter Five or Chapter Six for additional information.*

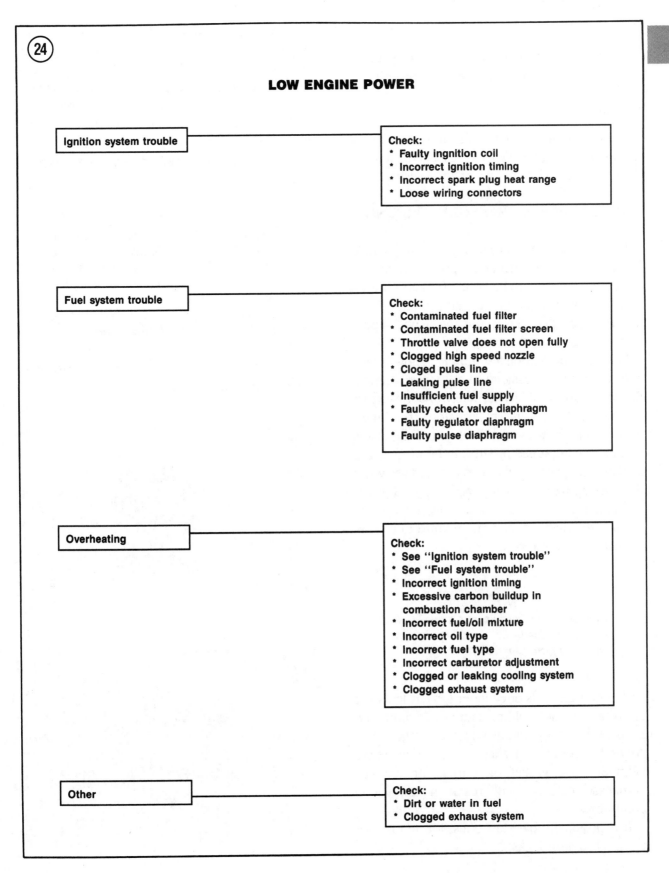

(24)

LOW ENGINE POWER

Ignition system trouble

Check:
* Faulty ingnition coil
* Incorrect ignition timing
* Incorrect spark plug heat range
* Loose wiring connectors

Fuel system trouble

Check:
* Contaminated fuel filter
* Contaminated fuel filter screen
* Throttle valve does not open fully
* Clogged high speed nozzle
* Cloged pulse line
* Leaking pulse line
* Insufficient fuel supply
* Faulty check valve diaphragm
* Faulty regulator diaphragm
* Faulty pulse diaphragm

Overheating

Check:
* See "Ignition system trouble"
* See "Fuel system trouble"
* Incorrect ignition timing
* Excessive carbon buildup in
  combustion chamber
* Incorrect fuel/oil mixture
* Incorrect oil type
* Incorrect fuel type
* Incorrect carburetor adjustment
* Clogged or leaking cooling system
* Clogged exhaust system

Other

Check:
* Dirt or water in fuel
* Clogged exhaust system

2

Excessive Vibrations

Excessive vibrations may be caused by loose engine, suspension or steering mount bolts.

Engine Noises

Experience is needed to diagnose accurately in this area (**Figure 27**). Noises are difficult to differentiate and even harder to describe.

TWO-STROKE PRESSURE TESTING

Many owners of 2-stroke engines are plagued by hard starting and generally poor running, for which there seems to be no cause. Carburetion and ignition may be good, and a compression test may show that all is well in the engine's upper end.

What a compression test does *not* show is lack of primary compression. In a 2-stroke engine, the crankcase must be alternately under pressure and vacuum. After the piston closes the intake port, further downward movement of the piston causes the entrapped mixture to be pressurized so that it can rush quickly into the cylinder when the scavenging ports are opened. Upward piston movement creates a lower vacuum in the crankcase, enabling fuel-air mixture to pass in from the carburetor.

> *NOTE*
> *The operational sequence of a two-stroke engine is illustrated in **Chapter One** under **Engine Operation**.*

If crankcase seals or cylinder gaskets leak, the crankcase cannot hold pressure or vacuum and proper engine operation becomes impossible. Any other source of leakage such as a defective cylinder base gasket or porous or cracked crankcase castings will result in the same conditions.

It is possible, however, to test for and isolate engine pressure leaks.

The test is simple but requires special equipment. A typical two-stroke pressure test kit is shown in **Figure 28**. Briefly, what is done is to seal off all natural engine openings, then apply air pressure. If the engine does not hold air, a leak or leaks is indicated. Then it is only necessary to locate and repair all leaks.

The following procedure describes a typical pressure test.

> *NOTE*
> *Because of the labyrinth seal on the crankshaft, the cylinders cannot be checked individually. When you pump up one cylinder you will also be pumping up the opposite cylinder. Thus both cylinders must be blocked before applying pressure during testing.*

1. Remove the carburetors as described in Chapter Seven.

2. Take a rubber plug and insert it tightly in the intake manifolds.

3. Remove the exhaust pipe and block off the exhaust ports, using suitable adaptors and fittings.

4. Plug the oil injection fittings on each cylinder block.

5. Plug the impulse hose fitting on the crankcase.

6. Remove one of the spark plugs and install the pressure gauge adaptor into the spark plug hole. Connect the pressurizing lever and gauge to the pressure fitting installed where the spark plug was, then squeeze the lever until the gauge indicates approximately 12 psi.

7. Observe the pressure gauge. If the engine is in good condition, the pressure should not drop more than 1 to 2 psi in several minutes. Any pressure loss of 1 psi in one minute indicates serious sealing problems.

Before condemning the engine, first be sure that there are no leaks in the test equipment or

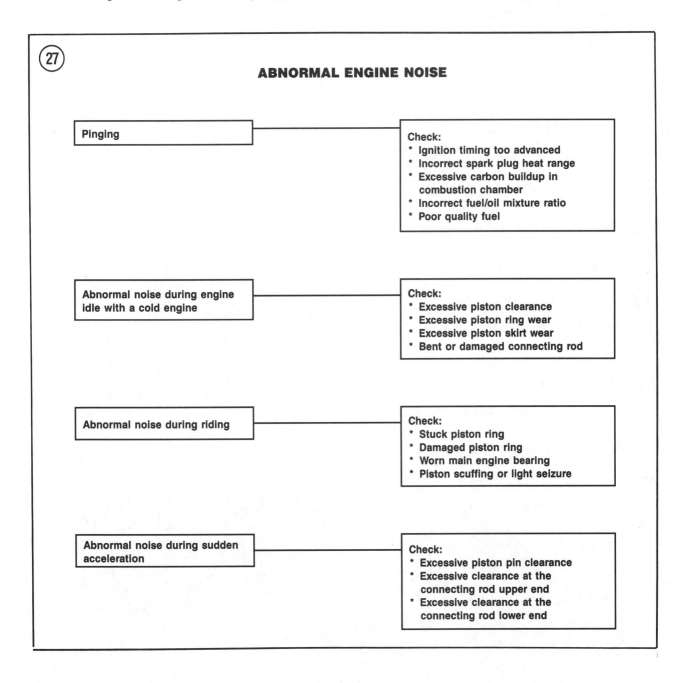

ABNORMAL ENGINE NOISE

Pinging

Check:
* Ignition timing too advanced
* Incorrect spark plug heat range
* Excessive carbon buildup in combustion chamber
* Incorrect fuel/oil mixture ratio
* Poor quality fuel

Abnormal noise during engine idle with a cold engine

Check:
* Excessive piston clearance
* Excessive piston ring wear
* Excessive piston skirt wear
* Bent or damaged connecting rod

Abnormal noise during riding

Check:
* Stuck piston ring
* Damaged piston ring
* Worn main engine bearing
* Piston scuffing or light seizure

Abnormal noise during sudden acceleration

Check:
* Excessive piston pin clearance
* Excessive clearance at the connecting rod upper end
* Excessive clearance at the connecting rod lower end

sealing plugs. If the equipment shows no signs of leakage, go over the entire engine carefully. Large leaks can be heard; smaller ones can be found by going over every possible leakage source with a small brush and soap suds solution. Possible leakage points are listed below:

a. Crankshaft seals.
b. Spark plug(s).
c. Cylinder head joint.
d. Cylinder base joint.
e. Carburetor base joint.
f. Crankcase joint.

POWER TRAIN

The following items provide a starting point from which to troubleshoot power train malfunctions. The possible causes for each malfunction are listed in a logical sequence.

Drive Belt Not Operating Smoothly in Primary Sheave

a. Drive sheave face is rough, grooved, pitted or scored.
b. Defective drive belt.

Uneven Drive Belt Wear

a. Misaligned primary and secondary sheaves.
b. Loose engine mounts.

Glazed Drive Belt

a. Excessive slippage caused by stuck or frozen track.
b. Engine idle speed too high.

Drive Belt Too Tight at Idle

a. Engine idle speed too high.
b. Incorrect sheave distance.
c. Incorrect belt length.
d. Incorrect drive belt tension.

Drive Belt Edge Cord Failure

a. Misaligned primary and secondary sheaves.
b. Loose engine mounts.

Brake Not Holding Properly

a. Worn brake pads.
b. Worn brake disc.
c. Oil saturated brake pads.
d. Air in brake line.
e. Sheared key on brake disc.

Brake Not Releasing Properly

a. Bent or damaged brake lever.
b. Incorrect brake adjustment.

Excessive Chaincase Noise

a. Incorrect chain tension.
b. Excessive chain stretch.
c. Worn sprocket teeth.
d. Damaged chain and/or sprockets.

Chain Slippage

a. Incorrect chain tension.
b. Excessive chain stretch.
c. Worn sprocket teeth.

Leaking Chaincase

a. Loose chaincase cover mounting bolts.
b. Damaged chaincase cover gasket.
c. Damaged chaincase oil seal(s).
d. Cracked or broken chaincase.

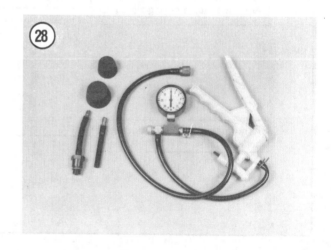

Rapid Chain and Sprocket Wear

a. Insufficient chaincase oil level.
b. Broken chain tensioner.
c. Misaligned sprockets.

Drive Clutch Engages Before Engagement RPM

a. Worn spring.
b. Incorrect weight.
c. Worn weight.

Drive Clutch Engages After Engagement RPM

a. Incorrect spring.
b. Worn or damaged secondary sheave buttons.

Erratic Shifting

a. Worn rollers and bushings.
b. Scuffed or damaged weights.
c. Dirty primary sheave assembly.
d. Worn or damaged secondary sheave buttons.

Engine Bogs During Engagement

a. Incorrect secondary sheave width adjustment.
b. Drive belt worn too thin.
c. Incorrect sheave distance.

Primary or Secondary Sheave Sticks

a. Damaged sheave assembly.
b. Sliding sheave damage.
c. Dirty sheave assembly.
d. Worn or damaged guide buttons.

SKIS AND STEERING

The following items provide a starting point from which to troubleshoot ski and steering malfunctions. The possible causes for each malfunction are listed in a logical sequence.

Loose Steering

a. Loose steering column.
b. Loose steering column fasteners.
c. Loose tie rod ends.
d. Worn spindle bushings.
e. Stripped spindle splines.

Unequal Steering

a. Improperly adjusted tie rods.
b. Improperly installed trailing arms.
c. Damaged steering components.

Rapid Ski Wear

a. Skis misaligned.
b. Worn out ski wear rods (skags).

TRACK ASSEMBLY

The following items provide a starting point from which to troubleshoot track assembly malfunctions. Also refer to *Track Wear Analysis* in Chapter Fourteen.

Frayed Track Edge

a. Incorrect track alignment.
b. Track contacts rivets in tunnel area.

Track Grooved on Inner Surface

a. Track too tight.
b. Frozen rear idler shaft bearing.

Track Drive Ratcheting

a. Track too loose.
b. Drive sprockets misaligned.
c. Damaged drive sprockets.

Rear Idlers Turning on Shaft

Frozen rear idler shaft bearings.

Table 1 EXCITER COIL TESTING

Model/wire color	Resistance reading
Sport	
Brown/white-to-black/red	225 ohms
Trail	
Brown/white-to-white	164 ohms
400/500	
Brown/white-to-white	164 ohms

Table 2 PULSER COIL TESTING

Model/wire color	Resistance reading
Sport	—
Trail	
1983-1985	
Brown/white-to-black/red	45 ohms
1986-on	
Brown/white-to-black/red	17 ohms
400/500	
Brown/white-to-black/red	45 ohms

Table 3 600/650 ELECTRICAL SPECIFICATIONS

Coil/wire color	Resistance reading
Exciter coil	
Black-to-white	261 ohms
Pulser coil	
Red-to-white	20 ohms
Control coil	
Green-to-blue	29.4 ohms
Ignition coil	
Primary	
Orange-to-black	0.106 ohms
Secondary	
Black-to-high tension lead	2,016 ohms

Chapter Three

Lubrication, Maintenance and Tune-up

This chapter covers all of the regular maintenance required to keep your snowmobile in top shape. Regular maintenance is the best guarantee of a trouble-free, long lasting vehicle. Because snowmobiles are high-performance vehicles, proper lubrication, maintenance and tune-ups have thus become increasingly important as ways in which you can maintain a high level of performance, extend engine life and extract the maximum economy of operation. You can do your own lubrication, maintenance and tune-ups if you follow the correct procedures and use common sense. Always remember that engine damage can result from improper tuning and adjustment. In addition, where special tools or testers are called for during a particular maintenance or adjustment procedure, the tool should be used or you should refer service to a qualified dealer or repair shop.

The following information is based on recommendations from Polaris that will help you keep your snowmobile operating at its peak level.

Maintenance intervals are listed in **Tables 1-4**. Recommended lubricants and fuels are listed in **Table 5**. **Tables 1-12** are at the end of the chapter.

NOTE
Be sure to follow the correct procedure and specifications for your specific model and year. Also use the correct quantity and type of fluid as indicated in the tables.

PRE-RIDE CHECKS

The machine should be checked before each ride. Refer to Chapter One.

FLUID CHECKS

Vital fluids should be checked daily or before each ride to assure proper operation and prevent severe component damage. Refer to **Table 1**.

BREAK-IN PROCEDURE

Following cylinder servicing (new pistons, new rings, honing, cylinder boring, etc.) and major lower end work, the engine should be broken in just as if it were new. The performance and service life of the engine depends greatly on a careful and sensible break-in.

For the first 3 hours of operation, no more than 3/4 throttle should be used and the speed should be varied as much as possible. Prolonged steady running at one speed, no matter how moderate, is to be avoided, as is hard acceleration. Wet snow and hard ice conditions should also be avoided during break-in.

To assure adequate protection to the engine during break-in, the first 5 gallons of fuel should be pre-mixed at a 40:1 ratio. This oil will be used *together* with the oil supplied by the injection system. Throughout the break-in period, check the oil injection reservoir tank to make sure the injection system is working (oil level diminishing). Refer to *Correct Fuel Mixing* in this chapter.

NOTE
The use of a 40:1 pre-mix mixture under normal operating conditions will lead to spark plug fouling and rapid carbon build-up.

After the initial 3 hour break-in, all engine and chassis fasteners should be checked for tightness. If the snowmobile is going to be used in extreme conditions, you may want to increase the break-in a few hours. After break-in, retighten the cylinder head nuts as described in this chapter.

NOTE
After the break-in is complete, install new spark plugs as described in this chapter.

Correct Fuel Mixing

When mixing oil and gasoline for engine break-in, note the following:

WARNING
Gasoline is an extreme fire hazard. Never use gasoline near sparks, heat or flame. Do not smoke while mixing fuel.

a. Mix the fuel and oil in a well-ventilated location. The oil should be at room temperature (68° F [20° C]) during mixing.

b. Mix the oil and gasoline thoroughly in a separate, clean, sealable container larger than the quantity being mixed to allow room for agitation. Always measure the quantities exactly. See **Table 6** for fuel tank capacity.

CAUTION
Do not mix oil and gasoline in the snowmobile gas tank. Incomplete mixing may result in engine seizure.

c. Always use fresh gasoline with an octane rating of 88 or higher. Gum and varnish deposits tend to form in gasoline stored in a tank for any length of time. Use of sour fuel can result in carburetor problems and spark plug fouling.

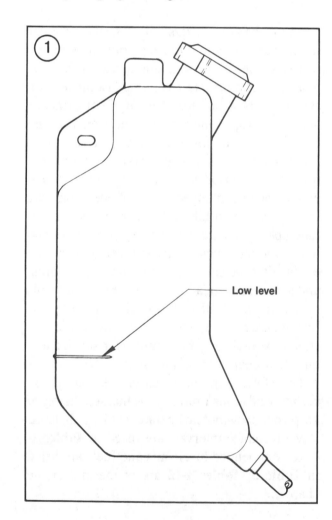

Low level

NOTE
Before adding pre-mixed fuel to the fuel tank, drain the fuel tank as described in **Chapter Seven**.

d. Pour approximately 2 1/2 gallons of gasoline into the mixing container and add 1 pint of Polaris Injection Oil. Agitate the mixture thoroughly, then add the remaining 2 1/2 gallons of gasoline and agitate again until all is mixed well.

e. To avoid any contaminants entering into the fuel system, use a funnel with a filter when pouring the fuel into the tank.

MAINTENANCE SCHEDULES

Polaris divides their maintenance schedules into 4 parts: weekly, initial 150 miles, 1,000 miles and 2,000 miles. Refer to **Table 1** (weekly), **Table 2** (initial 150 miles), **Table 3** (1,000 miles) or **Table 4** (2,000 miles).

NOTE
The engine inspection checks listed in the 150 mile maintenance schedule should be repeated whenever the engine top- or bottom-end has been overhauled or the engine removed from the frame. Likewise, chassis and steering inspection procedures should be performed after major service has been performed to these components.

LUBRICATION

WARNING
Serious fire hazards always exist around gasoline. Do not allow any smoking in areas where fuel is being mixed or while refueling your machine. Always have a fire extinguisher, rated for gasoline and electrical fires, within reach just to play it safe.

Proper Fuel Selection

Two-stroke engines are lubricated by mixing oil with the fuel. The various components of the engine are thus lubricated as the fuel-oil mixture passes through the crankcase and cylinders. All models are equipped with an oil injection system. Pre-mixing fuel is not required on any of the models covered in this manual *except* during engine break-in. See *Break-In* in this chapter.

Polaris recommends the use of gasoline with a minimum octane rating of 88. Use of a lower octane gasoline may cause engine damage.

CAUTION
Do not use gasoline additives, gasohol, white gas or gasoline containing methanol, ethanol or any other type of alcohol. The use of any of these additives or fuels will cause engine damage.

Engine Oil Tank

An oil injection system is used on all models. During engine operation, oil is automatically injected into the engine at a variable ratio depending on engine rpm.

Oil capacity in the reservoir tank (**Figure 1**) should be checked *daily* and during all fuel stops.

On some models, the oil tank is equipped with a oil level sensor that is wired to a low oil warning light on the instrument panel. When the oil level in the tank reaches a specified low point, the low oil warning light will light.

When the oil level is low, perform the following:

1. Open the shroud.
2. Wipe the area around the fill cap with a clean shop rag.
3. Remove the oil tank fill cap (**Figure 2**) and pour in the required amount of two-stroke injection oil specified in **Table 5**. Fill the tank.
4. Reinstall the fill cap and close the shroud.

3

Chaincase Oil

The oil in the chain housing lubricates the chain and sprockets.

Try to use the same brand of oil. Do not mix 2 brand types at the same time as they all vary slightly in their composition.

Oil level check

The chaincase oil level is checked at the check plug (A, **Figure 3**) on the chaincase cover.

1. Park the snowmobile on a level surface. Open the shroud.

2. Unscrew and remove the check plug at the bottom of the chaincase (A, **Figure 3**). Oil should drip from the check hole. If there is no evidence of oil at the check hole, perform Step 3.

3. If the oil level is low, top off with the chaincase oil recommended in **Table 5**. Add oil through the fill plug opening (B, **Figure 3**) until oil begins to leak from the check hole. Do not overfill as this will cause leakage.

4. Reinstall the check and fill plug(s). Tighten the plugs securely.

5. Close and secure the shroud.

Changing

The chaincase oil should be changed at the specified intervals. See **Tables 2-4**.

1. Park the snowmobile on a level surface.

2. Open the shroud.

3. Remove the exhaust pipe and muffler as described in Chapter Seven.

4. Place a number of shops rags underneath the chaincase cover.

> *NOTE*
> *The chaincase is filled with oil and the cover is not equipped with a drain plug. To drain the chaincase coil, the chaincase cover must be removed. Try to absorb as much of the oil on the rags as possible.*

5. Remove the bolts and lockwashers holding the chaincase cover to the chaincase. Remove the cover (**Figure 4**) and gasket.

> *NOTE*
> *If metallic particles are found in the oil residue or in the chaincase, inspect the gears, chain and chain tensioner assembly for damage. Refer to Chapter Fourteen. Repair or replace damaged parts as required.*

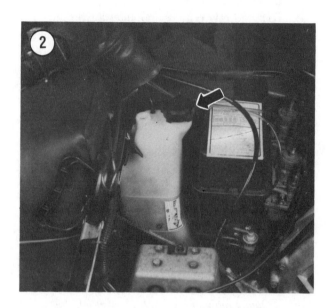

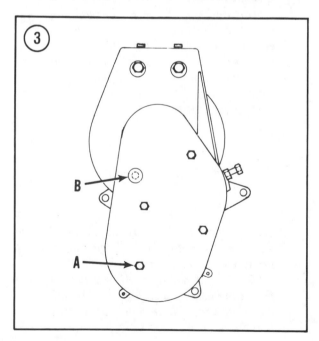

6. Soak up as much the oil with the rags and then remove the rags from the belly pan.

NOTE
Place the oil soaked rags into a suitable container until they can be cleaned.

7. Clean the chaincase cover thoroughly. Wipe up any remaining oil in the belly pan.
8. Replace the chaincase cover gasket if necessary.

NOTE
While the chaincase cover is removed, check drive chain tension as described in this chapter.

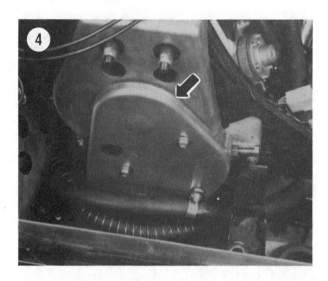

9. Align the slot in the gasket with the chaincase housing and fit the gasket (**Figure 5**) onto the chaincase housing. Make sure that the gasket fits snugly all the way around the housing.
10. Install the chaincase cover (**Figure 4**) onto the chain housing and hold it firmly in place. Install the mounting bolts and lockwashers. Tighten the bolts securely.
11. Remove the fill plug (B, **Figure 3**) at the top of the chaincase cover.
12. Insert a funnel into the fill plug hole and fill the chain housing with the correct type (**Table 5**) and quantity (**Table 6**) of chaincase oil.
13. Check the oil level as described in the previous procedure.
14. Reinstall the fill plug. Tighten the plug securely.
15. Check the belly pan for signs of oil leakage from the chaincase cover.

Suspension Lubrication

The following suspension components should be lubricated annually or every 1,000 miles with a low-temperature grease. If the snowmobile is operated under severe conditions or in wet snow, perform this service more frequently.

Spindles

NOTE
Raising the skis off of the ground will allow more grease penetration through the spindle housing.

Locate the grease fitting on the spindle housing (A, **Figure 6**). Wipe the grease fitting with a clean shop cloth to remove all traces of dirt and other trail residue. Fit the end of a hand-operated grease gun onto the grease fitting and pump grease into the spindle until it runs out of the top and bottom of the spindle housing. Remove the grease gun and wipe up all excess grease from the nipple and spindle housing. Repeat for the opposite side.

Steering post support brackets

Spray the upper and lower steering post brackets (**Figure 7**) with a low-temperature lubricant.

Radius rod end bushings

1. Disconnect the upper and lower radius rods (B, **Figure 6**) at the trailing arm.
2. Remove the bushing from the end of each rod.
3. Clean the bushing and radius rod thoroughly with solvent.
4. After the parts have dried thoroughly, lubricate the radius rod bushings and spacers with a low-temperature grease.
5. Reinstall the bushings and remount the radius rods onto the trailing arm. Tighten the radius rod bolts to the torque specification listed in **Table 7**.

Slide suspension pivot shafts

1. Remove the rear suspension mounting bolts as described in Chapter Sixteen.
2. Remove the 2 pivot shafts (**Figure 8**) from the rear suspension assembly.
3. Clean the pivot shafts with solvent and allow to thoroughly dry.
4. After the shafts have dried, lubricate the shafts with a low-temperature grease.
5. Insert the pivot shafts (**Figure 8**) into the rear suspension assembly.
6. Install the rear suspension assembly as described in Chapter Sixteen.

Throttle Cable Lubrication

The throttle cable should be periodically lubricated with Polaris Clutch and Cable Lubricant.
1. Turn the handlebars all the way to the left.
2. Open the throttle lever slightly to gain access to the throttle cable opening in the housing.
3. Insert the nozzle of the lubricant container into the cable opening and lubricate the cable.

4. Remove the lubricant container and wipe off all excessive lubricant from the throttle lever.

Choke Cable Lubrication

The choke cable and slide should be lubricated weekly with Polaris Clutch and Cable Lubricant.
1. Disconect the choke slide (**Figure 9**) at the console.
2. Insert the lubricant container into the choke cable slide and lubricate the cable assembly.
3. Remove the lubricant container and wipe off all excessive lubricant from the choke slide.
4. Reinstall the choke cable slide at the console. Make sure choke operates correctly.

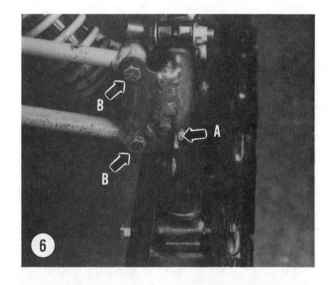

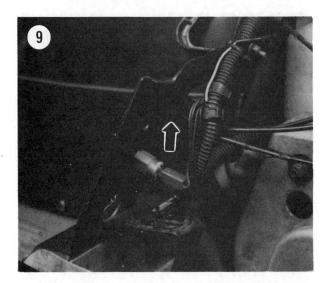

Drive Clutch Lubrication

The spider rollers and shift weight pins (**Figure 10**) should be lubricated with Polaris Clutch and Cable Lubricant at the intervals specified in **Table 1**. Remove the drive belt as described in Chapter Twelve. Then lubricate the rollers and pins. Wipe off all excessive lubricant with a rag before installing the drive belt.

COOLING SYSTEM

All 400, 500, 600 and 650 models are equipped with a closed liquid cooling system. This section describes scheduled and non-scheduled maintenance procedures for the cooling system.

Coolant Level

The coolant level should be checked at the beginning of each riding day.

> *WARNING*
> *Do not remove the pressure cap (**Figure 11**) when the engine is hot.*

1. Park the snowmobile on level ground.
2. Open the shroud.
3. Check the coolant level in the reserve tank (A, **Figure 12**); it should be 1/2 full (maintain between the maximum and minimum levels). If the tank level is low, perform Step 4. If the tank is empty, proceed to Step 5.
4. If the reserve tank level is below 1/2 full, remove the reserve tank cap and add antifreeze into the reserve tank to correct the level. Install the reserve tank cap.
5. If the reserve tank is empty, perform the following:
   a. Remove the pressure cap (**Figure 11**).
   b. Add a sufficient amount of antifreeze and water (in a 50:50 ratio) through the radiator cap opening as described under *Coolant*. See **Table 6** for coolant capacity.
   c. Reinstall the pressure cap with the pressure lever in its released position.

d. Remove the reserve tank cap (A, **Figure 12**) and fill the tank until it is 1/2 full.

e. Start and run the engine at a fast idle for 2-3 minutes to purge the cooling system of trapped air.

f. Lock the pressure cap lever and check the reserve tank level.

Coolant

Only a high-quality ethylene glycol-based coolant compounded for aluminum engines should be used. The coolant should be mixed with water in a 50:50 ratio. Coolant capacity is listed in **Table 6**. When mixing antifreeze with water, make sure to use only soft or distilled water. Never use tap or salt water as this will damage engine parts. Distilled water can be purchased at supermarkets in gallon containers.

> *CAUTION*
> *Always mix coolant in the proper ratio for the coldest temperature in your area. Do not use pure antifreeze; it will freeze without water.*

Coolant Change

The cooling system should be completely drained and refilled once a year (preferably before off-season storage).

> *CAUTION*
> *Use only a high-quality ethylene glycol antifreeze specifically labeled for use with aluminum engines. Do not use an alcohol-based antifreeze.*

The following procedure must be performed when the engine is *cold*.

> *CAUTION*
> *Be careful not to spill antifreeze on painted surfaces as it will destroy the surface. Wash immediately with soapy water and rinse thoroughly with clean water.*

1. Park the snowmobile on a level surface.
2. Open the shroud.

> *WARNING*
> *Do not remove the pressure cap (**Figure 11**) when the engine is hot. Scalding water will spew out and may cause severe burns and personal injury.*

3. Remove the pressure cap (**Figure 11**).
4. On 600 and 650 models, loosen the bleeder plug at the water pump (**Figure 13**).

> *WARNING*
> *Do not siphon coolant with your mouth and a hose. The coolant mixture is*

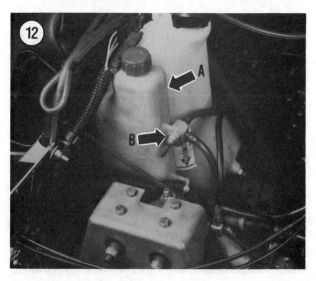

poisonous and may cause sickness. Observe warning labels on antifreeze containers. Animals are attracted to antifreeze so make sure you discard used antifreeze in a safe and suitable manner; do not store antifreeze in open containers.

CAUTION
Step 5 will make a mess. Spilled antifreeze is very slippery on cement floors. Wipe up spilled antifreeze as soon as possible.

5. Disconnect the 2 hoses at the engine. See **Figure 14** or **Figure 15**.

6. When the coolant stops draining, jack up the rear of the vehicle so that it is secure. This will allow the system to completely drain. When the coolant stops draining, lower the vehicle's track to the ground.

7. Reconnect the cooling hoses at the engine. Reposition the hose clamps and tighten securely.

8. Siphon coolant out of the reserve tank (A, **Figure 12**).

9. Using a mixture of antifreeze and distilled water that exceeds the anticipated temperature conditions in your riding area, add coolant through the pressure cap opening until it is even with the top of the filler neck. On 600 and 650 models, tighten the bleeder plug (**Figure 13**) when coolant flows freely from the plug hole.

10. Reinstall the pressure cap with the pressure lever in its released position.

11. Remove the reserve tank cap (A, **Figure 12**) and fill the tank until it is 1/2 full.

12. Start and run the engine at a fast idle for 2-3 minutes to purge the cooling system of trapped air.

13. Lock the pressure cap lever and check the reserve tank level.

14. Check all hose connections for leakage.

15. Turn the engine off and check the coolant freezing level with a coolant checker. Make sure freezing level exceeds the coldest operating temperatures in your area.

16. If the coolant level dropped, add coolant to maintain the correct level in the filler neck. If the coolant level dropped significantly, you may want to start the engine and allow it to idle again with the pressure cap removed. Continue to add coolant until the coolant level stabilizes. At this point, turn the engine off and install the pressure cap.

> *NOTE*
> *After flushing the cooling system, the coolant level may drop due to the displacement of entrapped air. To prevent from operating the engine with a low coolant level, check the level once again before starting the engine.*

17. Flush the engine compartment with clean water.
18. Close and secure the shroud.

Cooling System Inspection

> *WARNING*
> *When performing any service work to the engine or cooling system, never remove the pressure cap (**Figure 11**) or disconnect any hose while the engine is hot. Scalding fluid and steam may be blown out under pressure and cause serious injury.*

Once a year, or whenever troubleshooting the cooling system, the following items should be checked.
1. Remove the pressure cap (**Figure 11**).
2. Check the rubber washers on the pressure cap for tears or cracks. Check for a bent or distorted cap. Raise the vacuum valve and rubber seal and rinse the cap under warm tap water to flush away any loose rust or dirt particles.
3. Inspect the pressure cap neck seat on the filler neck for dents, distortion or contamination. Wipe the sealing surface with a clean cloth to remove any rust or dirt.

4. Check all cooling system hoses for damage or deterioration. Replace any hose that is questionable. Make sure all hose clamps are tight.
5. Check the heat exchangers (**Figure 16**) for cracks or damage. Replace if necessary, as described in Chapter Ten.

SCHEDULED MAINTENANCE

Refer to **Tables 1-4** for scheduled maintenance intervals. Refer to *Lubrication* in this chapter for all lubrication related service procedures. Scheduled maintenance service procedures are described as follows.

Carburetor Water Trap Service

The carburetor water traps (**Figure 17**) should be drained of all contamination weekly.

> *WARNING*
> *Fuel spillage will occur when draining the carburetor water traps. Work in a well-ventilated area, at least 50 feet from any open flame, sparks, lights or heaters. Do not smoke. Wipe up spills immediately.*

1. Open the shroud.
2. Turn the fuel tank supply valve (B, **Figure 12**) to its OFF position.

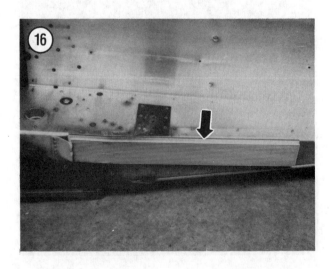

3. Remove air box as described in Chapter Seven.

4. Place shop rags underneath the carburetors.

5. Slide the clamp away from the drain plug on the end of the tube and remove the drain plug (**Figure 17**). Drain the tube contents onto the shop rags. When the tubes are clear, reinstall the drain plug and secure with the clamp.

6. Place the rags into a suitable container until they can be cleaned.

7. Reinstall the air box. See Chapter Seven.

8. Close and secure the shroud.

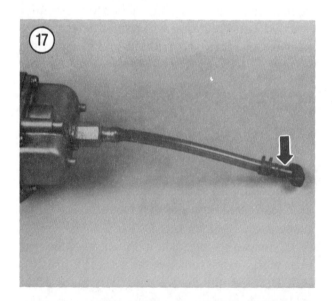

Drive Belt Check

A worn or damaged drive belt will reduce engine performance and may fail during engine operation. With the engine off, open and secure the shroud and remove the drive belt cover. Visually check the drive belt (**Figure 18**) for cracks, fraying or unusual wear as described in Chapter Twelve.

The drive belt should be removed once a month and checked for excessive drive belt width wear. See Chapter Twelve for procedures and specifications.

Drive Belt Tension

Check drive belt tension as described in Chapter Twelve.

Throttle Safety Switch and Kill Switch Testing

Test the throttle safety switch and the kill switch as described in Chapter Eight.

Headlight and Taillight Lens Check

Clean the headlight and taillight lens with a rag before heading out to ensure full light output for both assemblies. Replace damaged lens assemblies as required.

Headlight and Taillight Operation Check

The headlight and taillight must operate correctly for the safe operation of the snowmobile. Start the engine and make sure the headlight and taillight are on. Then operate the dimmer switch on the left-hand handlebar assembly and check that the headlight beam moves from low-beam to high-beam. Operate the brake lever and check that the brake light comes on. Replace damaged bulbs as required.

NOTE
If the headlight beam appears to be adjusted incorrectly, adjust it as described in Chapter Eight.

Steering Bolt Check

Steering and ski alignment cannot be maintained with loose or missing fasteners. Check the steering assembly for loose or missing fasteners as described in Chapter Fifteen.

Steering Check

After checking the steering bolts, check the steering assembly for proper play and adjustment. Perform the following:
1. Support the snowmobile so that the skis are off of the ground.
2. Turn the handlebars from side-to-side. The steering shaft should pivot smoothly with no signs of binding, noise or other abnormal conditions.
3. If the steering is tight or otherwise damaged, service the steering assembly as described in Chapter Fifteen.

Rear Suspension

Check the slide suspension mounting bolts as described in Chapter Sixteen.

Front Limiter Straps

The front limiter strap(s) (**Figure 19**) should be checked for fraying, tearing or other damage. The limiter strap bolts should be checked for tightness. Replace damaged limiter strap(s) as required. See Chapter Fifteen.

Track Adjustment

Because the track is made of rubber and subjected to high torque loads, track stretch and alignment should be monitored on a routine schedule. Failure to periodically adjust track tension and alignment will reduce track performance and eventually cause premature wear to the track. When maintaining the track, there are 2 adjustments—tension and alignment.

Tension adjustment

Correct track tension is important because a loose track will slap on the bottom of the tunnel and wear the track, tunnel and heat exchanger. A loose track can also ratchet on the drive sprockets and damage both the track and sprockets.

A too tight track will rapidly wear the slide runner material and the rubber on the idler wheels. This condition will reduce performance because of increased friction and drag on the system.

NOTE
The track should be warm when performing this procedure.

1. Raise the track off of the ground and secure the snowmobile with a suitable safety stand.
2. Clean ice, snow and dirt from the track and suspension.
3. Connect a spring scale 16 in. ahead of the rear idler wheel in the center of the track (**Figure 20**). Pull the track until the scale registers 10 lbs. Hold

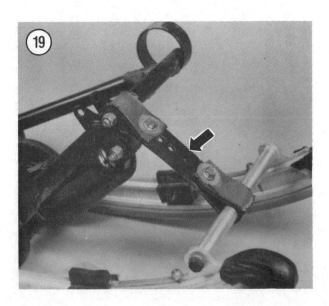

the track at this point and measure the distance between the hi-fax and the inner track clip surface. The distance should be 3/8-1/2 in. If track tension is incorrect, proceed to Step 4.

4. Loosen the rear idler wheel bolts (A, **Figure 21**) and the adjuster bolt (B, **Figure 21**) locknuts. Turn the adjuster bolts (B, **Figure 21**) in or out in equal amounts to adjust the track. Recheck track tension.

5. When track tension is correct, tighten the adjuster bolt (B, **Figure 21**) locknuts and the rear idler wheel bolts (A, **Figure 21**) securely.

6. After adjusting track tension, check track alignment as described in the following procedure.

Track alignment

Track alignment is related to track tension and should be checked and adjusted when the tension is checked and adjusted. If the track is misaligned, the rear idler wheels, driveshaft sprocket lugs and track lugs will wear rapidly. Because of resistance between the track and the sides of the wheels, performance will be greatly reduced.

1. Adjust the track tension as described in this chapter.

2. Position the machine on its skis so that the ski tips are placed against a wall or other immovable barrier. Elevate and support the machine with a shielded stand so that the track is completely clear of the ground and free to rotate.

WARNING
Don't stand behind or in front of the machine when the engine is running, and take care to keep hands, feet and clothing away from the track when it is turning.

3. Start the engine and apply just enough throttle to turn the track several complete revolutions. Then shut off the engine and allow the track to coast to a stop. Don't stop it with the brake.

NOTE
Step 3 ensures that the track will stop in its normal running position; this will insure an accurate alignment position for adjustment. Using the brake to stop track

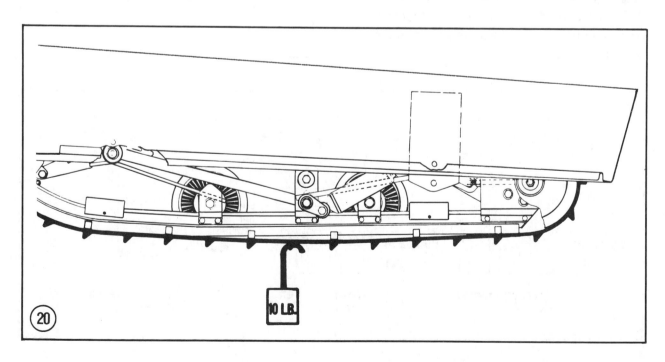

(20)   10 LB.

rotation may cause the track to stop out of its running position. This condition will cause the following adjustment to be inaccurate.

4. Check the alignment of the rear idler wheels and the track lugs (**Figure 22**). If the idlers are equal distance from the lugs and the openings in the track are centered with the hi-fax (**Figure 23**), the alignment is correct. However, if the track is offset to one side or the other, alignment adjustment is required.

5. Loosen the rear idler wheel adjuster bolts (A, **Figure 21**).

NOTE
Make sure to maintain track tension when adjusting track alignment.

6. If the track is offset to the left (**Figure 23**), tighten the left adjuster bolt and loosen the right one equal amounts until the track is centered (**Figure 23**). If the track is offset to the right (**Figure 23**), tighten the right adjuster bolt and loosen the left one in equal amounts.

7. Repeat until the track is properly aligned.

8. Tighten the adjuster bolt (B, **Figure 21**) locknuts securely.

9. Recheck track tension. If track tension is correct, tighten the rear idler wheel bolts securely (A, **Figure 21**).

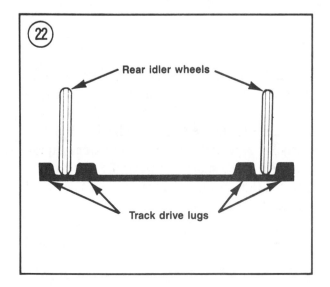

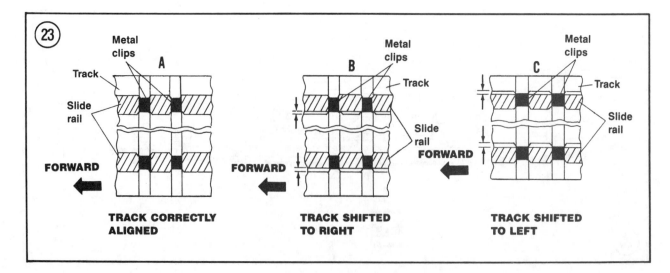

Brake Fluid Level Check

1. Turn the handlebars so that the front master cylinder is level. Wipe the master cylinder cover off with a clean rag.

2. Remove the screws holding the cover to the master cylinder and remove the cover (A, **Figure 24**) and diaphragm.

3. The brake fluid level should be 1/4-5/16 in. below the lip of the reservoir opening. If the level is low, add DOT 3 brake fluid as required.

NOTE
If the brake fluid was so low as to allow air in the hydraulic system, the brakes will have to be bled. Refer to Chapter Thirteen.

4. Reinstall the diaphragm and cover. Tighten the cover screws securely.

WARNING
Use only brake fluid clearly marked DOT 3 and specified for disc brakes. Others may vaporize and cause brake failure.

CAUTION
Be careful not to spill brake fluid on painted or plated surfaces as it will destroy the surface. Wash immediately with soapy water and thoroughly rinse it off.

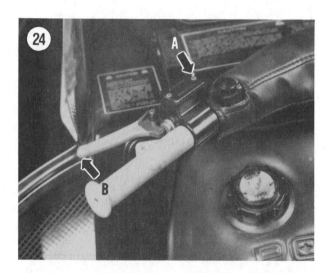

Brake Lever Check

Periodic adjustment of the front disc brake is not required because disc pad wear is automatically compensated. To check the front brake, apply the front brake and measure the distance from the brake lever (B, **Figure 24**) to the handlebar grip. If the distance is less than 1/2 in., check the brake pads for excessive wear as described in Chapter Thirteen. If the brake pads are okay, there may be air in the brake line. Bleed the brake as described in Chapter Thirteen.

Drive Chain
Check and Adjustment

1. Drain the chaincase oil as described in this chapter.

2. Check the drive chain and sprockets (**Figure 25**) for excessive wear or damage. Replace worn or damaged chain and sprockets as described in Chapter Fourteen. If the components are okay, check chain tension as described in Step 3.

3. Referring to **Figure 26**, adjust chain tension as follows:

   a. Turn brake disc a few degrees counterclockwise as indicated in A, **Figure 26**.

b. Turn the chain adjuster bolt (D) until the chain has approximately 1/4-3/8 in. free play at point B.

c. Tighten the adjuster locknut (C) and check free play. Readjust as required.

4. Reinstall the chaincase cover and refill the chaincase with the correct amount and type of gear oil specified in this chapter. See **Table 5** and **Table 6**.

Clutch Alignment

Refer to Chapter Twelve.

Engine Mounts and Fasteners

Loose engine mount bolts will cause incorrect clutch alignment. First check for signs of fraying, cracks or other abnormal conditions with the rubber engine mount dampers. If these are okay, check the front and rear engine mounting bolts to make sure they are tight (**Figure 27**, typical). Check all accessible engine assembly bolts for tightness. Tighten bolts to the torque specification in **Table 7**.

> *NOTE*
> *If the engine mount bolts were loose, check clutch alignment as described in Chapter Eleven.*

Cylinder Head Torque

Refer to *Tune-up* in this chapter.

Ignition Timing

Refer to *Tune-up* in this chapter.

Carburetor Adjustment

Refer to *Tune-up* in this chapter.

Choke Adjustment

Refer to *Tune-up* in this chapter.

Oil Pump Adjustment

Refer to Chapter Nine.

Fuel Filter Replacement

An inline fuel filter (**Figure 28**) is mounted between the fuel tank and fuel pump. To ensure adequate fuel flow, the fuel filter should be inspected monthly for contamination or damage. Perform the following to service the fuel filter:

1. Open the shroud.
2. Turn the fuel valve OFF.
3. *Electric start models*: Disconnect the negative battery cable.

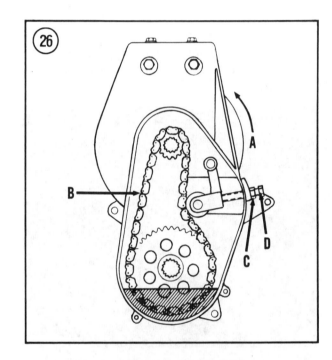

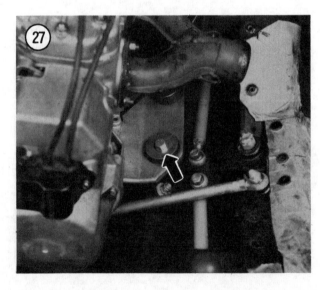

4. Locate the fuel filter in the engine compartment.

5. Disconnect the 2 hoses at the fuel filter. Plug the hoses to prevent fuel leakage and hose contamination.

6. Visually inspect the fuel filter for contamination or debris buildup.

7. If the filter is dirty, clean it by backflushing the filter with clean solvent. If the filter is still contaminated, install a new filter.

8. Install the fuel filter while noting the following:

   a. Reconnect the hoses at the fuel filter so that the arrow on the filter points *toward* the fuel pump.

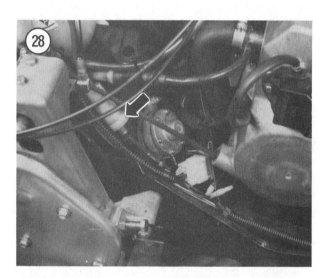

b. Make sure both hoses bottom out on the fuel filter fittings.

c. When installing new fuel filter hoses, make sure the hoses run fairly straight from the fuel tank to the filter and from the filter to the fuel pump; there should be no kinks or loops along any point of the hose(s). If necessary, cut one end of the hose to fit properly.

9. Reverse Steps 1 and 2 to complete installation.

Oil Filter Replacement

An inline oil filter (**Figure 29**) is mounted between the oil reservoir tank and engine. To ensure adequate oil flow, the oil filter should be inspected monthly for contamination or damage. Perform the following to service the oil filter.

1. Open the shroud.

2. Locate the oil filter in the engine compartment.

3. Disconnect the 2 hoses at the oil filter; discard the oil filter. Plug the hoses to prevent oil leakage and hose contamination.

4. Install the oil filter while noting the following:

   a. Reconnect the hoses at the oil filter so that the arrow on the filter (if so marked) points *toward* the engine.

   b. Make sure both hoses bottom out on the oil filter fittings.

   c. When installing new oil filter hoses, make sure the hoses run fairly straight from the reservoir tank to the filter and from the filter to the engine; there should be no kinks or loops along any point of the hose(s). If necessary, cut one end of the hose to fit properly.

   d. Bleed the oil pump as described in Chapter Nine.

5. Close and secure the shroud.

Water Pump Drive Belt And Water Pump Check

The drive belt should be checked for excessive wear, damage and tension. The water pump bearings and seals should be checked for damage or leakage. Refer to Chapter Ten.

Drive and Driven Clutch Inspection

At periodic intervals, the drive and driven clutches should be disassembled and all components inspected for excessive wear or damage. Refer to Chapter Twelve.

Battery Electrolyte

Refer to Chapter Eight.

Spark Plugs

The spark plugs should be checked periodically for firing tip condition and gap. Refer to spark plug service under *Tune-Up* in this chapter.

Ski and Ski Runner Check

Check the skis for cracks, bending or other signs of damage. Raise the front of the snowmobile so that the skis clear the ground. Check ski movement by pivoting both skis up and down; skis should pivot smoothly with no sign of binding. If a ski is tight, refer to Chapter Fifteen for ski removal, installation and ski pivot bolt tightness.

Excessively worn or damaged ski runners (skags) reduce handling performance and can cause wear to the bottom of the ski. Because track and snow conditions determine runner wear, they should be inspected often. Check the ski runners (**Figure 30**) for wear and replace them if they are more than half worn at any point or cracked. Refer to Chapter Fifteen.

Throttle Cable Inspection

On all models, the main throttle cable enters a junction block where it splits into 2 or 3 separate carburetor cables and an oil pump cable. Because the main throttle cable is pulling 3 or 4 additional cables, the cable assembly must be properly maintained to ensure correct throttle

lever, carburetor and oil pump operation. A malfunctioning or damaged throttle cable may cause severe engine damage.

1. Support the snowmobile so that the skis are off the ground.
2. Open and secure the shroud.
3. Disassemble the throttle cable at the throttle lever (**Figure 31**). Check the end of the cable for fraying or kinking. Replace the throttle cable assembly if only one inner cable strand has broken.
4. Check that the throttle cable junction box in the engine compartment for cracks or damage. Make sure the top of the junction box where the

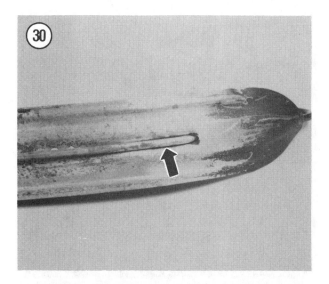

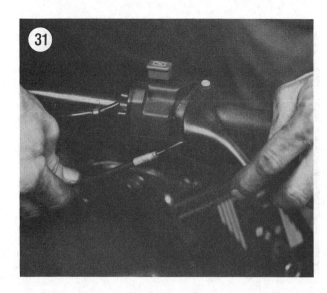

main throttle cable enters has not pulled off and that it is not damaged in anyway.

5. Check the throttle cables at the carburetors (**Figure 32**) for improper routing, kinking or other damage. Make sure the throttle cable enters the metal cable adjuster properly.

6. Check the oil pump cable from the junction box to the oil pump for the same conditions.

7. If the throttle cable assembly visually appears okay, reconnect the throttle cable at the throttle lever, then proceed to Step 8. If the throttle cable is damaged, replace it.

8. Turn the handlebar from side-to-side while checking the throttle cables for incorrect routing

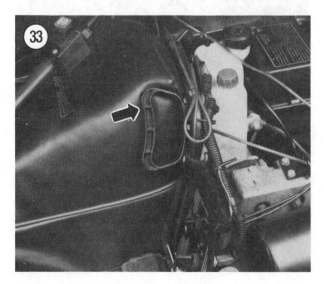

or other conditions that could cause the throttle lever to malfunction during operation.

9. Operate the throttle lever. The lever should move smoothly without any signs of roughness.

10. Turn the handlebar from side-to-side while operating the throttle lever. The lever should move smoothly as described in Step 9.

11. If the throttle lever failed to operate properly or if the throttle cable routing is incorrect, repair or correct the problem before starting the engine.

Throttle Cable Adjustment

If excessive play is felt at the throttle lever, adjust the carburetor as described under *Throttle Lever Free Play* in the *Tune-up* section in this chapter.

> *NOTE*
> *A throttle safety switch is installed in the throttle assembly. If any loss of spring tension occurs within the throttle lever assembly, the safety switch will turn the engine off (after the throttle lever is released). Because of the nature of this system, excessive throttle cable free play will cause the safety switch to override the ignition system and the engine will not start. Throttle cable adjustment should be maintained as described in this chapter.*

NON-SCHEDULED MAINTENANCE

The following service items should be checked frequently and serviced as required.

Recoil Starter

Pull out the starter rope (**Figure 33**) and inspect it for fraying. If its condition is questionable, replace the rope as described in Chapter Eleven.

Check the action of the starter. It should be smooth, and when the rope is released, it should return all the way. If the starter action is rough or if the rope does not return, service the starter as described in Chapter Eleven.

Exhaust System

The exhaust system is a vital link to engine performance and operation. Check the exhaust system from the cylinder exhaust port to the muffler for damaged gaskets, loose or missing fasteners or damaged components. Refer to Chapter Seven.

Shock Absorbers

The front and rear shock absorbers (**Figure 34**) should be checked for oil leakage, a bent shaft or housing damage. Because the front and rear shock absorbers are non-rebuildable, any sign of oil leakage or other damage requires replacement of the shock absorber.

Along with visual inspection, the damping of each shock absorber should be checked periodically. Remove the shock absorber(s) as described in Chapter Fifteen or Chapter Sixteen. Then remove the shock springs (if used) as described in Chapter Fifteen. With the spring removed, push the damper rod into the shock housing and then pull it out. There should be considerable resistance felt when pulling the damper rod out of the housing. If there is no resistance or if the resistance differs from a mating shock absorber, replace the shock absorber. Reinstall the springs and shock absorbers as described in Chapter Fifteen or Chapter Sixteen.

Rear Suspension
Rail Hi-fax Inspection

The hi-fax (**Figure 35**) mounted onto the bottom of the rear suspension rails should be checked weekly. Note the following:
1. Visually inspect the hi-fax (**Figure 35**) for cracks, severe wear or other damage.
2. Measure the thickness of the hi-fax (**Figure 36**) with a ruler or caliper. If the hi-fax thickness is 3/8 in. or less at any point along the rail, replace both hi-fax strips as described in Chapter Sixteen.

Track Inspection

Inspect the track as described in Chapter Sixteen.

Fasteners

All engine, steering and suspension fasteners should be checked weekly for tightness. Also replace damaged or missing fasteners or washers as required. Refer to the appropriate chapter for critical tightening torques.

Hose and Cable Inspection

The hoses and cables used on your Polaris snowmobile serve an important function. Check

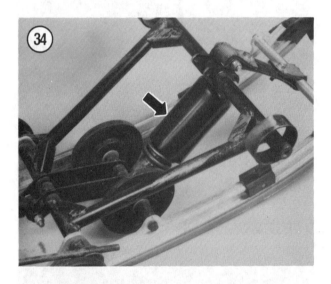

all hoses and cables weekly for wear or damage. Check also for proper routing. When installing new hoses or when reconnecting hoses, make sure hose ends bottom out on their fittings. This will allow proper installation of the hose clip to prevent the hose from slipping or being pulled off.

Electrical Connectors

Inspect the high-tension electrical leads to the spark plugs for cracks and breaks in the insulation and replace the leads if they are less than perfect; breaks in the insulation allow the spark to arc to ground and will impair engine performance.

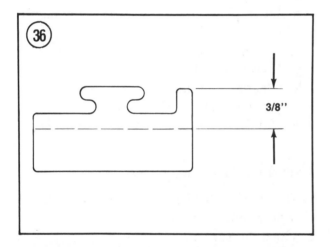

Check primary ignition wiring and lighting wiring for damaged insulation. Usually minor damage can be repaired by wrapping the damaged area with electrical insulating tape. If insulation damage is extensive, the damaged wires should be replaced.

Fuel System Inspection

Because gasoline is extremely flammable, the fuel system should be checked for loose or damaged connections. Note the following:
  a. Check all carburetor vent hoses for contamination, kinking or other damage.
  b. The fuel pump-to-engine pulse hose should be inspected monthly for cracks or other damage that results from age or secondary damage. Worn or damaged hoses will cause air leaks that will result in intermittent operating problems. Check the pulse hose from the fuel pump to the engine crankcase for loose connections or damage. Replace the pulse hose when necessary. **Figure 37** shows a typical pulse hose fitting.
  c. Check the fuel hoses at the carburetors for cracks, drying or other damage.
  d. Check the fuel tank vent hose for contamination or other damage.
  e. Replace worn or damaged hoses.

Body Inspection

Damaged body panels should be repaired or replaced.

Body Fasteners

Tighten any loose body bolts. Replace loose rivets by first drilling out the old rivet and then installing a new one with a pop-riveter. This tool, along with an assortment of rivets, is available through many hardware and auto parts stores. Follow the manufacturer's instructions for installing rivets.

Welded joints should be checked for cracks and damage. Damaged welded joints should be repaired by a competent welding shop.

Guide Wheel Inspection

Inspect the rubber on the guide wheels for wear and damage. Replace the wheels if they are in poor condition. Refer to Chapter Sixteen.

Driveshaft Sprockets

Inspect the teeth on the driveshaft sprockets for wear and damage. If the sprockets are damaged, replace them as described in Chapter Fourteen.

Fuel Tank Cleaning

The fuel tank should be removed and thoroughly flushed once a season. Refer to Chapter Seven.

Oil Tank

Inspect the oil tank for cracks and abrasions that are leaking or may soon be and replace the tank if its condition is in doubt.

Electrical System

All of the switches should be checked for proper operation. Refer to Chapter Eight.

Abnormal Engine Noise

> *WARNING*
> *Never lean into the snowmobile's engine compartment while wearing a scarf or other loose clothing when the engine is running or when attempting to start the engine. If the scarf or clothing should catch in the drive belt or clutch, severe injury could occur. Make sure the belt guard is in place.*

Open the shroud. Then start the engine and listen for abnormal noises. This could be a rattle indicating a loose fastener or a loud damaging engine sound. Periodic inspection for abnormal engine noises can prevent engine failure later on.

ENGINE TUNE-UP

The number of definitions of the term "tune-up" is probably equal to the number of people defining it. For the purposes of this book, a tune-up is general adjustment and maintenance to insure peak engine performance.

The following paragraphs discuss each facet of a proper tune-up which should be performed in the order given.

Have the new parts on hand before you begin.

To perform a tune-up on your snowmobile, you will need the following tools and equipment:

a. 14 mm spark plug wrench.
b. Socket wrench and assorted sockets.
c. Torque wrench.
d. Phillips head screwdriver.
e. Spark plug wire feeler gauge and gapper tool.
f. Dial indicator.
g. Flywheel puller.
h. Compression gauge.
i. Timing light.
j. Uni-Syn gauge for carburetor adjustment.

Cylinder and Cylinder Head Nuts

The engine must be at room temperature for this procedure.

1. Open the shroud.

2. *Air-cooled models*: Remove the air shroud assembly (**Figure 38**).

3. Using a torque wrench, tighten each cylinder head nut equally in a crisscross pattern to the torque specification in **Table 7**. Refer to the torque sequence shown in **Figure 39**, **Figure 40** or **Figure 41**.

4. *Air-cooled models*: Install the air shroud assembly (**Figure 38**).

5. Close and secure the shroud.

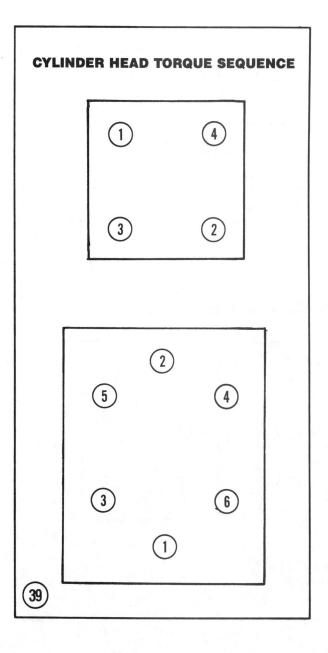

CYLINDER HEAD TORQUE SEQUENCE

Primary Compression Check

A cylinder cranking compression check is one of the quickest ways to check the internal condition of the engine: rings, head gasket, etc. It's a good idea to check compression at each tune-up, write it down, and compare it with the reading you get at the next tune-up. This will help you spot any developing problems.

1. Elevate and support the machine with a shielded stand so that the track is completely clear of the ground and free to rotate. Start and run the engine until it warms up to normal operating temperature. Turn the engine off.

> *WARNING*
> *Don't stand behind or in front of the machine when the engine is running, and take care to keep hands, feet and clothing away from the track when it is turning.*

> *CAUTION*
> *To prevent expensive engine damage, refer to **Caution** under **Spark Plug Removal** in this chapter before removing the spark plugs in Step 2.*

2. Using a spark plug socket, remove the spark plugs. Insert the plugs in the caps and ground the plugs to the cylinder head.

> *CAUTION*
> *If the plugs are not grounded during the compression test, the CDI ignition could be damaged.*

3. Screw a compression gauge into one spark plug hole or, if you have a press-in type gauge, hold it firmly in position.

4. Check that the kill switch is pushed down.

5. Hold the throttle wide open (**Figure 42**) and crank the engine several revolutions with the starter rope until the gauge gives its highest reading. Record the reading. Remove the pressure tester and relieve the pressure valve.

6. Repeat for the remaining cylinder(s).

40

CYLINDER HEAD TORQUE SEQUENCE

```
    (11)      (5)      (3)      (8)      (13)

(10)                (1)      (2)                (9)

    (14)      (7)      (4)      (6)      (12)
```

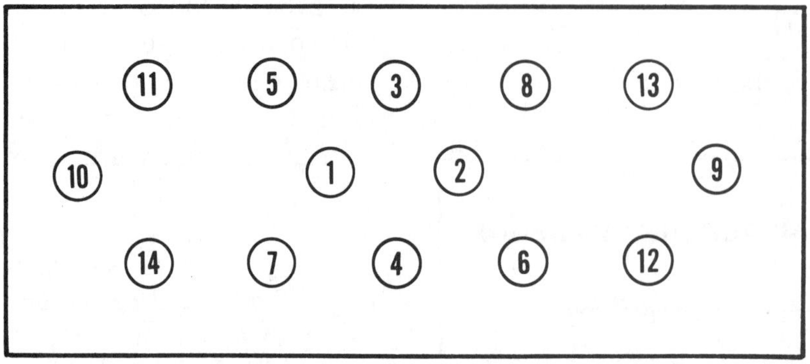

41

CYLINDER HEAD TORQUE SEQUENCE

```
 (1)  (7)  (4)      (1)  (7)  (4)      (1)  (7)  (4)

 (5)       (6)      (5)       (6)      (5)       (6)

 (3)  (8)  (2)      (3)  (8)  (2)      (3)  (8)  (2)
```

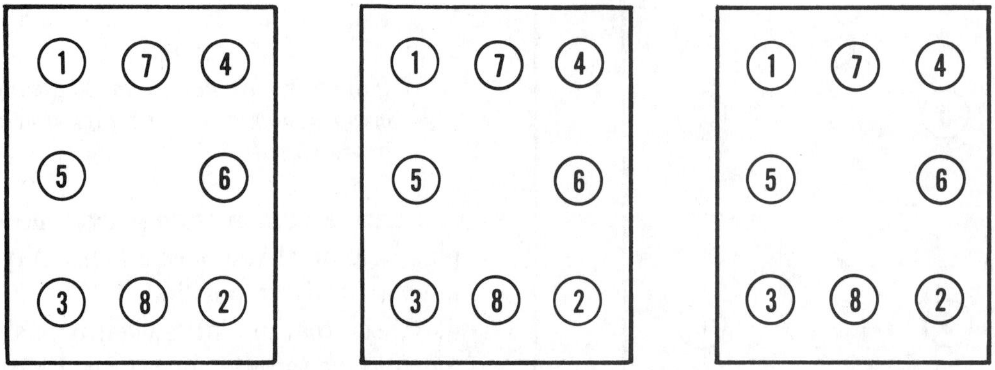

8mm: numbers 5, 6, 7 & 8
10mm: numbers 1, 2, 3 & 4

7. There should be no more than a 10% difference in compression between cylinders. For example, if the high reading is 140 in.-lbs., the low reading should not be below 126 in.-lbs.

8. If the compression is very low, it's likely that a ring is broken or there is hole in the piston.

9. Reinstall the spark plugs and reconnect the spark plug caps.

10. Close and secure the shroud.

Correct Spark Plug Heat Range

The proper spark plug is very important in obtaining maximum performance and reliability. The condition of a used spark plug can tell a trained mechanic a lot about engine condition and carburetion.

Select plugs of the heat range designed for the loads and conditions under which the snowmobile will be run. Use of incorrect heat ranges can cause a seized piston, scored cylinder wall, or damaged piston crown.

In general, use a hot plug for low speeds and low temperatures. Use a cold plug for high speeds, high engine loads and high temperatures. The plug should operate hot enough to burn off unwanted deposits, but not so hot that they burn themselves or cause preignition. A spark plug of the correct heat range will show a light tan color on the portion of the insulator within the

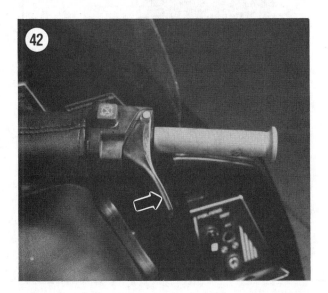

cylinder after the plug has been in service. See **Figure 43**.

The reach (length) of a plug is also important. A shorter than normal plug will cause hard starting, reduced engine performance and carbon buildup on the exposed cylinder head threads. A longer than normal plug could interfere with the piston or cause overheating; both conditions result with permanent and severe engine damage. Refer to **Figure 44**.

The standard heat range spark plug for the various models is is listed in **Table 8**.

Spark Plug Removal/Cleaning

1. Grasp the spark plug leads as near the plug as possible and pull it off the plug. If it is stuck to the plug, twist it slightly to break it loose. See **Figure 45**.

2. Blow away any dirt that has accumulated next to the spark plug base.

CAUTION
The dirt could fall into the cylinder when the plug is removed, causing serious engine damage.

3. Remove the spark plug with a 14 mm spark plug socket.

NOTE
If the plug is difficult to remove, apply penetrating oil, like WD-40 or Liquid Wrench, around the base of the plug and let it soak in about 10-20 minutes.

4. Inspect the plug carefully. Look for a broken center porcelain, excessively eroded electrodes, and excessive carbon or oil fouling. See **Figure 43**.

Gapping and Installing the Plug

A new spark plug should be carefully gapped to ensure a reliable, consistent spark. You must use a special spark plug gapping tool and a wire feeler gauge.

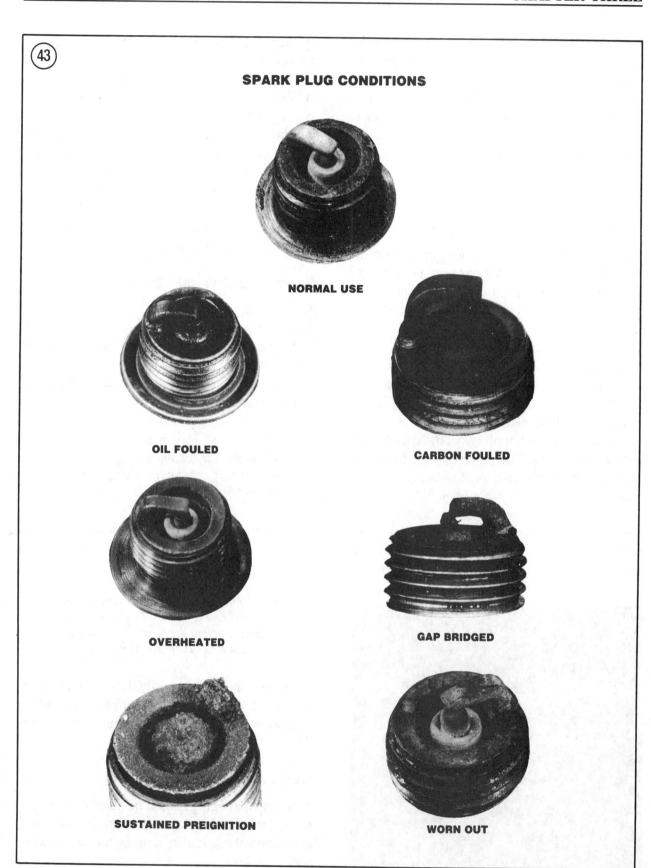

SPARK PLUG CONDITIONS

NORMAL USE

OIL FOULED

CARBON FOULED

OVERHEATED

GAP BRIDGED

SUSTAINED PREIGNITION

WORN OUT

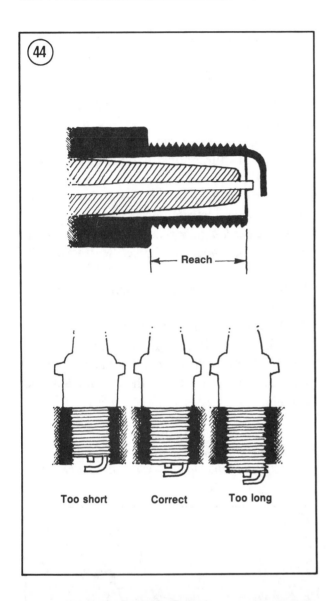

Reach

Too short — Correct — Too long

1. Insert a wire feeler gauge between the center and side electrode (**Figure 46**). The correct gap is listed in **Table 8**. If the gap is correct, you will feel a slight drag as you pull the wire through. If there is no drag, or the gauge won't pass through, bend the side electrode with a gapping tool (**Figure 47**) to set the proper gap.

> *NOTE*
> *Never try to close the spark plug gap by tapping the spark plug on a solid surface. This can damage the plug internally. Always use the special tool to open or close the gap.*

2. Apply anti-seize to the plug threads before installing the spark plug.

> *NOTE*
> *Anti-seize can be purchased at most automotive parts stores.*

3. Screw the spark plug in by hand until it seats. Very little effort is required. If force is necessary, you have the plug cross-threaded. Unscrew it and try again.

4. If you are using a torque wrench, tighten the spark plugs to the torque specification listed in **Table 7**. If you are not using a torque wrench, tighten the plug an additional 1/4 to 1/2 turn after the gasket has made contact with the head. If you

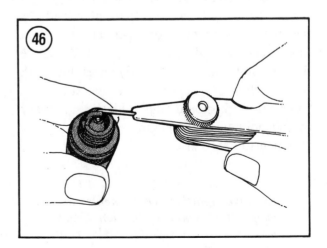

are installing an old, regapped plug and reusing the old gasket, only tighten an additional 1/4 turn.

> *CAUTION*
> *Do not overtighten. This will only squash the gasket and destroy its sealing ability. This could cause compression leakage around the base of the plug.*

5. Install the spark plug wires. Make sure they snap onto the top of the plugs tightly.

> *CAUTION*
> *Make sure the spark plug wires are pulled away from the exhaust pipe.*

Reading Spark Plugs

Much information about engine and spark plug performance can be determined by careful examination of the spark plugs. Refer to Chapter Four.

Ignition Timing

These models are equipped with a capacitor discharge ignition (CDI). This system uses no breaker points and greatly simplifies ignition timing and makes the ignition system much less susceptible to failures caused by dirt, moisture and wear.

Dynamic Timing Check

Dynamic engine timing uses a timing light connected to the MAG side spark plug lead. As the engine is cranked or operated, the light flashes each time the spark plug fires. When the light is pointed at the moving flywheel, the mark on the flywheel appears to stand still. The flywheel mark should align with the stationary timing pointer on the engine.

1. Open the shroud.

> *NOTE*
> *Because ignition components are temperature sensitive, ignition timing must be checked when the engine is at*

its normal operating temperature; start and run engine for approximately 4 to 5 minutes.

2. Warm the engine to normal operating temperature.

3. Hook up a stroboscopic timing light according to the manufacturer's instructions on the MAG side spark plug lead.

4. Connect a tachometer according to the manufacturer's instructions.

5. *Liquid-cooled models*: Remove the timing plug from the upper crankcase. See **Figure 48**, typical.

6A. Remove the drive belt (Chapter Twelve) or perform Step 6B.

6B. Position the machine, on its skis, so the tips of the skis are against a wall or other immovable barrier. Elevate and support the machine so the track is completely clear of the ground and free to rotate.

> *WARNING*
> *Don't allow anyone to stand behind or in front of the machine when the engine*

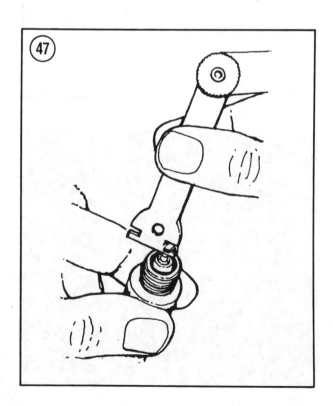

(47)

is running, and take care to keep hands, feet and clothing away from the track when it is running.

7. Start the engine.

8. When the engine has reached its normal operating temperature, run the engine idle up to 3,000 rpm and point the timing light at the crankcase inspection hole (**Figure 48**) or to the timing mark on the fan housing (**Figure 49**). The light will flash as the flywheel timing mark aligns with the pointer. See **Figures 50-53** for timing mark identification. Note the ignition timing and

turn the engine off. Lower the snowmobile to the ground. Note the following:

   a. Dynamic ignition timing specifications are listed in **Table 9**.

   b. If the ignition timing is incorrect, perform the *Static Timing Check* in this chapter and then repeat this procedure. If the ignition timing is still incorrect, proceed to Step 9.

   c. If the ignition timing is correct, proceed to Step 11.

9. Remove the recoil starter assembly and starter pulley as described in Chapter Eleven.

NOTE
Some of the following photographs are shown with the engine removed for clarity.

10. Ignition timing is changed by moving the stator plate. Note the following:

   a. Each mark on either side of the center flywheel mark represents different degrees of crankshaft rotation. See **Figures 50-53**. See **Table 9** for dynamic timing specifications. Ignition timing is changed by moving the stator plate.

   b. If the ignition timing was retarded, rotate the stator plate counterclockwise. If the ignition timing was advanced, rotate the stator plate clockwise.

   c. To change the ignition timing, insert a Phillips screwdriver through the cutouts in the flywheel face (**Figure 54**) and loosen the stator plate screws (A, **Figure 55**). Then turn the stator plate counterclockwise to advance or clockwise to retard ignition timing.

NOTE
Turn the stator plate by inserting a screwdriver into the stator plate notch and pivoting the screwdriver against the crankcase tabs. See B, Figure 55.

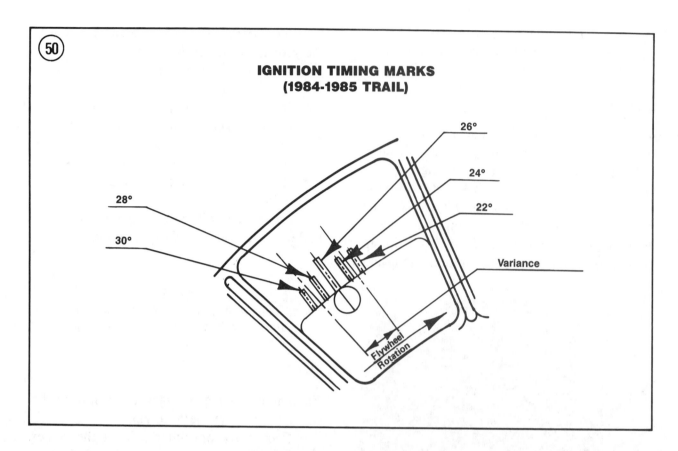

(50)

IGNITION TIMING MARKS
(1984-1985 TRAIL)

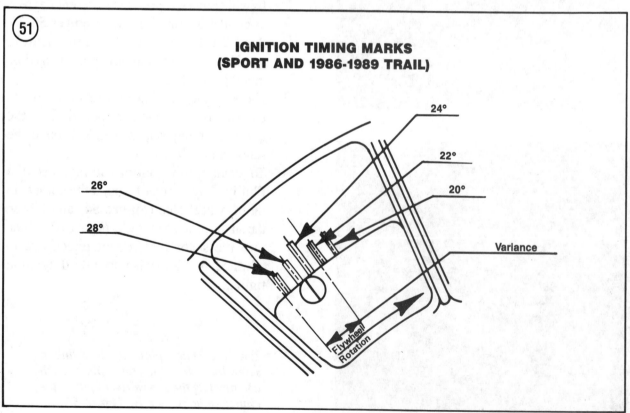

(51)

IGNITION TIMING MARKS
(SPORT AND 1986-1989 TRAIL)

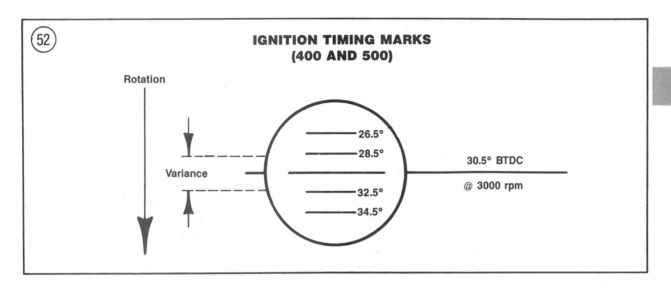

**IGNITION TIMING MARKS
(400 AND 500)**

Rotation

Variance

26.5°
28.5°
30.5° BTDC
@ 3000 rpm
32.5°
34.5°

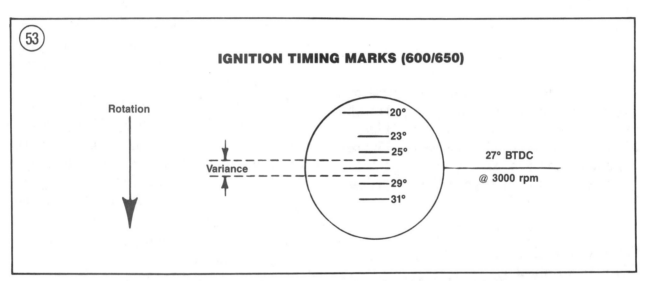

IGNITION TIMING MARKS (600/650)

Rotation

Variance

20°
23°
25°
27° BTDC
@ 3000 rpm
29°
31°

d. Tighten the stator plate set screws.

e. Reinstall the starter pulley and recoil starter assembly.

f. Repeat Steps 6-8 to recheck ignition timing. Repeat procedure until ignition timing is correct. If ignition timing cannot be corrected, troubleshoot the ignition system as described in Chapter Two.

11. Remove the timing light and tachometer.

12. *Liquid-cooled models*: Install the timing hole plug (**Figure 48**).

13. Reinstall the drive belt or lower the snowmobile to the ground and close the shroud.

Static Timing Check (Without Dial Indicator)

1. Remove the flywheel as described in Chapter Eight.

2. Static timing can be set on all models by aligning the index mark on the stator plate with the crankcase parting line as shown in **Figure 56**. Tighten the stator plate screws (A, **Figure 55**) after aligning marks.

3. Reinstall the flywheel as described in Chapter Eight.

Static Timing Check (With Dial Indicator)

An accurate dial indicator will be required to determine top dead center (TDC) of the piston when performing this procedure. TDC is determined by removing a spark plug and installing the dial indicator in the plug opening. The static timing method is used to verify or determine the following:

a. Verify factory timing marks.

b. Detect a broken or missing flywheel Woodruff key.

c. Detect a twisted crankshaft.

d. To scribe timing marks on a new flywheel.

1. Open and secure the shroud.

2. Remove the drive belt as described in Chapter Twelve.

3. Remove the spark plugs as described in this chapter.

4. *Liquid-cooled models*: Remove the timing hole plug from the upper crankcase. See **Figure 48**, typical.

5. Install and position a dial indicator as follows:

a. Screw the extension onto a dial indicator and insert the dial indicator into the adaptor (**Figure 57**).

b. Screw the dial indicator adaptor into the cylinder head on the MAG side. Do not lock the dial indicator in the adaptor at this time.

c. Rotate the flywheel (by turning the primary sheave) until the dial indicator rises all the way up in its holder (piston is approaching top dead center). Then slide the indicator far enough into the holder to obtain a reading.

d. Lightly tighten the set screw on the dial indicator adaptor to secure the dial gauge.

e. Rotate the flywheel until the dial on the gauge stops and reverses direction. This is top dead center. Zero the dial gauge by aligning the zero with the dial needle (**Figure 58**).

f. Tighten the set screw on the dial indicator adaptor securely.

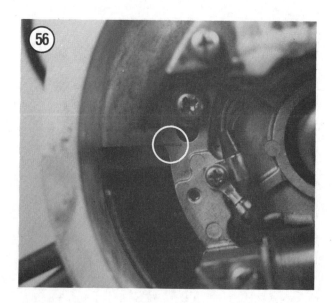

6. Rotate the crankshaft counterclockwise (viewed from the right-hand side) until the gauge needle has made approximately 3 revolutions. Then carefully turn the crankshaft clockwise until

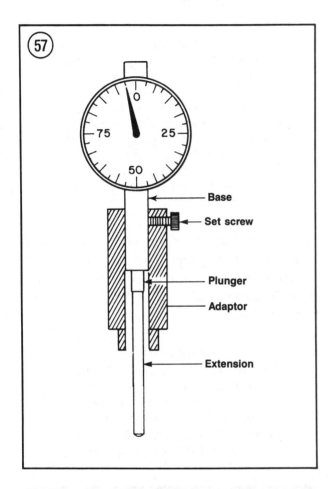

the gauge indicates the timing shown in **Table 10**. View the timing marks through the hole in the crankcase (**Figure 48**) or on the fan housing (**Figure 49**). See **Figures 50-53** for timing marks. The center timing mark should align. If the timing is incorrect, note the following:

  a. Remove the flywheel and check for a broken or missing flywheel Woodruff key.

  b. The crankshaft may be twisted; refer to Chapter Five or Chapter Six for crankshaft service.

  c. If the above checks do not point to the cause of the inaccuracy, the flywheel timing marks may be inaccurate.

  d. Scribe a new mark on the flywheel that aligns with the crankcase center mark. This mark will be used as the reference when using the timing light.

  e. Repeat the timing procedure to check the accuracy of the new mark.

7. Remove the gauge and adaptor. Install the spark plugs and connect the high-tension leads.

8. *Liquid-cooled models*: Install the timing hole plug (**Figure 48**).

9. Reinstall the drive belt (Chapter Twelve).

Carburetor Adjustment

For maximum engine performance, the cylinders must work equally. If one cylinder's throttle valve opens earlier, that cylinder will be required to work harder. This will cause poor acceleration, rough engine performance and engine overheating. For proper carburetor synchronization, the throttle valves must lift at the same time.

Because of throttle cable stretch, carburetor synchronization should be checked at each tune-up or whenever the engine suffers from reduced performance.

When adjusting the carburetors with the engine running, the air silencer must be connected to the carburetors with the mating air boots. Failure to have this assembly properly installed will cause a lean fuel mixture, resulting in engine seizure.

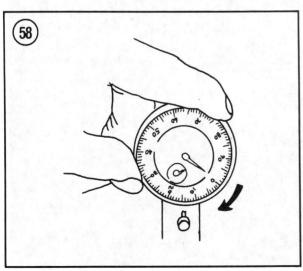

NOTE
A throttle safety switch is installed in the throttle assembly to monitor throttle spring tension. If any loss of spring tension occurs within the throttle lever assembly, the safety switch will turn the engine off (after the throttle lever is released). Because of the nature of this system, excessive throttle cable free play will cause the safety switch to override the ignition system and the engine will not start. If the engine will not start after performing this procedure, perform the **Throttle Lever Free Play** *adjustment in this chapter.*

1. With the engine turned off, open and secure the shroud.
2. Check the choke cable adjustment as described in this chapter.
3. Remove the air box (**Figure 59**) as described in Chapter Seven.
4. Back out the carburetor idle speed screws (A, **Figure 60**) so that the carburetor throttle valves (B, **Figure 60**) drop all the way to the bottom.
5. Loosen the throttle cable locknut (C, **Figure 60**) and turn the throttle cable adjuster (D, **Figure 60**) until the throttle valves (B, **Figure 60**) at each carburetor lift at the same time when the throttle lever is pulled in. When the throttle valves lift simultaneously, tighten the cable locknuts.
6. Turn one carburetor idle speed screw (A, **Figure 60**) clockwise until it just touches the carburetor slide. Then turn the screw 2 turns clockwise. Repeat for each carburetor.
7. Check throttle lever free play as described in this chapter.
8. Operate the throttle lever and with a finger lightly placed on the bottom of each throttle valve, gauge how quickly the throttle valves reach the top of their carburetor bore in relation to each other. For proper synchronization, the throttle valves must reach the top at the same time. If necessary, readjust the throttle cables so that the valves arrive at the same time by turning one of the cable adjusters at the top of the carburetor cap (D, **Figure 60**). When the throttle valves are

moving at the same time, tighten the throttle cable locknuts (C, **Figure 60**).
9. Operate the throttle lever a few times, then recheck synchronization as described in Step 8. Readjust as necessary.

WARNING
When operating the throttle lever in Step 9, make sure the throttle valves move smoothly and return to their closed position against the idle speed screw. If the throttle valves do not move smoothly or if they fail to return properly, check the throttle cables for kinks or other damage. Do not start the engine until the throttle valves and throttle lever are operating correctly.

10. Turn both pilot air screws (E, **Figure 60**) in until they lightly seat. Then back the screws out the number of turns specified in **Table 11**.
11. Reinstall the air box (**Figure 59**). See Chapter Seven.

WARNING
Never lean into the snowmobile's engine compartment while wearing a scarf or other loose clothing when the engine is running or when attempting to start the engine. If the scarf or clothing should catch in the drive belt or clutch, severe

59

injury or death could occur. Make sure the pulley guard is in place.

12. Start the engine and warm to operating temperature. Check idle speed (see **Table 12**). If necessary, turn idle speed adjustment screws (A, **Figure 60**) in equal amounts to obtain specified idle speed.

CAUTION
*Do not use the pilot air screws to attempt to set engine idle speed. Pilot air screws must be set as specified in **Table 11** or a "too lean" mixture and subsequent engine damage may result.*

13. After adjusting the carburetors, check the oil pump cable adjustment as described in Chapter Nine.

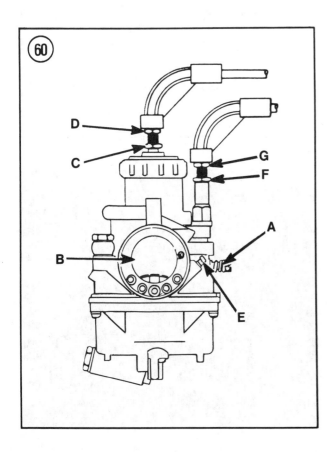

Uni-Syn Carburetor Adjustment

To obtain a precise synchronization of multiple carburetor models, the Uni-Syn air flow meter can be used.

WARNING
Never lean into the snowmobile's engine compartment while wearing a scarf or other loose clothing when the engine is running or when the driver is attempting to start the engine. If the scarf or clothing should catch in the drive belt or clutch, severe injury or death could occur. Make sure the pulley guard is in place.

1. Position the machine, on its skis, so the tips of the skis are against a wall or other immovable barrier. Elevate and support the machine so the track is completely clear of the ground and free to rotate.

WARNING
Don't allow anyone to stand behind or in front of the machine when the engine is running, and take care to keep hands, feet and clothing away from the track when it is running.

2. Perform *Carburetor Adjustment* as described in this chapter.
3. Remove the air box (**Figure 59**) as described in Chapter Seven.
4. Start the engine. Bind and wedge the throttle lever to maintain engine speed at 3,000 rpm (**Figure 61**).
5. Place the Uni-Syn air flow meter over the PTO side carburetor throat (**Figure 62**). The tube on meter must be vertical.
6. Slowly close the air flow control until float in tube aligns with centered mark on sight tube.
7. Without changing the adjustment of the air flow control, place the meter on the opposite carburetor(s). If the carburetors are equal, no adjustment is necessary.
8. If adjustment is required, loosen the locknut on the carburetor with the lowest float level and

turn the cable adjuster (D, **Figure 60**) until the float level matches that of the other carburetor(s).

9. Return the engine to idle and repeat this procedure. Turn the idle speed screw (A, **Figure 60**) as required to obtain a balanced idle.

10. Reinstall the air box (Chapter Seven).

11. Lower the track to the ground and close the shroud.

Throttle Lever Free Play

Measure the throttle lever clearance at the point indicated in **Figure 63**. The clearance should be 0.010-0.030 in. If clearance is incorrect, loosen the throttle cable locknuts (C, **Figure 60**) and turn the cable adjusters (D, **Figure 60**) to obtain the correct throttle lever clearance. Turn the throttle cable adjusters in *equal* amounts. Tighten the locknuts and recheck the clearance.

NOTE
Turning the throttle cable adjusters will affect carburetor synchronization. If necessary, check synchronization by performing the **Carburetor Adjustment** *as described in this chapter.*

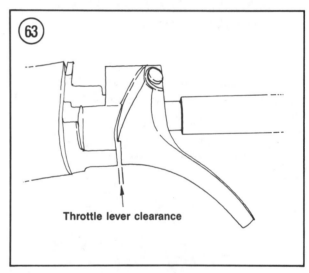

Throttle lever clearance

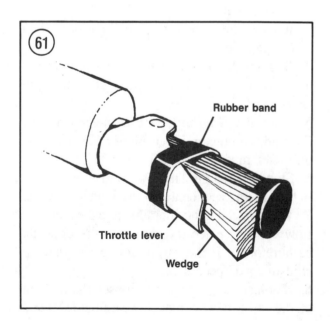

Rubber band

Throttle lever

Wedge

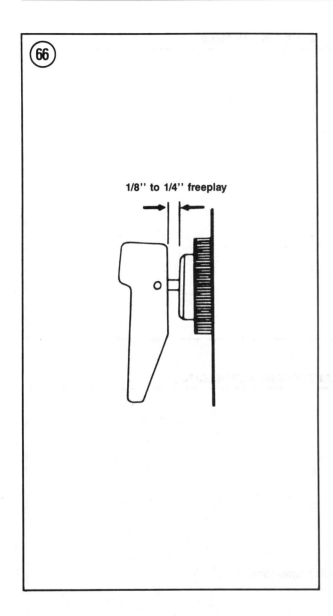

1/8'' to 1/4'' freeplay

Choke Adjustment

Incorrect choke adjustment can cause engine flooding (not enough cable slack) or hard starting (too much cable slack).

1. Open the shroud.

2. Check the choke cables from the toggle switch to the carburetors (**Figure 64**) for damage. If the cables are okay, perform the following.

3. Flip the choke control toggle switch to its OFF position (**Figure 65**).

4. Loosen the choke cable adjuster locknut on each carburetor (F, **Figure 60**). Then turn the choke cable adjusting nut (G, **Figure 60**) clockwise to provide 1/4 in. of free play at the toggle switch (**Figure 66**). Repeat for each carburetor.

5. When the toggle switch free play is correct (Step 4), turn the choke cable adjusting nut (G, **Figure 60**) counterclockwise until the toggle switch has no free play. Then rotate the choke cable adjusting nut (G, **Figure 60**) clockwise to provide 1/8-1/4 in. toggle switch free play (**Figure 66**). Repeat for each carburetor.

6. When the choke cable adjustment is correct, tighten each choke cable adjuster locknut (F, **Figure 60**).

7. Check choke adjustment by operating the toggle switch in the half-on and full-on positions (**Figure 65**). Toggle switch should move freely with no sign of binding.

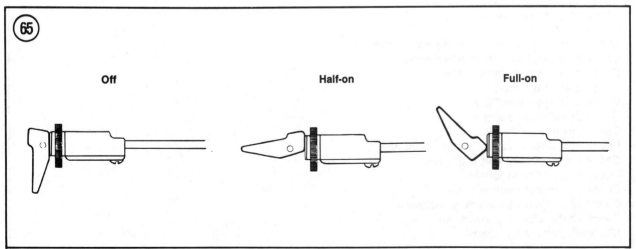

Off   Half-on   Full-on

Table 1 WEEKLY MAINTENANCE SCHEDULE

Check oil injection tank level
Check chaincase oil level
Check coolant level
Check for coolant system leakage
Check carburetor water trap (if so equipped)
Check drive belt condition
Check drive belt tension
Lubricate choke slide and cables
Lubricate primary sheave (bi-weekly)
Check kill switch and throttle safety switch operation
Check headlight and taillight lenses
Check headlight and taillight operation
Check steering bolt tightness
Check steering operation
Check rear suspension mounting bolt tightness
Check front limiter strap(s) for excessive wear or damage
Check front limiter strap bolt tightness
Check track alignment and adjustment
Check brake operation and adjustment

Table 2 INITIAL 150 MILE MAINTENANCE SCHEDULE

Check chain and sprockets for wear
Check and adjust drive chain tension
Flush chaincase housing and reinstall cover
Refill chaincase with new oil
Check all hoses for leakage and tightness
Check heat exchangers for damage
Check coolant level
Check drive belt condition
Check drive belt tension
Check clutch offset adjustment
Check clutch center distance adjustment
Check battery electrolyte level (if so equipped)
Check emergency shut off switch and throttle safety switch operation
Check headlight and taillight operation
Check brake light operation
Check steering bolt tightness
Check steering operation
Check ski alignment
Check rear suspension mounting bolt tightness
Check track alignment and adjustment
Check brake system for leakage
Check brake fluid level
Check brake pad condition
Check brake disc condition
Check brake lever travel
Check condition of rubber engine mounts
Check engine mount bolt tightness
Check cylinder head torque
Check and adjust ignition timing
Check and adjust carburetor adjustment
Check choke plunger adjustment
Check oil pump adjustment

Table 3 1,000 MILE MAINTENANCE SCHEDULE

Check and adjust drive chain tension
Flush chaincase housing and reinstall cover
Refill chaincase with new oil
Replace fuel filter
Replace oil filter
Lubricate front suspension components
Check all hoses for leakage and tightness
Check heat exchangers for damage
Check coolant level
Check drive belt condition
Check drive belt tension
Check clutch offset adjustment
Check clutch center distance adjustment
Check battery electrolyte level (if so equipped)
Check emergency shut off switch and throttle safety switch operation
Check headlight and taillight operation
Check brake light operation
Check steering bolt tightness
Check steering operation
Check ski alignment
Check ski wear bar condition
Check rear suspension mounting bolt tightness
Check track alignment and adjustment
Check brake system for leakage
Check brake fluid level
Check brake pad condition
Check brake disc condition
Check brake lever travel
Check condition of rubber engine mounts
Check engine mount bolt tightness
Check and adjust carburetor adjustment
Check oil pump adjustment

Table 4 2,000 MILE MAINTENANCE SCHEDULE

Check and adjust drive chain tension
Flush chaincase housing and reinstall cover
Refill chaincase with new oil
Replace fuel filter
Replace oil filter
Lubricate front suspension components
Remove water pump drive belt; check water pump bearings and seals for damage; reinstall drive belt and
  check alignment
Check all hoses for leakage and tightness
Check heat exchangers for damage
Check coolant level
Disassemble drive clutch; inspect all components
Disassemble driven clutch; inspect all components
Check drive belt condition
Check drive belt tension
Check clutch offset adjustment
Check clutch center distance adjustment

(continued)

Table 4 2,000 MILE MAINTENANCE SCHEDULE (continued)

Check battery electrolyte level (if so equipped)
Check emergency shut off switch and throttle safety switch operation
Check headlight and taillight operation
Check brake light operation
Check steering bolt tightness
Check steering operation
Check ski alignment
Check ski wear bar condition
Check rear suspension mounting bolt tightness
Check track alignment and adjustment
Check brake system for leakage
Check brake fluid level
Check brake pad condition
Check brake disc condition
Check brake lever travel
Check condition of rubber engine mounts
Check engine mount bolt tightness
Decarbonize engine top end; retorque cylinder heads
Install new spark plugs
Check and adjust carburetor adjustment
Check oil pump adjustment

Table 5 RECOMMENDED LUBRICANTS AND FUEL

Engine oil	Polaris Injection Oil
Coolant	Glycol-based automotive type antifreeze compounded for aluminum engines
Chaincase	Polaris Chaincase Oil
Grease	Low-temperature grease
Fuel	88 minimum octane leaded*
Brake fluid	DOT 3
Throttle cable	Polaris Clutch and Cable Lubricant
Choke cable	Polaris Clutch and Cable Lubricant
Clutch and drive system	Polaris Clutch and Cable Lubricant

* Unleaded gasoline with a minimum 88 octane rating can be used in place of leaded gasoline. However, do not use gasoline containing ethanol or methanol. In addition, Polaris specifies that gasohol, white gas or gasoline containing additives must not be used.

Table 6 APPROXIMATE REFILL CAPACITY

Chaincase housing	3 oz. (88.8 cc)
Oil tank	2.5 qts. (2.4 l)
Fuel tank	
1984-1987	7 gal. (26.5 l)
1988-1989	7.3 gal. (27.7 l)
Cooling system	
Twins	4 quarts
Triples	5 quarts

3

Table 7 MAINTENANCE TIGHTENING TORQUES

	ft.-lb.	N·m
Radius rods	25	34
Engine mounting bolts		
3/8 in.	34-38	46-52
7/16 in.	55-60	75-82
Cylinder head		
Sport and Trail	17-18	22-23
All other models		
8 mm	17-18	22-23
10 mm	24-26	33-35

Table 8 SPARK PLUGS

Year	NGK	Champion	Gap
1984-1986	BR9ES	RN-2C	0.5 mm (0.020 in.)
1987-on	BR9ES	RN-2C	0.6 mm (0.025 in.)

Table 9 IGNITION TIMING—DYNAMIC

Model	Peak timing @ rpm
Trail	
1984-1985	26.5° @ 3,000 rpm
1986-1989	24° @ 3,000 rpm
Sport	25.5° @ 3,000 rpm
400	30.5° @ 3,000 rpm
500	30.5° @ 3,000 rpm
600/650	27° @ 3,000 rpm

Table 10 IGNITION TIMING CHECK—WITH DIAL INDICATOR

Model	BTDC		Range BTDC	
	mm	in.	mm	in.
Sport	3.41	0.134	2.90-3.94	0.114-0.160
Trail				
1984-1985	3.93	0.155	3.39-4.56	0.133-0.179
1986-on	3.26	0.128	2.75-3.53	0.108-0.150
400	5.19	0.204	4.55-5.85	0.179-0.230
500	5.19	0.204	4.55-5.85	0.179-0.230
600/650	4.10	0.162	3.81-4.40	0.150-0.173

Table 11 PILOT AIR SCREW ADJUSTMENT

Model	Turns out
Trail	1.0
Sport	1.0
400	
1984-1986	1.0
1987-1989	1 1/2
500	1.0
600	
1984	1.0
1985-1986	3/4
1987	1.0
650	1.0

Table 12 IDLE SPEED ADJUSTMENT

Model	RPM
Trail	
1984-1985	2,300
1986	2,000
1987-1989	1,900
Sport	2,100
400	
1984-1986	2,200
1987-1989	1,900
500	1,900
600	
1984-1986	2,000
1987	1,900
650	1,900

Chapter Four

High-Altitude and Rear Suspension Adjustment

If the snowmobile is to deliver its maximum efficiency and peak performance, the engine and chassis must be properly adjusted. This chapter describes high altitude tuning and rear suspension adjustment.

Basic tune-up procedures are described in Chapter Three.

Tables 1-11 are found at the end of the chapter.

HIGH-ALTITUDE CARBURETOR TUNING

Polaris snowmobiles are tuned at the factory for sea level conditions. However, when the snowmobile is operated at a higher altitude, engine performance will drop because of a change in air density. At sea level, the air is quite a bit denser than the air at 10,000 feet. You should figure on a 3% loss of power output for every 1000 feet of elevation change (increase). This decrease in power is caused by a drop in cylinder pressure and a change in the fuel:air ratio. For

example, an engine that produces 40 horsepower at sea level will produce approximately 38.8 horsepower at 1,000 feet. At 10,000 feet, the engine produces 29.5 horsepower. With sea level jetting, the engine would run extremely rich at 10,000 feet.

Air temperature must also be considered when jetting the carburetor. For example, the carburetors are set at the factory to run at temperatures of 32 to minus 4° F (0 to minus 20° C) at sea level. If the snowmobile is to be operated under conditions other than those specified, the carburetors must be adjusted accordingly.

Figure 1 illustrates the different carburetor circuits and how they overlap during engine operation. **Tables 1-6** show carburetor specifications for sea level conditions.

Before adjusting the carburetor, make the following adjustments:

    a. Carburetor adjustment (including synchronization). See Chapter Three.

    b. Oil pump adjustment. See Chapter Nine.

NOTE
Changes in port shape and smoothness, expansion chamber, carburetor, etc., will also require jetting changes because these factors alter the engine's ability to breathe. When installing aftermarket equipment or when the engine has been modified, equipment manufacturers often include a tech sheet listing suitable jetting changes to correspond to their equipment or modification. This information should be taken into account along with altitude and temperature conditions previously mentioned.

NOTE
It is important to note that the following jetting guidelines should be used as guidelines only. Individual adjustments will vary because of altitude, temperature and snow conditions. The condition of the spark plugs should be used as the determining factor when changing jets and adjusting the carburetor.

Carburetor Adjustment

When the snowmobile is going to be run at a higher altitude, tune carburetor as follows. Refer to Chapter Seven for carburetor removal and disassembly.

Low speed tuning

The pilot jet and pilot air screw setting control the fuel mixture from 0 to about 1/4 throttle (**Figure 1**). In addition, the pilot air screw (A, **Figure 2**) controls mixture adjustment when the throttle is opened from idle to the full open position quickly and when the engine is run at half-throttle. Note the following when adjusting the pilot air screw:

a. Turning the pilot air screw clockwise enriches the fuel mixture.

b. Turning the pilot air screw counterclockwise leans the fuel mixture.

c. Pilot jets are identified by number (**Figure 3**). As the pilot numbers increase, the fuel mixture richens.

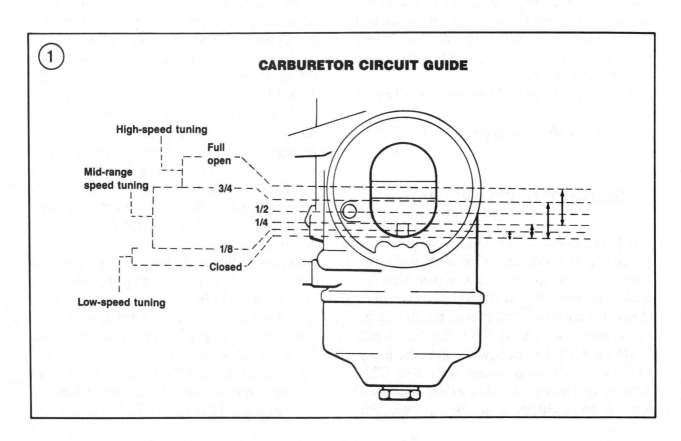

d. When operating the snowmobile in relatively warm weather or at a higher altitude, turn the pilot air screw clockwise. When operating the snowmobile in excessively cold weather conditions, turn the pilot air screw counterclockwise.

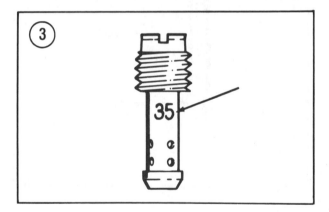

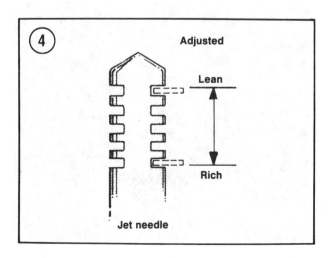

1. Open the shroud.

NOTE
Figure 2 shows the carburetor removed for clarity.

2. Locate the pilot air screws on the side of each carburetor (A, **Figure 2**).
3. Turn both pilot air screws in until they lightly seat. Then back the screws out the number of turns specified in **Table 7** for your model.
4. Start the engine and allow it to warm up to normal operating temperature.
5. Adjust the throttle stop screw (B, **Figure 2**) until the engine is idling at the rpm listed in **Table 8** for your model. Then slowly turn the pilot air screws (A, **Figure 2**) counterclockwise. (Turning the pilot air screws counterclockwise will richen the fuel mixture.) Continue to turn the pilot air screws counterclockwise until the highest engine rpm can be reached. When the highest engine rpm is reached, adjust the throttle stop screw (B, **Figure 2**) to set the idle speed to the specification listed in **Table 8**.
6. Operate the engine and check performance. If the engine performance is off at high altitudes or in extremely cold areas or if the off-idle pickup is poor, install larger pilot jets. Refer to Chapter Seven for carburetor removal and disassembly.
7. After replacing the pilot jets, repeat Steps 2-6.

Mid-range tuning

The jet needle controls the mixture at medium speeds, from approximately 1/4 to 3/4 throttle (**Figure 1**). The jet needle has 2 operating ends. The top of the needle has 5 evenly spaced circlip grooves (**Figure 4**). The bottom half of the needle is tapered; this portion extends into the needle jet. While the jet needle is fixed into position by the circlip, fuel cannot flow through the space between the needle jet and jet needle until the throttle valve is raised approximately 1/4 open.

As the throttle valve is raised, the jet needle's tapered portion moves out of the needle jet (**Figure 5**). The grooves permit adjustment of the mixture ratio. If the clip is raised (thus dropping the needle deeper into the jet), the mixture will be leaner; lowering the clip (raising the needle) will result in a rich mixture.

1. Open the shroud.

> *NOTE*
> *Prior to removing the top cap, thoroughly clean the area around it so that no dirt can fall into the carburetor.*

2. Unscrew and remove the carburetor top cap (**Figure 6**) and pull the throttle valve assembly (**Figure 7**) out of the carburetor.

3. Remove the jet needle from the throttle valve.

> *NOTE*
> *Some models have a washer installed on the jet needle. Don't loose it when removing the jet needle.*

4. Note the position of the clip (**Figure 4**) before removing it. Remove the clip and reposition it on the jet needle. Make sure the clip seats in the needle groove completely.

5. Reverse to install. Make sure the O-ring in the cap is positioned correctly (**Figure 8**) before installing the cap.

High-speed tuning

The main jet controls the mixture from 3/4 to full throttle and has some effect at lesser throttle openings (**Figure 1**). Each main jet is stamped with a number (**Figure 9**). Larger numbers provide a richer mixture, smaller numbers a leaner mixture. Refer to **Tables 1-6** for main jet sizes at sea level conditions. When operating the snowmobile in relatively warm weather or at a higher altitude, a smaller main jet should be

used. When operating the snowmobile in excessively cold weather conditions, install a larger main jet.

> *CAUTION*
> *The information given in Step 1 for determining main jet sizes should be used*

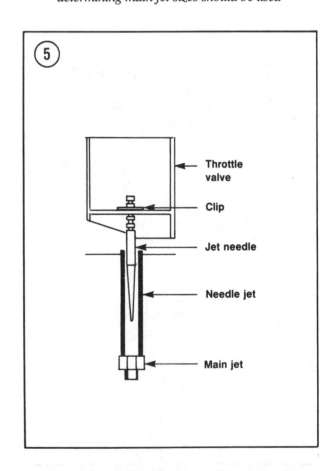

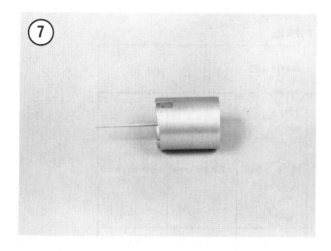

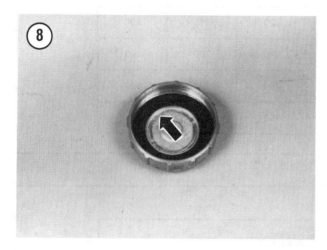

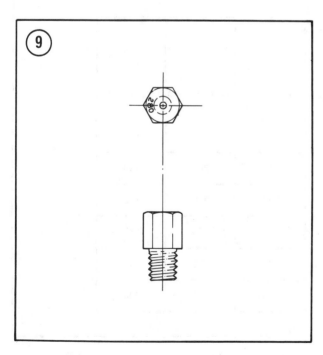

as a guideline only. Because of variables that exist with each individual machine, the spark plug condition should be used as the determining factor. When in doubt, always jet on the rich side.

4

1. After determining the altitude and temperature range that the snowmobile will be operated in, compute the necessary jet changes as follows:

    a. Refer to **Figures 10-19** for your model and find the temperature and altitude range the snowmobile will be operated in. Then cross reference the altitude and temperature to obtain the new main jet size.

    b. For example, refer to **Figure 10**. When the ambient temperature is 0° F (-20° C) at an altitude of 5000 ft., the new main jet size would be 210.

2. Refer to Chapter Seven for carburetor removal and installation. Then replace main jets as required.

3. Reinstall the carburetor as described in Chapter Seven.

CAUTION
Do not run the engine without the air box installed as engine seizure may result.

WARNING
If you are taking spark plug readings, the engine will be HOT! Use caution as the fuel in the float bowl will spill out when the main jet cover is removed from the bottom of the float bowl. Have a fire extinguisher and an assistant standing by when performing this procedure.

4. Start and run the snowmobile at high speed and then stop the engine.

5. Open the shroud and remove the spark plugs. Read the spark plugs as described in this chapter.

6. Reinstall the spark plugs.

7. If it is necessary to change main jets, perform the following:

    a. Remove the carburetor as described in Chapter Seven.

⑩

1984-1985 TRAIL

TEMPERATURE °F	MAIN JET				
ALTITUDE FEET	−40 to −20	−20 to 0	0 to +20	+20 to +40	+40 to +60
1000-3000	240	230	230	220	210
3000-5000	230	220	220	210	200
5000-7000	220	210	210	200	190
7000-9000	210	200	200	190	180
9000-11000	200	190	190	180	170

⑪

1986 TRAIL

TEMPERATURE °F	MAIN JET				
ALTITUDE FEET	−40 to −20	−20 to 0	0 to +20	+20 to +40	+40 to +60
1000-3000	230	220	210	200	190
3000-5000	220	210	200	190	180
5000-7000	210	200	190	180	170
7000-9000	200	190	180	170	160
9000-11000	190	180	170	160	160

(12)

1987-1989 TRAIL

	MAIN JET				
TEMPERATURE °F **ALTITUDE FEET**	**−40 to −20**	**−20 to 0**	**0 to +20**	**+20 to +40**	**+40 to +60**
1000-3000	240	230	220	210	200
3000-5000	230	220	210	200	190
5000-7000	210	210	200	190	180
7000-9000	200	190	180	180	170
9000-11000	190	180	170	170	160

4

(13)

1985-1986 400

	MAIN JET				
TEMPERATURE °F **ALTITUDE FEET**	**−40 to −20**	**−20 to 0**	**0 to +20**	**+20 to +40**	**+40 to +60**
1000-3000	230	220	220	210	200
3000-5000	220	210	210	200	190
5000-7000	210	200	200	190	180
7000-9000	200	190	190	180	170
9000-11000	190	180	180	170	170

⑭

1987-1989 400

	MAIN JET				
TEMPERATURE °F ALTITUDE FEET	−40 to −20	−20 to 0	0 to +20	+20 to +40	+40 to +60
1000-3000	240	230	220	210	200
3000-5000	230	220	210	200	190
5000-7000	210	200	200	190	180
7000-9000	200	190	190	180	170
9000-11000	190	180	170	160	160

⑮

1987-1989 SPORT

	MAIN JET				
TEMPERATURE °F ALTITUDE FEET	−40 to −20	−20 to 0	0 to +20	+20 to +40	+40 to +60
1000-3000	190	180	170	160	160
3000-5000	180	170	160	160	150
5000-7000	170	160	150	150	140
7000-9000	160	150	140	140	130
9000-11000	150	140	130	130	125

⑯

1989 500

TEMPERATURE °F	MAIN JET				
	−40 to −20	−20 to 0	0 to +20	+20 to +40	+40 to +60
ALTITUDE FEET					
1000-3000	290	280	270	260	250
3000-5000	280	270	250	240	230
5000-7000	260	250	240	230	220
7000-9000	240	240	230	220	210
9000-11000	230	220	210	200	190

4

⑰

1984 INDY 600

TEMPERATURE °F	MAIN JET				
	−40 to −20	−20 to 0	0 to +20	+20 to +40	+40 to +60
ALTITUDE FEET					
1000-3000	260	250	240	240	230
3000-5000	250	240	230	230	220
5000-7000	240	230	220	220	210
7000-9000	220	210	200	200	190
9000-11000	210	200	190	190	180

⑱

1985-1986 600

TEMPERATURE °F ALTITUDE FEET	MAIN JET				
	−40 to −20	−20 to 0	0 to +20	+20 to +40	+40 to +60
1000-3000	270	260	250	240	230
3000-5000	260	250	240	230	220
5000-7000	250	240	230	220	210
7000-9000	230	220	210	200	190
9000-11000	220	210	200	190	180

⑲

1987 600; 1988-1989 650

TEMPERATURE °F ALTITUDE FEET	MAIN JET				
	−40 to −20	−20 to 0	0 to +20	+20 to +40	+40 to +60
1000-3000	280	270	260	250	240
3000-5000	270	260	250	240	230
5000-7000	250	240	230	220	210
7000-9000	240	230	220	210	200
9000-11000	220	210	200	200	190

b. Remove the float bowl (**Figure 20**).

c. Remove and replace the main jet (**Figure 21**).

d. Reinstall the float bowl. Make sure the float bowl gasket is in place and not torn or damaged.

e. Repeat for the remaining carburetor(s).

f. Reinstall the carburetors as described in Chapter Seven.

g. Make sure the throttle cables work smoothly before starting the engine.

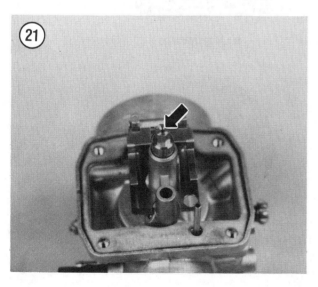

Reading Spark Plugs

Because the firing end of a spark plug operates in the combustion chamber, it reflects the operating condition of the engine. Much information about engine and spark plug performance can be determined by careful examination of the spark plug. Refer to **Figure 22**.

Normal condition

If the plug has a light tan- or gray-colored deposit and no abnormal gap wear or erosion, good engine, carburetion and ignition condition are indicated. The plug in use is of the proper heat range and may be serviced and returned to use.

Carbon fouled

Soft, dry, sooty deposits covering the entire firing end of the plug are evidence of incomplete combustion. Even though the firing end of the plug is dry, the pug's insulation decreases. An electrical path is formed that lowers the voltage from the ignition system. Engine mis-firing is a sign of carbon fouling. Carbon fouling can be caused by one or more of the following:

a. Too rich fuel mixture (incorrect jetting).

b. Spark plug heat range too cold.

c. Over-retarded ignition timng.

d. Ignition component failure.

e. Low engine compression.

Oil fouled

The tip of an oil fouled plug has a black insulator tip, a damp oily film over the firing end and a carbon layer over the entire nose. The electrodes will not be worn. Common causes for this condition are:

a. Too much oil in the fuel (incorrect jetting or incorrect oil pump adjustment).

b. Wrong type of oil.

c. Ignition component failure.

4

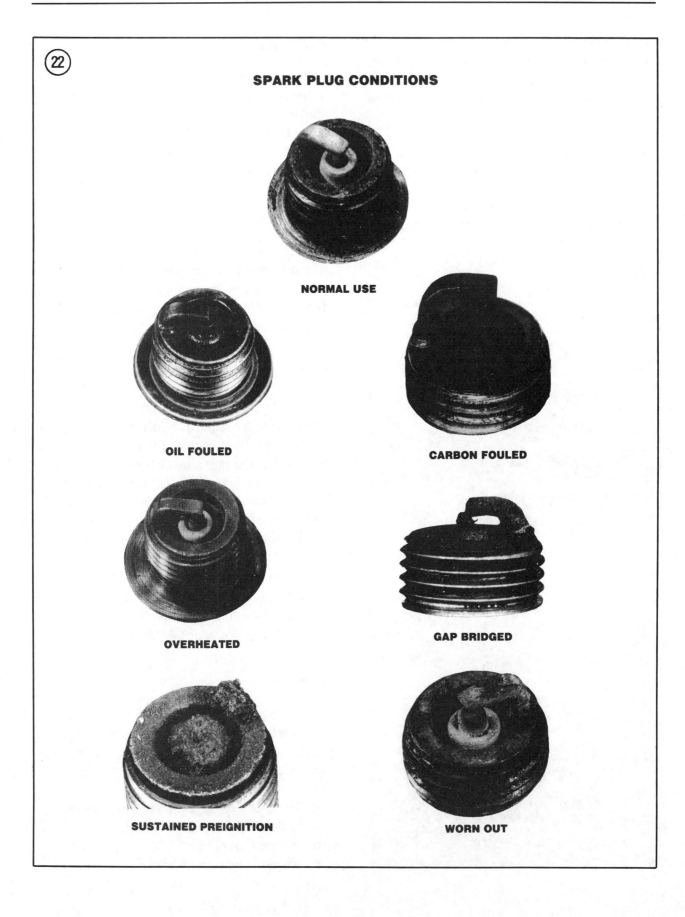

(22)

SPARK PLUG CONDITIONS

NORMAL USE

OIL FOULED

CARBON FOULED

OVERHEATED

GAP BRIDGED

SUSTAINED PREIGNITION

WORN OUT

d. Spark plug heat range too cold.

e. Engine still being broken in.

Oil fouled spark plugs may be cleaned in an emergency, but it is better to replace them. It is important to correct the cause of fouling before the engine is returned to service.

Gap bridging

Plugs with this condition exhibit gaps shorted out by combustion deposits between the electrodes. If this condition is encountered, check for an improper oil type, excessive carbon in combustion chamber or a clogged exhaust port and pipe. Be sure to locate and correct the cause of this condition.

Overheating

Badly worn electrodes and premature gap wear are signs of overheating, along with a gray or white "blistered" porcelain insulator surface. The most common cause for this condition is using a spark plug of the wrong heat range (too hot). If you have not changed to a hotter spark plug and the plug is overheated, consider the following causes:

a. Lean fuel mixture (incorrect main jet or incorrect oil pump adjustment).

b. Ignition timing too advanced.

c. Cooling system malfunction.

d. Engine air leak.

e. Improper spark plug installation (overtightening).

f. No spark plug gasket.

Worn out

Corrosive gases formed by combustion and high voltage sparks have eroded the electrodes. Spark plugs in this condition require more voltage to fire under hard acceleration. Replace with a new spark plug.

Preignition

If the electrodes are melted, preignition is almost certainly the cause. Check for carburetor mounting or intake manifold leaks and overadvanced ignition timing. It is also possible that a plug of the wrong heat range (too hot) is being used. Find the cause of the preignition before returning the engine to service.

HIGH-ALTITUDE CLUTCH TUNING

When the snowmobile is operated at an altitude of more than 5,000 feet (1,200 meters), the clutch should be changed to the specifications listed in **Table 9** or **Table 10** (for your model) to compensate for engine power loss (see *High-Altitude Carburetor Tuning* in this chapter). If the clutch is not adjusted for altitude, engine bogging at the point of belt engagement will occur. In addition, the engine may bog when running in deep snow. Both conditions can cause premature drive belt failure.

Refer to Chapter Twelve for complete clutch service procedures.

GEARING

Depending upon altitude, snow and track conditions, a different gear ratio may be required. Snow conditions that offer few rough sections require a smaller gear ratio. Less optimum snow conditions or more rugged terrain require a higher gear ratio. Refer to **Table 11** for your model. Replacement sprockets and chains can be purchased through Polaris dealers. Refer to Chapter Fourteen for sprocket and chain replacement procedures.

SUSPENSION ADJUSTMENT

The suspension can be adjusted to accommodate rider weight and snow conditions.

Correct suspension adjustment is arrived at largely through a matter of trial-and-error "tuning." There are several fundamental points that must be understood and applied before the suspension can be successfully adjusted to your needs.

Ski pressure—the load on the skis relative to the load on the track—is the primary factor controlling handling performance. If the ski pressure is too light, the front of the machine tends to float and steering control becomes vague, with the machine tending to drive straight ahead rather than turn, and wander when running straight at steady throttle.

On the other hand, if ski pressure is too heavy, the machine tends to plow during cornering and the skis dig in during straight line running rather than stay on top of the snow.

Ski pressure for one snow condition is not necessarily good for another condition. For instance, if the surface is very hard and offers little steering traction, added ski pressure—to permit the skis to "bite" into the snow—is desirable. Also, the hard surface will support the skis and not allow them to penetrate when the machine is running in a straight line under power.

On the other hand, if the surface is soft and tacky, lighter ski pressure is desirable to prevent the skis from sinking into the snow. Also, the increased traction afforded by the snow will allow the skis to turn with light pressure.

It's apparent, then, that good suspension adjustment involves a thorough analysis relating to ski pressure versus conditions. The Polaris suspension has been set at the factory to work in most conditions encountered by general riding. However, when the snowmobile is operated in varying or more difficult conditions, suspension adjustment may be required. It is important to remember that suspension tuning is a compromise. An adjustment that works well in one situation may not work as well in another.

Rear Torque Arm Spring Adjustment

The rear torque arm springs can be set to best suit the weight of the individual rider (and passenger). This will provide optimum rider comfort when riding the snowmobile. Adjustment is made by comparing height measurements with and without a rider (and passenger) seated on the snowmobile. Because of the weight factor, the rider (and passenger) should be wearing their complete riding gear when performing the following procedure.

NOTE
When making the measurements in Step 1 and Step 2, measurements should be made on both sides of the snowmobile.

1. With an assistant, lift the rear of the snowmobile so that the track is clear of the ground. Then carefully lower the snowmobile until the track is firmly on the ground—do not place any weight on the snowmobile after sitting the track onto the ground. Measure the distance from the snowmobile running board to the ground.

2. The rider (and passenger) should mount the snowmobile and assume their normal riding positions. After the rider (and passenger) are seated, have someone measure the distance from the running board to the ground as in Step 1.

3. The difference between the 2 measurements should be 1 1/2 in. If the measured distance is incorrect, perform the following:

   a. Turn the locknut (**Figure 23**) on the rear torque arm eyebolt to decrease or increase spring tension.

b. Repeat Steps 1 and 2 until the spring adjustment is correct for both sides of the snowmobile.

c. If the ride is too hard after making this adjustment, the rear shock can be repositioned to soften the ride. Refer to *Rear Shock Adjustment* in this chapter.

NOTE
Adjustment of the rear torque springs also affects ski pressure. If the ski pressure is too light after making this adjustment, tighten the rear torque springs to increase ski pressure. You will have to experiment to arrive at a suitable

compromise between spring adjustment and ski pressure.

Rear Shock Adjustment

The rear shock lower mounting position (**Figure 24**) can be moved to the upper mount hole in the rear torque arm to soften the ride.

Rear Suspension/Tunnel Adjustment Guide

Refer to the chart in **Figure 25** or **Figure 26** when adjusting the rear suspension and the front limiter strap(s). Use the charts as a guideline as to the number of possible adjustment positions.

Front limiter strap adjustment

The front torque arm limiter strap(s) (**Figure 27**) can be shortened to increase ski pressure or lengthened to decrease pressure.

Front torque arm spring pressure

The front shock absorber has 2 mounting positions (**Figure 28**) in the front torque arm. By changing the position of the upper shock mount, you can increase or decrease rear suspension travel firmness. The upper mount position will increase travel firmness whereas the bottom shock mount position will soften it.

Figures 25-28 and tables are on the following pages.

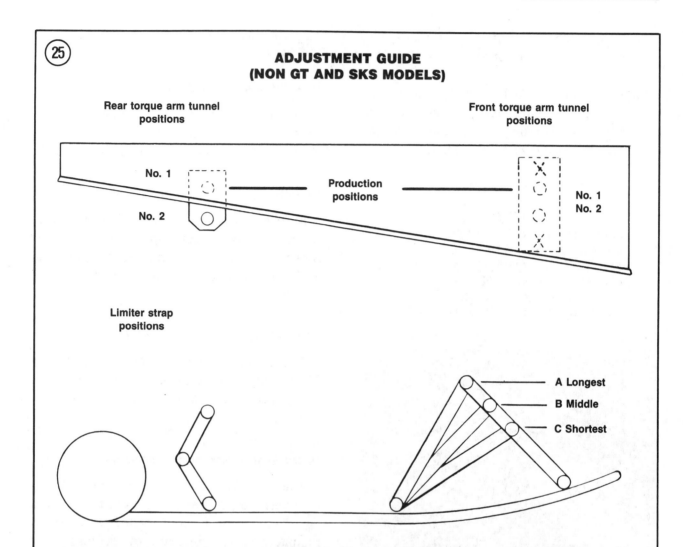

㉕

ADJUSTMENT GUIDE
(NON GT AND SKS MODELS)

Rear torque arm tunnel
positions

Front torque arm tunnel
positions

No. 1

Production
positions

No. 1
No. 2

No. 2

Limiter strap
positions

A Longest

B Middle

C Shortest

Rear Torque Arm Tunnel Location	Front Torque Arm Tunnel Location	Limiter Strap Position	Mounting Condition	
			Go	No Go (Do Not Use)
1	1	A,B,C	X	
1	2	A,B		X
1	2	C	X	
2	1	A,B,C	X	
2	2	A		X
2	2	B,C	X	

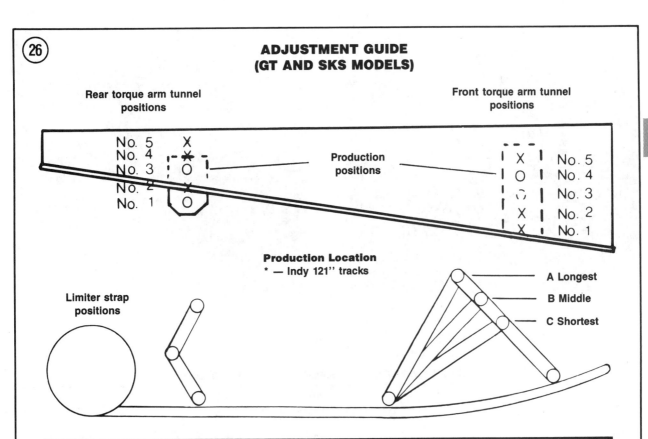

**ADJUSTMENT GUIDE
(GT AND SKS MODELS)**

Rear torque arm tunnel positions

Front torque arm tunnel positions

Production positions

Production Location
* — Indy 121'' tracks

Limiter strap positions

A Longest
B Middle
C Shortest

Rear Torque Arm Tunnel Location	Front Torque Arm Tunnel Location	Limiter Strap Position	Indy 121" Mounting Condition		GT SKS Supertrak Mounting Condition	
			Go	No Go (Do Not Use)	Go	No Go (Do Not Use)
o 3	o 4	o A,B,C	X			X
3	3	A,B		X		X
3	3	C	X			X
*1	*4	*A,B,C	X		X	
1	3	A		X		X
1	3	B,C	X		X	
5	5	A,B,C		X		X

⚠ Using any other location or track tension than recommended will result in severe tunnel and/or drive shaft damage.

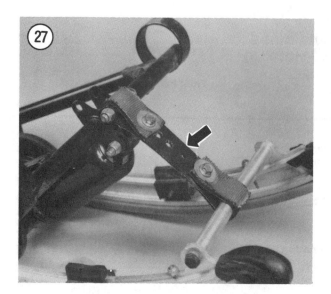

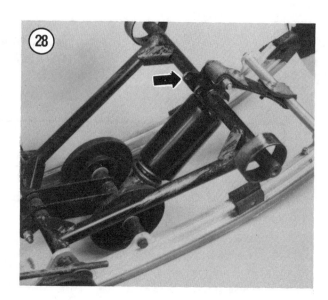

Table 1 CARBURETOR SPECIFICATIONS—TRAIL

Type	Mikuni
Model	VM34SS
Main jet	
1984-1985	230
1986	210
1987-1988	220
1989	230
Pilot jet	
1984-1985	35
1986-1987	30
1988-on	35
Jet needle/clip position	
1984-1986	6DH7/2
1987	6DH7/3
1988-on	6DH7/2
Needle jet	
1984-1985	P-6
1986-on	P-8
Throttle valve cutaway	
1984-on	3.0

Table 2 CARBURETOR SPECIFICATIONS—SPORT

Type	Mikuni
Size	VM30SS
Main jet	180
Pilot jet	35
Jet needle/clip position	5DP7/3
Needle jet	0-8
Throttle valve cutaway	3.0

Table 3 CARBURETOR SPECIFICATIONS—400

Type	Mikuni
Size	VM345SS
Main jet	
1985-1988	220
1989	230
Pilot jet	
1985	35
1986-on	30
Jet needle/clip position	
1985	6DH7/2
1986	6DP17/2
1987-on	6DP17/3
Needle jet	
1985	P-8
1986-on	Q-2
Throttle valve cutaway	
1985	3.0
1986-on	2.0

Table 4 CARBURETOR SPECIFICATIONS—500

Type	Mikuni
Size	VM38SS
Main jet	280
Pilot jet	40
Jet needle/clip position	6F9/3
Needle jet	Q-0
Throttle valve cutaway	3.0

Table 5 CARBURETOR SPECIFICATIONS—600

Type	Mikuni
Size	VM38SS
Main jet	
1984	250
1985-1987	260
Pilot jet	
1984	35
1985-1986	40
1987	35
Jet needle/clip position	
1984-1985	6F4/3
1986-1987	6F9/3
Needle jet	
1984	Q-2
1985-1987	P-8
Throttle valve cutaway	
1984	3.5
1985-1987	3.0

Table 6 CARBURETOR SPECIFICATIONS—650

Type	Mikuni
Size	VM38SS
Main jet	260
Pilot jet	50
Jet needle/clip position	
1988	6F9-3
1989	6DH4/3
Needle jet	P-8
Throttle valve cutaway	3.0

Table 7 PILOT AIR SCREW ADJUSTMENT

Model	Turns out
Trail	1.0
Sport	1.0
400	
1984-1986	1.0
1987-1989	1 1/2
500	1.0
600	
1984	1.0
1985-1986	3/4
1987	1.0
650	1.0

Table 8 IDLE SPEED ADJUSTMENT

Model	RPM
Trail	
1984-1985	2,300
1986	2,000
1987-1989	1,900
Sport	2,100
400	
1984-1986	2,200
1987-1989	1,900
500	1,900
600	
1984-1986	2,000
1987	1,900
650	1,900

Table 9 HIGH-ALTITUDE CLUTCH ADJUSTMENT—5,000-8,000 FEET

Model	Shift Weight	Spring	Helix
Trail			
1984	M-1 Mod*	Purple	36°
1985	N-1 Mod	Gold	34°
1986-1987	M-1 Mod	Orange	36°
1988-1989			
SKS	M-1 Mod or 10M	Orange	36°
All other	M-1 Mod or 10M	Red/white	36°
Sport	10	Gold	34°
400			
1984	**	**	**
1985	P-1	Red	34°
1986-1987	M-1 Mod	Gold	34°
1988	10M Red	Blue	34°
500	10M Blue	Blue	36°
600			
1984	06	Red	34°
1985	06	Blue	34°
1986-1987	06	Blue/gold	34°
650	10M Blue	Red	34°

* Use 1321372 spider roller with 2 washers per side.
** Specifications not listed.

Table 10 HIGH-ALTITUDE CLUTCH ADJUSTMENT—8,000-12,000 FEET

Model	Shift Weight	Spring	Helix
Trail			
1984	M-1 Mod*	Purple	36°
1985	N-1 Mod	Gold	34°
1986-1987	M-1 Mod	Red/white	36°
1988-1989	10M Red	Red/white	36°
Sport			
1987-1988	10	Blue	34°
1989	10	Blue	34°
1989	10M Blue	Gold	34°
400			
1984	**	**	**
1985	P-1	Red	34°
1986	P-1	Red	34°
1986	M-1 Mod	Blue/gold	34°
1987	P-1	Red	34°
1987	M-1 Mod	Blue/gold	34°
1988-1989	10M Red	Blue	34°
500	10M Blue	Blue	36°
600			
1984***	M-1 Mod	Red	34°
1985	06	Blue	34°
1986-1987	06	Blue/gold	34°
650	10M Blue	Blue	34°

* Use 1321372 spider roller with 2 washers per side.
** Specifications not listed.
*** Use J weight and red spring when operating above 10,000 feet and at 32° F.

Table 11 HIGH-ALTITUDE GEARING ADJUSTMENT

	5,000-8,000 feet	8,000-12,000
Trail		
1984	19/35	18/35
1985	19/35	19/35
1986-1987	21/35	21/35
1988-1989		
Trail	21/35	21/35
SKS	19/35	19/35
Sport	17/35	17/35
400		
1984	*	*
1985-1989	19/35	19/35
500		
SKS	18/35	18/35
All other models	19/35	19/35
600**		
1984-1985	19/35	19/35
1986-1987	21/35	21/35
650		
1988-1989		
SKS	19/35	19/35
All other models	21/25	21/35

* Specifications not listed.
** On 1984 600 models, optional 19/39 gearing can be used when altitude exceeds 10,000 feet and a temperature of 32° F.

Chapter Five

Engine—Two Cylinder

This chapter describes service procedures for the Polaris 2 cylinder engines installed on Indy models from 1984-on. The engine is equipped with ball-type main crankshaft bearings and needle bearings on both ends of the connecting rods. Crankshaft components are available as individual parts. However, other than to replace the outer seals, it is recommended that all crankshaft work be entrusted to a dealer or other competent engine specialist.

This chapter covers information to provide routine top-end service as well as crankcase disassembly and crankshaft inspection. Work on the snowmobile engine requires considerable mechanical ability. You should carefully consider your own capabilities before attempting any operation involving major disassembly of the engine.

Much of the labor charge for dealer repairs involves the removal and disassembly of other parts to reach the defective component. Even if you decide not to tackle the entire engine overhaul after studying the text and illustrations in this chapter, it can be cheaper to perform the preliminary operations yourself and then take the engine to your dealer. Since dealers have lengthy waiting lists for service (especially during the fall and winter season), this practice can reduce the time your unit is in the shop. If you have done much of the preliminary work, your repairs can be scheduled and performed much quicker. General engine specifications are listed in **Table 1**. Engine service specifications are listed in **Table 2** and **Table 3**. **Tables 1-4** are found at the end of the chapter.

ENGINE LUBRICATION

Engine lubrication is provided by the fuel/oil mixture used to power the engine.

SERVICE PRECAUTIONS

Whenever you work on your Polaris, there are several precautions that should be followed to help with disassembly, inspection and reassembly.

1. In the text there is frequent mention of the left-hand and right-hand side of the engine. This refers to the engine as it is mounted in the frame, not as it sits on your workbench. See **Figure 1**.

2. Always replace a worn or damaged fastener with one of the same size, type and torque requirements. Make sure to identify each bolt before replacing it with another. Bolt threads should be lubricated with engine oil, unless otherwise specified, before torque is applied. If a tightening torque is not listed in **Table 4** (end of this chapter), refer to the torque and fastener information in Chapter One.

3. Use special tools where noted. In some cases, it may be possible to perform the procedure with makeshift tools, but this procedure is not recommended. The use of makeshift tools can damage the components and may cause serious personal injury. Where special snowmobile tools are required, these may be purchased through any Polaris snowmobile dealer. Other tools can be purchased through your dealer, or from a snowmobile, motorcycle or automotive accessory store. When purchasing tools from an automotive accessory dealer or store, remember that all threaded parts must have metric threads.

4. Before removing the first bolt and to prevent frustration during assembly, get a number of boxes, plastic bags and containers and store the parts as they are removed (**Figure 2**). Also have on hand a roll of masking tape and a permanent, waterproof marking pen to label each part or assembly as required. If your snowmobile was purchased second hand and it appears that some of the wiring may have been changed or replaced, label each electrical connection before disconnecting it.

5. Use a vise with protective jaws to hold parts. If protective jaws are not available, insert wooden blocks on each side of the part(s) before clamping them in the vise.

6. Remove and install pressed-on parts with an appropriate mandrel, support and hydraulic press. **Do not** *try to pry, hammer or otherwise force them on or off.*

7. Refer to **Table 4** at the end of the chapter for torque specifications. Proper torque is essential to assure long life and satisfactory service from snowmobile components.

8. Discard all O-rings and oil seals during disassembly. Apply a small amount of heat durable grease to the inner lips of each oil seal to prevent damage when the engine is first started.

9. Keep a record of all shims and where they came from. As soon as the shims are removed, inspect them for damage and write down their thickness and location.

10. Work in an area where there is sufficient lighting and room for component storage.

SERIAL NUMBERS

Polaris snowmobiles are identified by frame and engine identification numbers. Refer to *Serial Numbers* in Chapter One. These numbers identify your snowmobile and should always be used when ordering replacement parts.

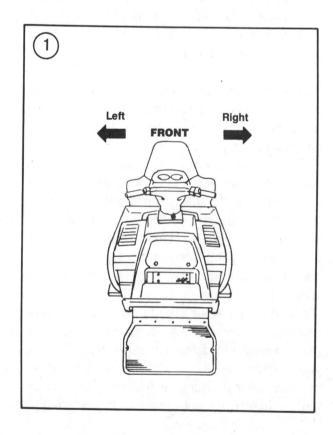

SERVICING ENGINE IN FRAME

Some of the components can be serviced while the engine is mounted in the frame:

a. Cylinder heads.
b. Cylinders.
c. Pistons.
d. Carburetors.
e. Magneto.
f. Oil pump.
g. Starter motor.
h. Recoil starter.
i. Drive assembly.

ENGINE REMOVAL

Engine removal and crankcase separation is required for repair of the "bottom end" (crankshaft, connecting rod and bearings).

1. Open the shroud all the way. Unplug the harness from the headlight and free the harness from the shroud. Disconnect the shroud restraining cable and with assistance, support the shroud and remove the bolts that attach the shroud hinge to the chassis. Remove the shroud and set it well out of the way to prevent it from being damaged.

2. Refer to **Figure 3** or **Figure 4** and remove the exhaust system as follows:

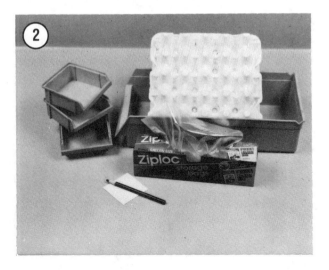

a. Disconnect the springs that secure the muffler to the exhaust pipe.
b. Remove the nut and washers or spring securing the exhaust pipe to the bracket. Account for the spacers and grommet, if used.
c. Remove the spring securing the muffler to the exhaust pipe and remove the exhaust pipe.
d. Remove the rubber ball (A, **Figure 5**) from the bracket.

3. Remove the air box (**Figure 6**) as described in Chapter Seven.

4. Remove the carburetors as described in Chapter Seven.

5. Remove the drive belt as described in Chapter Twelve.

6. Remove the primary sheave as described in Chapter Twelve.

7. *Liquid cooled models*—perform the following:
   a. Drain the cooling system as described under *Coolant Change* in Chapter Three.
   b. *Liquid cooled models*: Disconnect the coolant hose at the engine. See A, **Figure 7**, typical.
   c. Disconnect the lower coolant hose at the water pump. See B, **Figure 7**, typical.
   d. Disconnect the electrical connector at the thermosensor.

8. Label and disconnect the oil hoses at the oil pump. Plug the hoses to prevent oil from spilling into the belly pan.

9. Remove the recoil starter as described in Chapter Eleven.

10. Disconnect the CDI electrical connectors.

11. Disconnect the pulse hose (**Figure 8**) at the rear of the crankcase.

12. If necessary, the engine top end (cylinder heads, pistons, and cylinder blocks) can be removed before removing the engine from the frame. Refer to *Cylinder* in this chapter.

NOTE
Because of the number of bushings, washers and rubber dampers used on the

5

③

EXHAUST SYSTEM
(1984-ON TRAIL; 1984-1987 400; 1988-ON SPORT)

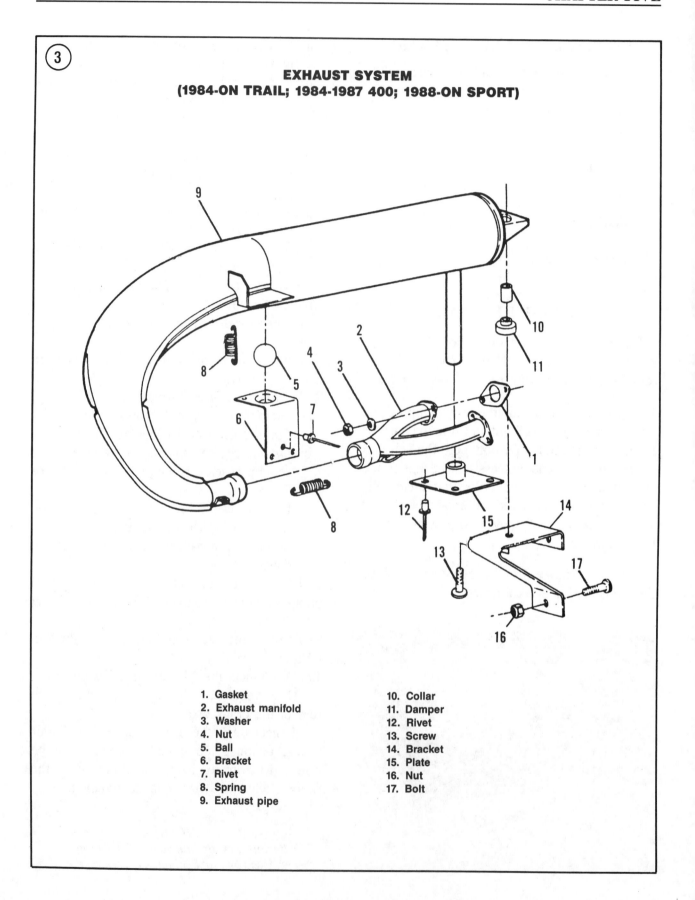

1. Gasket
2. Exhaust manifold
3. Washer
4. Nut
5. Ball
6. Bracket
7. Rivet
8. Spring
9. Exhaust pipe
10. Collar
11. Damper
12. Rivet
13. Screw
14. Bracket
15. Plate
16. Nut
17. Bolt

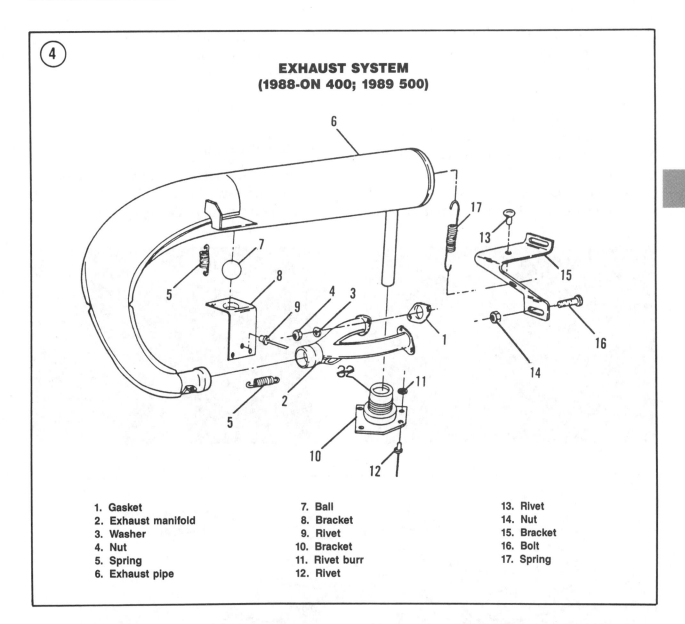

EXHAUST SYSTEM
(1988-ON 400; 1989 500)

1. Gasket
2. Exhaust manifold
3. Washer
4. Nut
5. Spring
6. Exhaust pipe
7. Ball
8. Bracket
9. Rivet
10. Bracket
11. Rivet burr
12. Rivet
13. Rivet
14. Nut
15. Bracket
16. Bolt
17. Spring

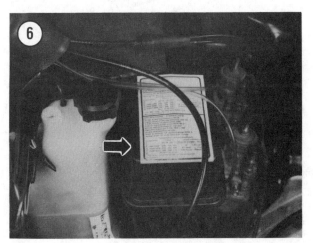

engine bracket assemblies, refer to **Figures 9-13** *for your model when performing Step 13.*

13. Loosen the front (B, **Figure 5**) and rear engine bracket nuts or bolts. Remove the nut and washer from the rear bracket. On the 1988 Trail Special, disconnect the torque stop assembly (**Figure 11**).

14. Check to make sure all of the wiring and hoses have been disconnected from the engine.

15. With at least one assistant, lift the engine up and remove it from the frame. Take it to a workbench for further disassembly.

ENGINE PLATE

Removal/Installation

The engine plate is a critical part of the snowmobile drive train. A damaged or loose engine plate will allow the engine to shift or pull

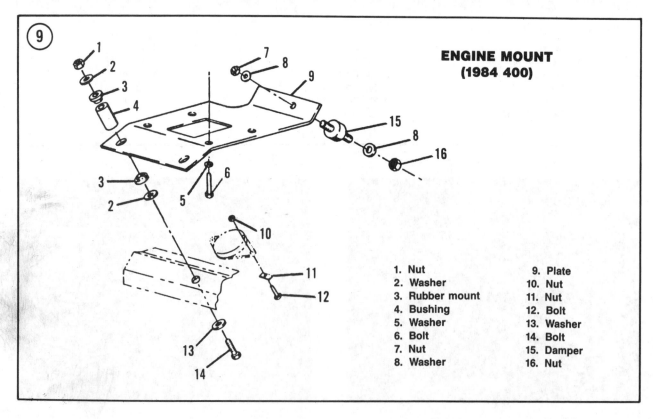

**ENGINE MOUNT
(1984 400)**

1. Nut	9. Plate
2. Washer	10. Nut
3. Rubber mount	11. Nut
4. Bushing	12. Bolt
5. Washer	13. Washer
6. Bolt	14. Bolt
7. Nut	15. Damper
8. Washer	16. Nut

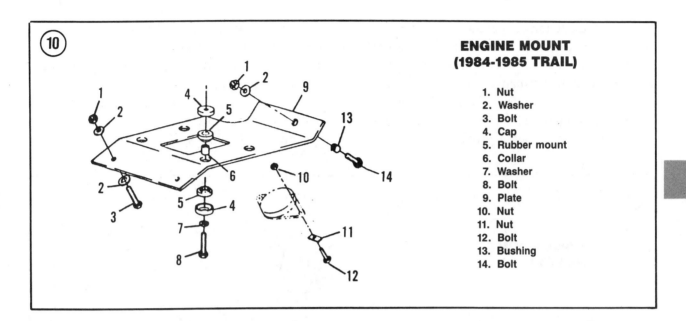

ENGINE MOUNT
(1984-1985 TRAIL)

1. Nut
2. Washer
3. Bolt
4. Cap
5. Rubber mount
6. Collar
7. Washer
8. Bolt
9. Plate
10. Nut
11. Nut
12. Bolt
13. Bushing
14. Bolt

5

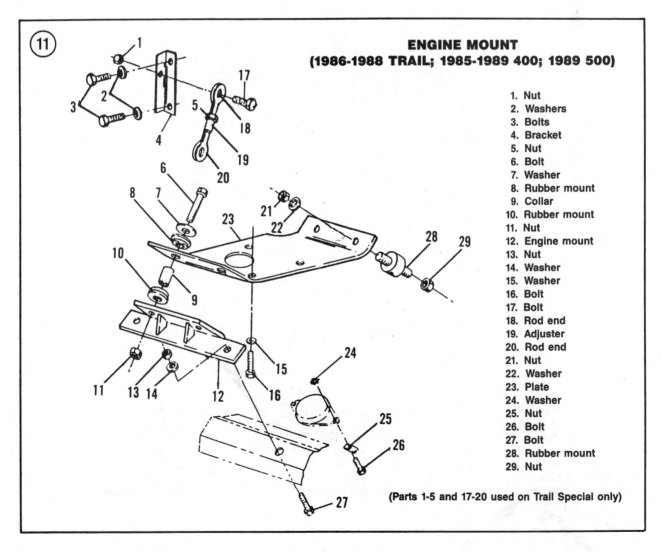

ENGINE MOUNT
(1986-1988 TRAIL; 1985-1989 400; 1989 500)

1. Nut
2. Washers
3. Bolts
4. Bracket
5. Nut
6. Bolt
7. Washer
8. Rubber mount
9. Collar
10. Rubber mount
11. Nut
12. Engine mount
13. Nut
14. Washer
15. Washer
16. Bolt
17. Bolt
18. Rod end
19. Adjuster
20. Rod end
21. Nut
22. Washer
23. Plate
24. Washer
25. Nut
26. Bolt
27. Bolt
28. Rubber mount
29. Nut

(Parts 1-5 and 17-20 used on Trail Special only)

out of alignment during operation. This condition will cause primary and secondary sheave misalignment that will result in drive belt wear and reduced performance. The engine plate assembly should be inspected carefully whenever the engine is removed from the frame or if clutch misalignment becomes a problem.

Refer to **Figures 9-13** for your model when performing this procedure.

1. Carefully turn the engine over to gain access to the engine plate.

2. Remove the engine-to-engine plate bolts and washers and remove the plate. Account for the spacers and bushings, if used.

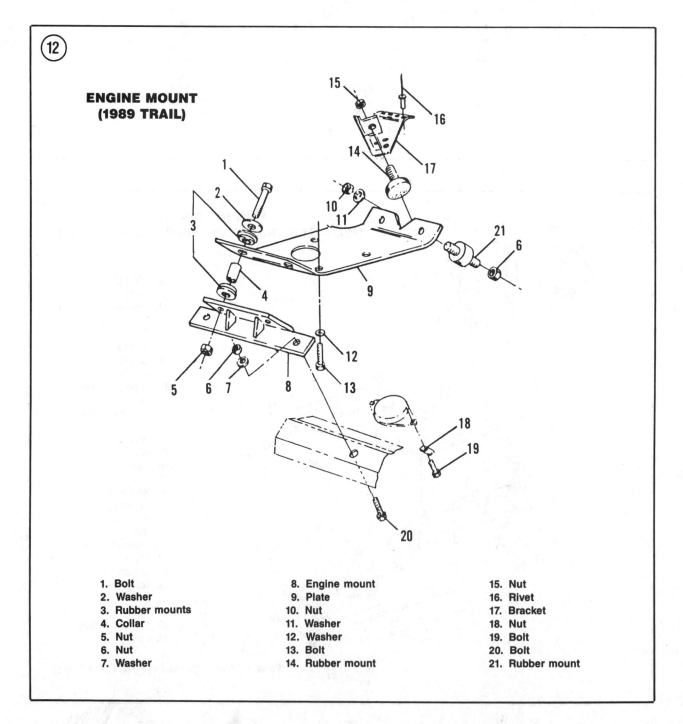

⑫

ENGINE MOUNT (1989 TRAIL)

1. Bolt	8. Engine mount	15. Nut
2. Washer	9. Plate	16. Rivet
3. Rubber mounts	10. Nut	17. Bracket
4. Collar	11. Washer	18. Nut
5. Nut	12. Washer	19. Bolt
6. Nut	13. Bolt	20. Bolt
7. Washer	14. Rubber mount	21. Rubber mount

3. Check the engine plate for cracks or other damage.

4. Check all of the rubber dampers for separation or other damage. Check the dampers still mounted in the engine compartment.

5. Check all of the engine plate bolts for thread damage. Replace damaged bolts with the same

grade bolt. Weaker bolts will loosen and allow the engine to slip.

6. Check the engine plate tapped holes in the lower crankcase for stripped threads or debris buildup. Clean threads with a suitable size metric tap. If Loctite was previously used, make sure to remove all traces of Loctite residue before reinstalling bolts.

7. Replace worn or damaged parts as required.

8. Installation is the reverse of these steps. Tighten the engine plate bolts to the torque specification listed in **Table 4**.

ENGINE INSTALLATION

1. Wash the engine compartment with clean water.

2. Spray all of the exposed electrical connectors with a spray electrical contact cleaner.

3. Before installing the engine, now is a good time to check components (**Figure 14**, typical)

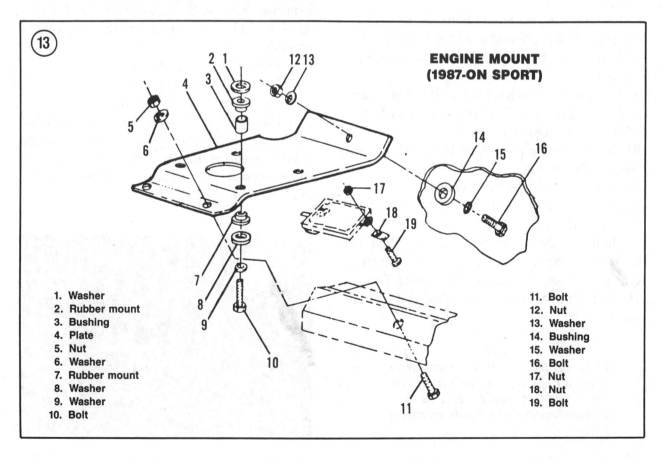

ENGINE MOUNT (1987-ON SPORT)

1. Washer
2. Rubber mount
3. Bushing
4. Plate
5. Nut
6. Washer
7. Rubber mount
8. Washer
9. Washer
10. Bolt
11. Bolt
12. Nut
13. Washer
14. Bushing
15. Washer
16. Bolt
17. Nut
18. Nut
19. Bolt

which are normally inaccessible for visual inspection. Note the following:

   a. *Liquid cooled models*: Check all of the coolant hoses for damage. Replace damaged hoses as required.

   b. Check all of the steering component fasteners for loose or missing parts. See Chapter Fifteen.

   c. Make sure the engine pads or dampers are mounted on the firewall, if used.

   d. Check that the CDI unit is secured tightly.

4. Bolt the engine plate onto the bottom of the crankcase as described in this chapter.

5. Check inside the engine compartment for tools or other objects that may interfere with engine installation. Make sure all wiring harnesses are routed and secured properly.

6. Before installing the engine in the frame, reconnect the 2 oil pump-to-check valve oil lines (**Figure 15**).

7. With an assistant, place the engine into the frame.

8. Install the engine mount bolts finger-tight at this time. Refer to **Figures 9-13** for your model when installing the engine mounts, washers and dampers.

9. Reconnect the oil hose at the oil pump. Then bleed the oil pump as described in Chapter Nine. Wipe up all excess oil.

10. Reconnect the pulse hose at the crankcase nozzle (**Figure 8**).

11. Reconnect all electrical connectors.

12. Reinstall the recoil starter housing. See Chapter Eleven.

13. Reinstall the carburetors as described in Chapter Seven. Check carburetor synchronization as described in Chapter Three.

14. Install the air box as described in Chapter Seven.

> *NOTE*
> *Make sure the hoses from the carburetor to the air box are not pinched or torn. This condition would cause a lean air/fuel mixture and result in engine seizure.*

15. *Liquid cooled models*: Perform the following:

   a. Reconnect the coolant hoses at the engine and water pump.

   b. Refill the cooling system and bleed it as described in Chapter Three.

16. Reinstall the primary sheave and drive belt as described in Chapter Twelve.

17A. Check clutch alignment as described in Chapter Twelve.

17B. Tighten engine mount bolts to the torque specification listed in **Table 4**.

18. Reinstall the exhaust pipe assembly. Secure the muffler with the springs and dampers.

19. Reinstall the engine shroud. Reconnect the headlight electrical connector.

ENGINE TOP END

The engine "top end" consists of the cylinder heads, cylinder blocks, pistons, piston rings, piston pins and the connecting rod small-end bearings. Refer to the following figure for your model when performing the following procedures:

   a. **Figure 16**: 1984-on Trail.

   b. **Figure 17**: 400 and 500

   c. **Figure 18**: 1987-on Sport.

   d. **Figure 19**: Piston assembly (all models).

The engine top end can be serviced with the engine installed in the frame. However, the following service procedures are shown with the engine removed for clarity.

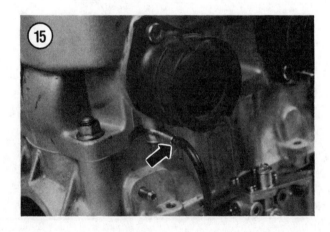

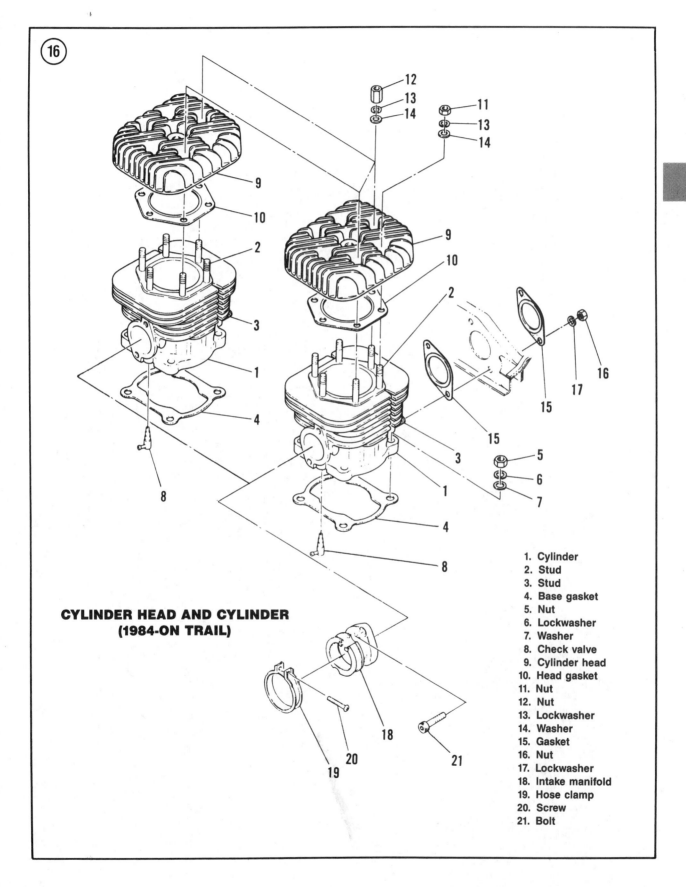

**CYLINDER HEAD AND CYLINDER
(1984-ON TRAIL)**

1. Cylinder
2. Stud
3. Stud
4. Base gasket
5. Nut
6. Lockwasher
7. Washer
8. Check valve
9. Cylinder head
10. Head gasket
11. Nut
12. Nut
13. Lockwasher
14. Washer
15. Gasket
16. Nut
17. Lockwasher
18. Intake manifold
19. Hose clamp
20. Screw
21. Bolt

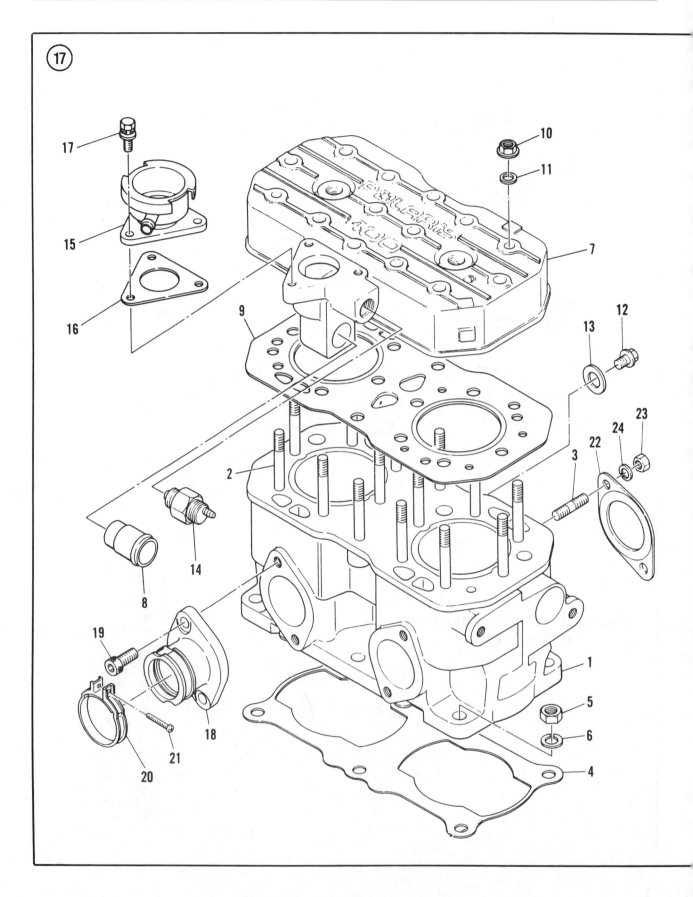

5

CYLINDER HEAD AND CYLINDER
(400 AND 500)

1. Cylinder
2. Stud
3. Stud
4. Base gasket
5. Nut (500 models use a flange nut)
6. Lockwasher
7. Cylinder head
8. Water pipe
9. Head gasket
10. Flange nut
11. Washer
12. Flange bolt
13. Gasket
14. Thermoswitch
15. Filler neck assembly
16. Gasket
17. Bolt
18. Intake manifold
19. Bolt
20. Hose clamp
21. Screw
22. Gasket
23. Nut
24. Lockwasher

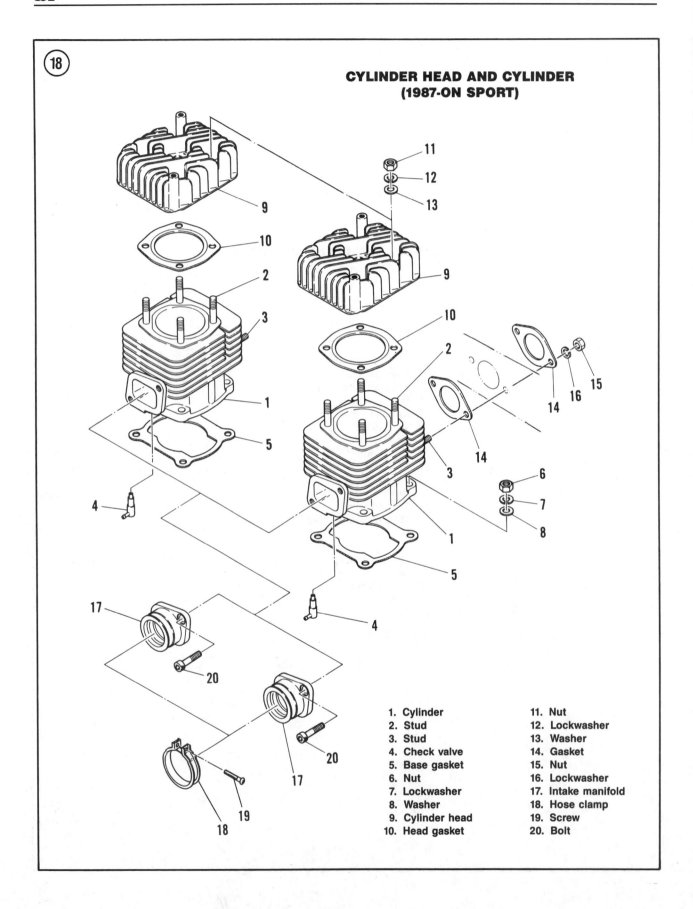

**CYLINDER HEAD AND CYLINDER
(1987-ON SPORT)**

1. Cylinder
2. Stud
3. Stud
4. Check valve
5. Base gasket
6. Nut
7. Lockwasher
8. Washer
9. Cylinder head
10. Head gasket
11. Nut
12. Lockwasher
13. Washer
14. Gasket
15. Nut
16. Lockwasher
17. Intake manifold
18. Hose clamp
19. Screw
20. Bolt

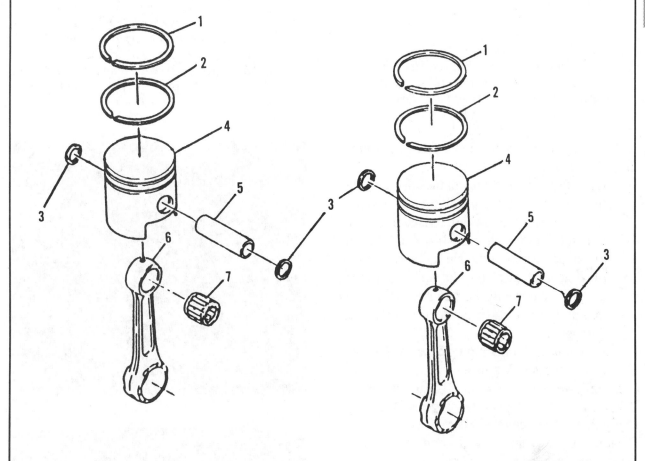

PISTON ASSEMBLY

5

1. Piston ring
2. Piston ring
3. Wrist pin clips
4. Piston
5. Wrist pin
6. Connecting rod
7. Bearing

Cylinder Heads
Removal/Installation
(Air Cooled Models)

CAUTION
To prevent warpage and damage to any component, remove the cylinder heads only when the engine is at room temperature.

NOTE
If the engine is being disassembled for inspection procedures, check the compression before removing the cylinder heads. See Chapter Three.

1. If the engine is mounted in the frame, perform the following:
   a. Open the engine shroud all the way. Unplug the harness from the headlight and free the harness from the shroud. Disconnect the shroud restraining cable and with assistance, support the shroud and remove the bolts that attach the shroud hinge to the chassis. Remove the shroud and set it well out of the way to prevent it from being damaged.
   b. Disconnect the spark plug caps (A, **Figure 20**) at the spark plugs.
   c. Remove the air shroud (B, **Figure 20**).
2. Loosen the spark plugs (A, **Figure 21**) if they are going to be removed later.
3. Referring to **Figure 22**, loosen the cylinder head nuts in a crisscross pattern.
4. Remove the nuts, lockwashers and flat washers. On Trail models, note the location of the long nuts that the air shroud is bolted to.

NOTE
Identify the cylinder heads before removal so that they can be reinstalled in their original position.

5. Loosen the cylinder heads by tapping around the perimeter with a rubber or plastic mallet. Remove the cylinder heads (B, **Figure 21**).

6. Remove and discard the cylinder head gaskets.
7. Lay a rag over the cylinder blocks to prevent dirt from falling into the cylinders.
8. Inspect the cylinder heads as described in this chapter.

NOTE
While the cylinder heads are removed, check the cylinder studs for thread strippage, looseness or other damage. If necessary, remove the cylinder as described in this chapter and replace damaged studs as described in Chapter One.

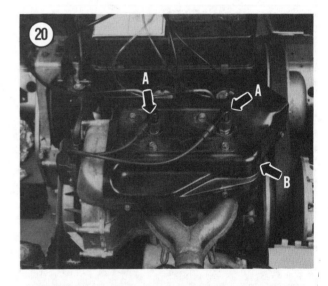

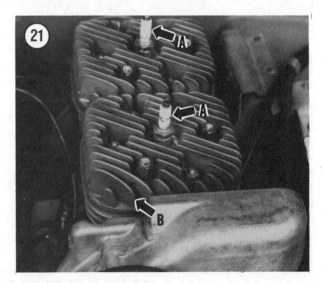

9. Remove all gasket residue from each cylinder head and cylinder mating surface with a gasket scraper.

10. Install new cylinder head gaskets.

11. Install the cylinder heads on their respective cylinders. Install the flat washers, lockwashers and nuts. On Trail models, install the long nuts in their original positions.

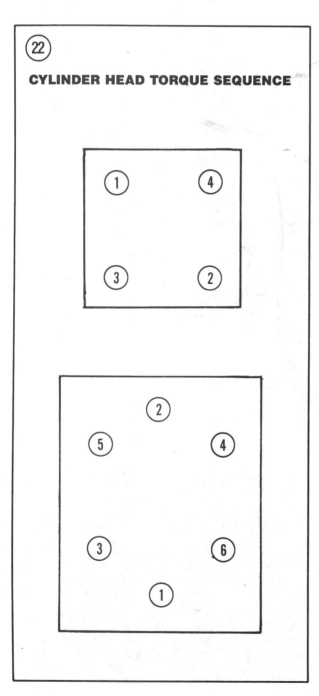

22

CYLINDER HEAD TORQUE SEQUENCE

12. Tighten the cylinder head nuts in the numerical sequence shown in **Figure 22** to the torque specification listed in **Table 4**.

13. Install the spark plugs and tighten securely.

14. If the engine is installed in the frame, note the following:

   a. Install the air shroud (B, **Figure 20**).

   b. Connect the spark plug caps (A, **Figure 20**).

   c. Reinstall the engine shroud.

Cylinder Heads
Removal/Installation
(Liquid Cooled Models)

CAUTION
To prevent warpage and damage to any component, remove the cylinder heads only when the engine is at room temperature.

NOTE
If the engine is being disassembled for inspection procedures, check the compression before removing the cylinder head. See Chapter Three.

1. If the engine is mounted in the frame, perform the following:

   a. Open the engine shroud all the way. Unplug the harness from the headlight and free the harness from the shroud. Disconnect the shroud restraining cable and with assistance, support the shroud and remove the bolts that attach the shroud hinge to the chassis. Remove the shroud and set it well out of the way to prevent it from being damaged.

   b. Drain the cooling system as described in Chapter Three.

   c. Disconnect the coolant hose at the cylinder head (A, **Figure 7**).

   d. Disconnect the spark plug caps at the spark plugs.

2. Loosen the spark plugs if they are going to be removed later.

3. Loosen the cylinder head nuts in the numerical order shown in **Figure 23**.

4. Remove the cylinder head nuts (A, **Figure 24**) and washers and remove the cylinder head (B, **Figure 24**). If the cylinder head is stuck, pry the head loose with a screwdriver at the pry point shown in **Figure 25**. *Do not* pry between the cylinder head-to-cylinder mating surfaces.

5. Remove and discard the cylinder head gasket (**Figure 26**).

6. Lay a rag over the cylinder block to prevent dirt from falling into the cylinders.

7. Inspect the cylinder head as described in this chapter.

NOTE
While the cylinder head is removed from the engine, check the cylinder studs for thread strippage, looseness or other damage. If necessary, replace damaged studs as described in Chapter One.

8. Remove all gasket residue from the cylinder head and cylinder mating surfaces with a gasket scraper.

9. Install the cylinder head gasket so that all of the cylinder head stud and coolant passage holes align with the cylinder block. See **Figure 26**.

10. Install the cylinder head (B, **Figure 24**).

11. Install the cylinder head washers and nuts and tighten the nuts in the numerical order shown in **Figure 23** in 2 stages; tighten to the torque specification listed in **Table 4**.

12. Install the spark plugs and tighten securely.

13. If the engine is installed in the frame, note the following:

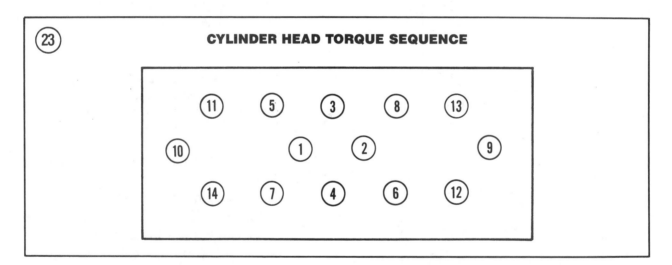

24 **CYLINDER HEAD TORQUE SEQUENCE**

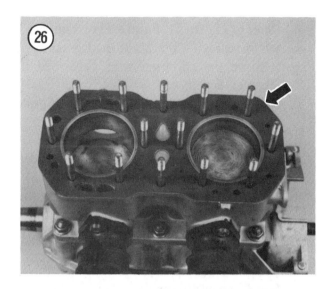

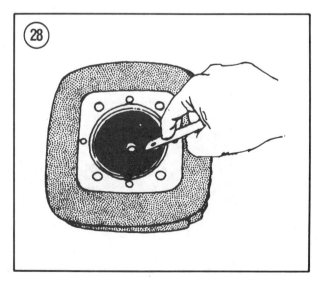

a. Connect the spark plug caps.

b. Connect the coolant hose at the cylinder head (A, **Figure 7**). Tighten the hose clamp securely.

c. Refill and bleed the cooling system as described in Chapter Three.

d. Reinstall the engine shroud.

Inspection (All Models)

1. Wipe away any soft deposits on the cylinder head (**Figure 27**) combustion chamber and gasket surface. Hard deposits should be removed with a wire brush mounted in a drill or drill press or with a soft-metal scraper (**Figure 28**). Be careful not to gouge the aluminum surfaces. Burrs created from improper cleaning will cause preignition and heat erosion.

> *NOTE*
> *Always use an aluminum thread fluid or kerosene on the thread chaser and cylinder head threads when performing Step 2.*

2. With the spark plug removed, check the spark plug threads in the cylinder head (**Figure 27**) for any sign of carbon buildup or cracking. The carbon can be removed with a 14 mm spark plug chaser.

3. Wash the cylinder head in hot soapy water and rinse thoroughly.

4. Check for cracks in the combustion chamber or along the spark plug thread hole. A cracked head must be replaced.

> *NOTE*
> *Polaris does not list wear limits for cylinder head flatness. The procedure given in Step 5 should be used as a guideline when determining cylinder head flatness.*

5. After the head has been thoroughly cleaned, place a straightedge across the cylinder head

gasket surface at several points (**Figure 29**). Measure the warp by attempting to insert a flat feeler gauge between the straightedge and the cylinder head at each location. There should be no warpage; if a small amount is present, it can be resurfaced as follows:

 a. Tape a piece of 400-600 grit wet emery sandpaper onto a piece of thick plate glass or surface plate (**Figure 30**).

 b. Slowly resurface the head by moving it in figure-eight patterns on the sandpaper.

 c. Rotate the head several times to avoid removing too much material from one side. Check progress often with the straightedge and feeler gauge.

 d. If there is still evidence of warpage, it will be necessary to have the head resurfaced by a machine shop familiar with snowmobile and motorcycle service. Note that removing material from the cylinder head mating surface will change the compression ratio. Consult with the machinist on how much material was removed.

6. *Liquid cooled models*: Check the coolant passages through the cylinder head for residue buildup. Flush coolant passages thoroughly.

CYLINDER

An aluminum cylinder block is used with a cast iron liner pressed into the block. When severe wear is experienced, the cylinder liner can be bored to 0.25 mm (0.10 in.) and 0.50 mm (0.20 in.) over and new pistons and rings installed.

Removal
(Air Cooled Models)

Refer to **Figure 16** or **Figure 18** for this procedure.

1. Remove the cylinder heads as described in this chapter.

2. If the engine is installed in the frame, perform the following:

 a. Refer to **Figure 31**, **Figure 32** or **Figure 33** when removing the air shroud assembly.

 b. Remove the carburetors as described in Chapter Seven.

 c. Remove the intake manifolds (A, **Figure 34**).

 d. Remove the rear air shroud (B, **Figure 34**).

 e. Remove the exhaust pipe as described under *Engine Removal* in this chapter.

 f. Remove the nuts and washers securing the exhaust manifold to the cylinder blocks. Remove the exhaust manifold (**Figure 35**).

 g. Remove front air shroud.

 h. Remove the recoil starter housing assembly. See Chapter Eleven.

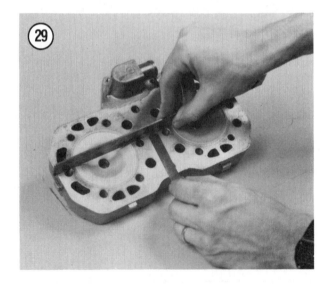

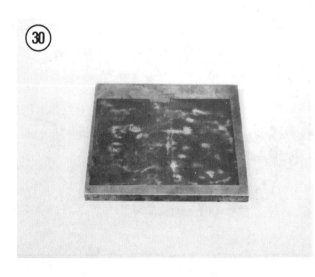

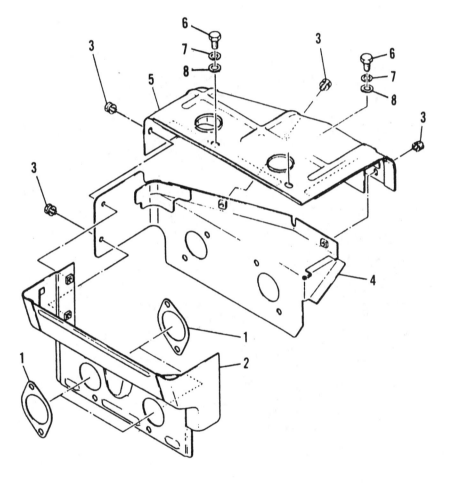

AIR SHROUD
(1984-1985 TRAIL)

1. Gasket
2. Shroud
3. Bolt
4. Shroud
5. Shroud
6. Bolt
7. Lockwasher
8. Washer

5

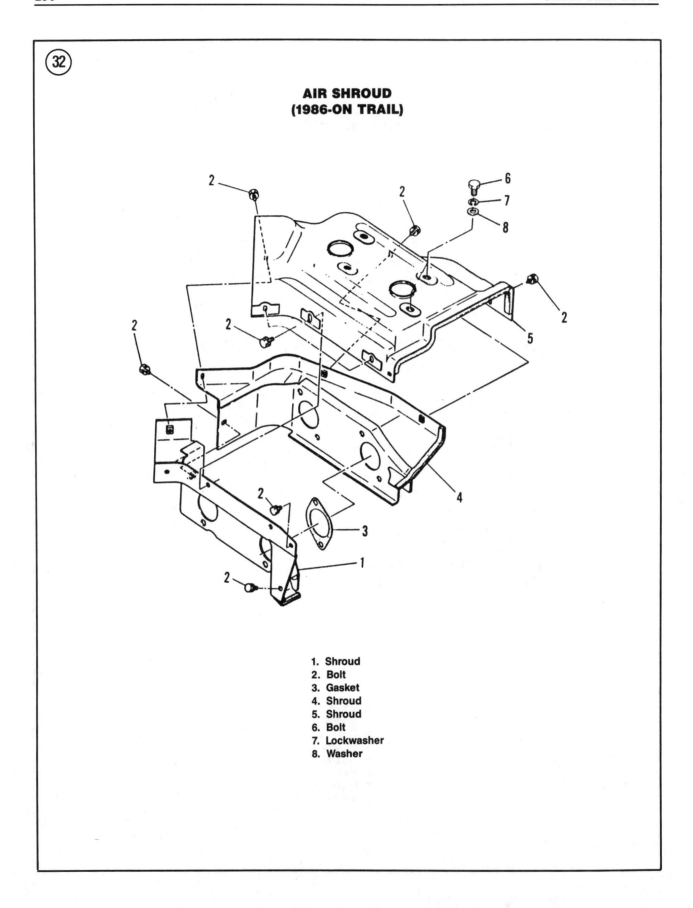

**AIR SHROUD
(1986-ON TRAIL)**

1. Shroud
2. Bolt
3. Gasket
4. Shroud
5. Shroud
6. Bolt
7. Lockwasher
8. Washer

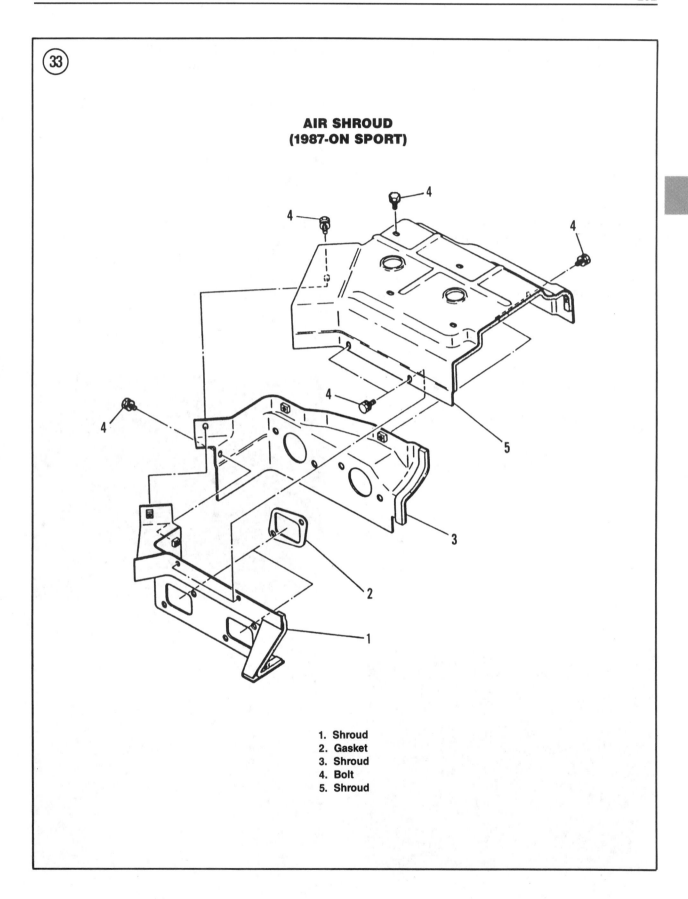

**AIR SHROUD
(1987-ON SPORT)**

5

1. Shroud
2. Gasket
3. Shroud
4. Bolt
5. Shroud

i. Disconnect the oil lines at the check valves on the cylinder blocks. Plug the oil lines to prevent leakage and contamination.

NOTE
Mark the cylinders for location before removal.

3. Remove the nuts, flat washers and lockwashers securing the cylinder blocks to the crankcase.

4. Loosen the cylinder by tapping around the perimeter with a rubber or plastic mallet. Repeat for each cylinder.

CAUTION
When removing the cylinders in Step 5, do not twist the cylinder block so far that the piston rings could snap into the intake port. This would cause the cylinder block to bind and cause piston and ring damage. If this should happen, push the rings back into position.

5. Rotate the engine so that the piston is at the bottom of its stroke. Pull the cylinder straight up and off the crankcase studs.

6. Remove the cylinder base gasket and discard it.

7. Repeat for the opposite cylinder.

8. Stuff clean rags around the connecting rods to keep dirt and loose parts from entering the crankcase.

9. Clean and inspect the cylinder as described in this chapter.

Installation
(Air Cooled Models)

1. Clean the cylinder bore as described under *Inspection* in this chapter.

2. Check that the top surface of the crankcase and the bottom cylinder surface are clean prior to installation.

3. Install a new base gasket.

4. Make sure the end gaps of the piston rings are lined up with the locating pins in the ring

grooves (**Figure 36**). Lightly oil the piston rings and the inside of the cylinder bore.

5. Place a piston holding tool under one piston and turn the crankshaft until the piston is down firmly against the tool. This will make cylinder installation easier. You can make this tool out of wood as shown in **Figure 37**.

6. Install the cylinder block on the crankcase studs so that the intake port faces the rear of the engine. Compress each piston ring with your fingers as the cylinder starts to slide over it. Remove the piston holding tool. Make sure the cylinder block is fully seated on the crankcase.

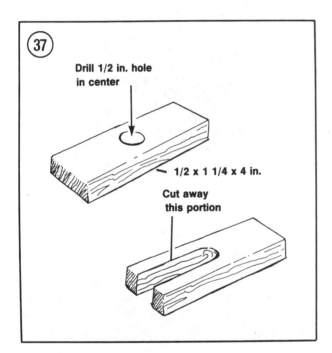

Drill 1/2 in. hole in center

1/2 x 1 1/4 x 4 in.

Cut away this portion

7. Install the cylinder flat washers, lockwashers and nuts. Tighten the nuts securely in a crisscross pattern.

8. Repeat for the opposite cylinder.

9. Reverse Step 1 and Step 2 under *Removal* to complete assembly.

10. Bleed the oil pump as described in Chapter Nine.

11. If new components were installed or if the cylinders were bored or honed, the engine must be broken in as if it were new. Refer to *Break-In Procedure* in Chapter Three.

Removal
(Liquid Cooled Models)

Refer to **Figure 17** for this procedure.

1. Remove the cylinder head as described in this chapter.

2. If the engine is installed in the frame, perform the following:

   a. Remove the carburetors as described in Chapter Seven.

   b. Disconnect the oil lines at the check valves on the cylinder blocks (A, **Figure 38**).

   c. Remove the exhaust pipe as described under *Engine Removal* in this chapter.

3. Remove the nuts and lockwashers holding the exhaust manifold (**Figure 39**) to the cylinders. Remove the exhaust pipe.

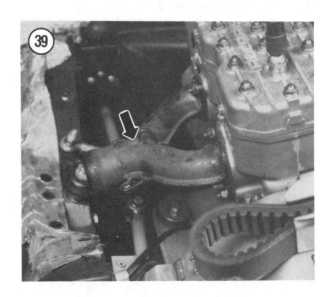

5

4. Remove the recoil starter housing as described in Chapter Eleven.

5. Remove the water pump as described in Chapter Ten.

6. Loosen the cylinder nuts (B, **Figure 38**) in a crisscross pattern.

7. Remove the cylinder nuts and washers (400) or the flange nuts (500).

8. Loosen the cylinder (**Figure 40**) by tapping around the perimeter with a rubber or plastic mallet. Repeat for each cylinder.

9. Pull the cylinder block (**Figure 40**) straight up and off the crankcase studs.

10. Remove the cylinder base gasket and discard it.

11. Stuff clean rags around the connecting rods to keep dirt and loose parts from entering the crankcase.

12. If necessary, remove the bolts holding the intake manifold to the engine. Remove the manifold.

13. Inspect the cylinder block as described in this chapter.

Installation
(Liquid Cooled Models)

1. Clean the cylinder bore as described under *Inspection* in this chapter.

2. Install the intake manifold, if removed. Tighten the bolts securely.

3. Check that the top surface of the crankcase and the bottom cylinder surface are clean prior to installation.

4. Install a new base gasket.

5. Make sure the end gaps of the piston rings are lined up with the locating pins in the ring grooves (**Figure 41**). Lightly oil the piston rings and the inside of the cylinder bores.

6. Place a piston holding tool under one piston and turn the crankshaft until the piston is down firmly against the tool (**Figure 42**). This will make cylinder installation easier. You can make this tool out of wood as shown in **Figure 37**.

7. Install the cylinder block onto the crankcase studs so that the intake ports face the rear of the engine. Compress each piston ring with your fingers as the cylinder starts to slide over it. After one piston is installed in its cylinder, lower the

cylinder and install the remaining piston. Make sure the cylinder block is fully seated on the crankcase. See **Figure 40**.

8. Install the cylinder lockwashers and nuts (400) or the flange nuts (500). Tighten the nuts securely in a crisscross pattern.

9. Reverse Steps 1-5 under *Removal* to complete assembly.

10. Bleed the oil pump as described in Chapter Nine.

11. Refill and bleed the cooling system as described in Chapter Three.

12. If new components were installed or if the cylinders were bored or honed, the engine must be broken in as if it were new. Refer to *Break-In Procedure* in Chapter Three.

Inspection (All Models)

Cylinder measurement requires a precision inside micrometer or bore gauge. If you don't have the right tools, have your dealer or a machine shop take the measurements. The following procedures are shown on a liquid cooled model. Service procedures for the air-cooled Trail and Sport models are the same.

1. Remove all gasket residue from the top and bottom (**Figure 43**) gasket surfaces. If the cylinder head surface appears warped or excessively contaminated with exhaust deposits, resurface as follows:

    a. Remove the cylinder studs (**Figure 44**) as described in Chapter One.

    b. Tape a piece of 400-600 grit wet emery sandpaper onto a piece of thick plate glass or surface plate (**Figure 30**).

    c. Slowly resurface the cylinder by moving it in figure-eight patterns on the sandpaper.

    d. Rotate the cylinder several times to avoid removing too much material from one side.

    e. Reinstall the cylinder studs as described in Chapter One.

2. Using a soft scraper or a wire wheel mounted on a drill or hand grinder, remove all carbon deposits from the exhaust ports (**Figure 45**).

3. Wash the cylinder in solvent to remove any oil and carbon particles. The cylinder bore must be cleaned thoroughly before attempting any measurement as incorrect readings may be obtained.

4. Measure the cylinder bore diameter as described under *Piston/Cylinder Clearance* in this chapter.

5. If the cylinder is not worn past the service limit, check the bore carefully for scratches or gouges. The bore may require reconditioning.

6. Check the threaded holes in the cylinder block for thread damage. Minor damage can be cleaned up with a suitable metric tap. Refer to Chapter One for information pertaining to threads, fasteners and repair tools. If damage is severe, a thread insert should be installed.

7. Check the cylinder studs (**Figure 44**) for thread damage. Replace studs as described in Chapter One.

8. After the cylinder has been serviced, wash the bore in hot soapy water. This is the only way to clean the cylinder wall of the fine grit material left from the boring or honing job. After washing the cylinder wall, run a clean white cloth through it. The cylinder wall should show no traces of grit or other debris. If the rag shows any sign of debris, the cylinder wall is not clean and must be rewashed. After the cylinder is thoroughly cleaned, lubricate the cylinder walls with clean engine oil to prevent the cylinder liners from rusting.

> *CAUTION*
> *A combination of soap and water is the only solution that will completely clean the cylinder wall. Solvent and kerosene cannot wash fine grit out of cylinder crevices. Grit left in the cylinder will act as a grinding compound and cause premature wear to the new rings.*

9. After cleaning the cylinder, check that the oil line check valve (**Figure 46**) in each cylinder is clear. A plugged check valve will cause piston and cylinder damage. Clean check valves with compressed air if available.

PISTON, WRIST PIN AND PISTON RINGS

The piston is made of an aluminum alloy. The wrist pin is a precision fit and is held in place by a clip at each end. A caged needle bearing is used on the small end of the connecting rod. See **Figure 19**.

Piston and Piston Ring Removal

1. Remove the cylinders as described in this chapter.

2. Identify the pistons by scratching PTO or MAG on the piston crowns. In addition, keep each piston together with its own pin, bearing and piston rings to avoid confusion during reassembly.

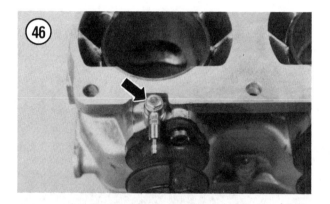

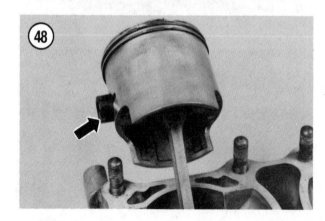

3. Before removing the piston, hold the rod tightly and rock the piston. Any rocking motion (do not confuse with the normal sliding motion) indicates wear on the wrist pin, needle bearing, wrist pin bore, or more likely a combination of all three.

NOTE
Wrap a clean shop cloth under the piston so that the clip will not fall into the crankcase.

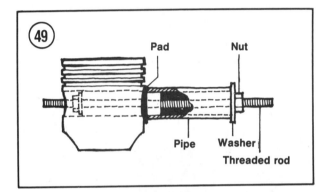

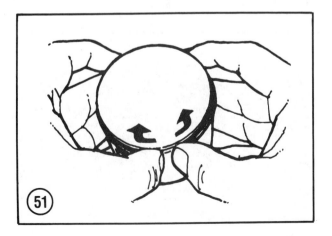

WARNING
Safety glasses should be worn when performing Step 4.

4. Remove the wrist pin clips (**Figure 47**) from the outside of the piston with an awl, scribe or other sharp pointed tool. Hold your thumb over one edge of the clip when removing it to prevent it from springing out.

5. Use a proper size wooden dowel or socket extension and push out the wrist pin (**Figure 48**).

CAUTION
If the engine ran hot or seized, the wrist pin may be difficult to remove. However, do not drive the wrist pin out of the piston. This will damage the piston, needle bearing and connecting rod. If the wrist pin will not push out by hand, remove it as described in Step 6.

6. If the wrist pin is tight, fabricate the tool shown in **Figure 49**. Assemble the tool onto the piston and pull the wrist pin out of the piston. Make sure to install a pad between the piston and piece of pipe to prevent from scoring the side of the piston.

7. Lift the piston off the connecting rod.

8. Remove the needle bearing (**Figure 50**).

9. Repeat for each piston.

10. If the pistons are going to be left off for some time, place a piece of foam insulation tube, or shop cloth, over the end of each rod to protect it.

NOTE
On pistons which use 2 piston rings, always remove the top piston ring first.

11. The top and bottom piston rings are identical. However, if they are going to be reused, mark each ring after removal to ensure correct installation. Remove the upper ring by spreading the ends with your thumbs just enough to slide it up over the piston (**Figure 51**). Repeat for the lower ring.

**Wrist Pin and
Needle Bearing Inspection**

1. Clean the needle bearing (A, **Figure 52**) in solvent and dry it thoroughly. Use a magnifying glass and inspect the bearing cage for cracks at the corners of the needle slots (**Figure 53**) and inspect the needles themselves for cracking. If any cracks are found, the bearing must be replaced.

2. Check the wrist pin (B, **Figure 52**) for severe wear, scoring or chrome flaking. Also check the wrist pin for cracks along the top and side. Replace the wrist pin if necessary.

3. Oil the needle bearing and pin and install them into the connecting rod. Slowly rotate the pin and check for radial and axial play (**Figure 54**). If any play exists, the pin and bearing should be replaced, providing the rod bore is in good condition. If the condition of the rod bore is in question, the old pin and bearing can be checked with a new connecting rod.

4. Oil the wrist pin and install it in the wrist pin hole (**Figure 55**). Check for up and down play between the pin and piston. There should be no noticeable play. If play is noticeable, replace the wrist pin and/or piston.

CAUTION
If there are signs of piston seizure or overheating, replace the wrist pins and bearings. These parts have been weakened from excessive heat and may fail later.

5. Check the needle bearing spacers for cracks or other damage; replace if necessary.

Connecting Rod Inspection

1. Wipe the wrist pin bore in the connecting rod with a clean rag and check it for galling, scratches, or any other signs of wear or damage. If any of these conditions exist, replace the connecting rods as described in this chapter.

2. Check the connecting rod big end bearing play. You can make a quick check by simply rocking the connecting rod back and forth (**Figure 56**). If there is more than a very slight rocking motion (some side-to-side sliding is normal), the connecting rods may require replacement.

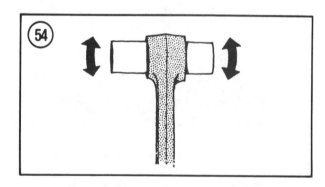

Piston and Ring Inspection

1. Carefully check the piston for cracks at the top edge of the transfer cutaways (**Figure 57**) and replace if found. Check the piston skirt (**Figure 58**) for brown varnish buildup. More than a slight amount is an indication of worn or sticking rings which should be replaced.

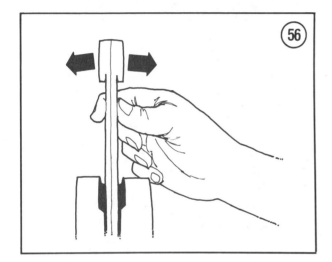

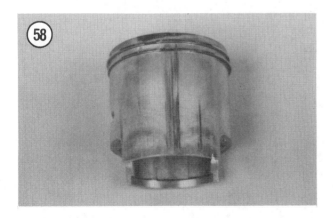

2. Check the piston skirt for galling and abrasion which may have resulted from piston seizure. If light galling is present, smooth the affected area with No. 400 emery paper and oil or a fine oilstone. However, if galling is severe or if the piston is deeply scored, replace it.

3. Check the condition of the piston crown (**Figure 59**). Normal carbon buildup can be removed with a wire wheel mounted on a drill press. If the piston is damaged, it is important to pinpoint the cause so that the failure will not repeat after engine assembly. Note the following when checking damaged pistons:

    a. If the piston damage is contained to the area above the wrist pin bore, the engine is probably overheating. Seizure or galling conditions contained to the area below the wrist pin bore is usually caused by a lack of lubrication, rather than overheating.

    b. If the piston has seized and appears very dry (apparent lack of oil or lubrication on the piston), a lean fuel mixture probably caused the overheating. Overheating can result from incorrect jetting, air leaks or over advanced ignition timing.

    c. Preignition will cause a sand-blasted appearance on the piston crown. This condition is discussed in Chapter Two.

    d. If the piston damage is confined to the exhaust port area on the front of the piston, look for incorrect jetting (too lean) or over advanced ignition timing.

e. If the piston has a melted pocket starting in the crown or if there is a hole in the piston crown, the engine is running too lean. This can be due to incorrect jetting, an air leak or over advanced ignition timing. A spark plug that is too hot will also cause this type of piston damage.

f. If the piston is seized around the skirt but the dome color indicates proper lubrication (no signs of dryness or excessive heat), the damage may result from a condition referred to as cold seizure. This condition typically results from running a water-cooled engine too hard without first properly warming it up. A lean fuel mixture can also cause skirt seizure.

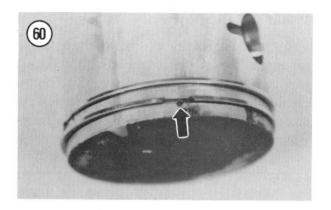

4. Check the piston ring locating pins in the piston (**Figure 60**). The pins should be tight and the piston should show no signs of cracking around the pins. If a locating pin is loose, replace the piston. A loose pin will fall out and cause severe engine damage. Do not attempt to repair a loose locating pin.

5. Check the wrist pin clip grooves (**Figure 61**) in the piston for cracks or other damage that could allow a clip to fall out. This would cause severe engine damage. Replace the piston if any one groove shows signs of wear or damage.

NOTE
Maintaining proper piston ring end gap helps to ensure peak engine performance. Excessive ring end gap reduces engine performance and can cause overheating. Insufficient ring end gap will cause the ring ends to butt together and cause the ring to break, resulting in severe engine damage. So that you don't have to wait for parts, always order extra cylinder head and base gaskets to have on hand for routine top end inspection and maintenance.

6. Measure piston ring end gap. Place a ring into the cylinder approximately 9.5 mm (3/8 in.) from the top; square the ring in the cylinder by pushing

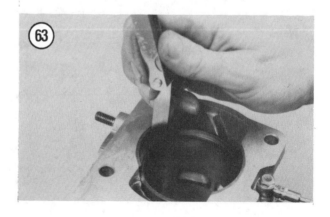

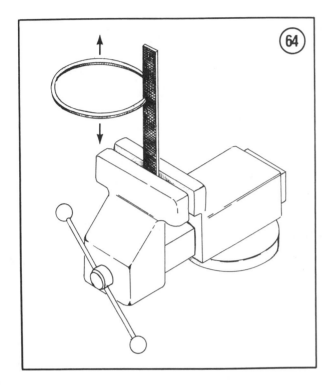

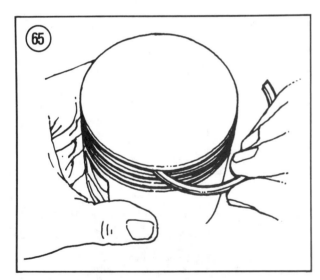

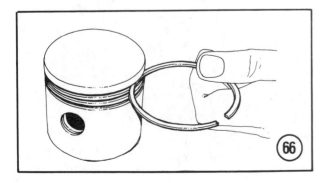

it with the piston (**Figure 62**). This ensures that the ring is square in the cylinder bore. Measure the gap with a flat feeler gauge (**Figure 63**) and compare to the wear limit in **Table 2**. If the gap is greater than specified, the rings should be replaced as a set.

NOTE
*When installing new rings, measure the end gap in the same manner as for old ones. If the gap is less than specified, make sure you have the correct piston rings. If the replacement rings are correct but the end gap is too small, carefully file the ends with a fine cut file until the gap is correct (**Figure 64**). Insufficient ring end gap will allow the rings to butt together. This can result in ring breakage, ring seizure or rapid cylinder wear. These conditions will reduce engine performance and may cause seizure or severe damage to the cylinder wall.*

7. Carefully remove all carbon buildup from the ring grooves with a broken ring (**Figure 65**). Inspect the grooves carefully for burrs, nicks, or broken and cracked lands. Recondition or replace the piston if necessary.

8. Check piston ring side clearance by rolling each ring around its piston groove as shown in **Figure 66**. Check the ring for ease of movement and squareness. If you are checking side clearance with old rings, and severe wear is indicated, recheck with new rings. If the new rings still show wear, replace the piston.

9. If the piston checked out okay after performing these inspection procedures, measure the piston outside diameter as described under *Piston/Cylinder Clearance* in this chapter.

10. If new piston rings are required, the cylinders should be honed before assembling the engine. Refer to *Cylinder Honing* in this chapter.

Piston/Cylinder Clearance

The following procedure requires the use of highly specialized and expensive measuring tools. If such equipment is not readily available, have the measurements performed by a dealer or machine shop. Always replace pistons as a set.

1. Measure the outside diameter of the piston with a micrometer at the point indicated in **Figure 67** above the bottom of the piston skirt, at a 90° angle to the piston pin (**Figure 68**). Record the measurement.

NOTE
Always install new rings when installing a new piston.

2. Wash the cylinder block in solvent to remove any oil and carbon particles. The cylinder bore must be cleaned thoroughly before attempting any measurement as incorrect readings may be obtained.

3. Measure the cylinder bore with a bore gauge or telescoping gauge (**Figure 69**). Then measure the bore gauge or telescoping gauge with a micrometer to determine the bore diameter. Measure the cylinder bore at the points shown in **Figure 70**. Measure in 3 axes—in line with the piston pin and at 90° to the pin. Record the measurement.

4. Piston clearance is the difference between the maximum piston diameter and the minimum cylinder diameter. For a run-in (used) piston and cylinder, subtract the dimension of the piston from the cylinder dimension. If the clearance exceeds the dimension in **Table 3**, the piston should be replaced.

NOTE
Polaris does not list specifications for the piston outside diameter or the cylinder bore inside diameter. Thus piston and cylinder wear is monitored by checking piston-to-cylinder clearance; when clearance is excessive, the piston is replaced to renew the clearance specification. If you find that the clearance is still excessive with a new piston, the cylinder bore is worn. A bore job and oversize pistons are required.

NOTE
The new pistons should be obtained first before the cylinders are bored so that the pistons can be measured. The cylinders must be bored to match the pistons. Piston-to-cylinder clearance is specified in Table 3.

Cylinder Honing

The surface condition of a worn cylinder bore is normally very shiny and smooth. When installing new piston rings for use in a cylinder with minimum wear, they would not seat properly and engine performance would suffer from compression losses. Cylinder honing, often

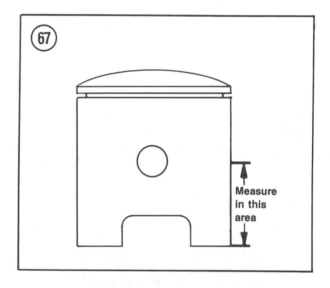

Measure in this area

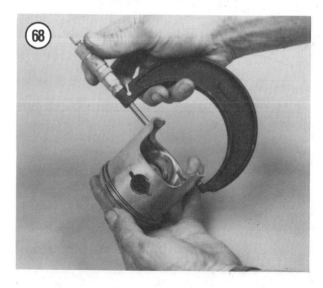

referred to as bead breaking or deglazing, is required whenever new piston rings are installed. When a cylinder bore is honed, the surface is slightly roughed up to provide a textured or crosshatched surface. This surface finish controls

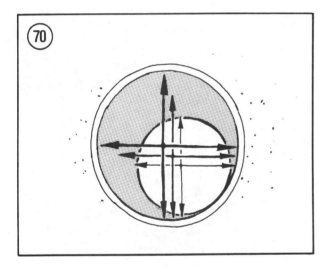

wear of the new rings and helps them to seat and seal properly. *Whenever* new rings are installed, the cylinder surface should be honed. This service can be performed by a Polaris dealer or independent repair shop. The cost of having the cylinder honed by a dealer is usually minimal compared to the cost of purchasing a hone and doing the job yourself. If you choose to hone the cylinder yourself, follow the hone manufacturer's directions closely.

5

CAUTION
*After a cylinder has been reconditioned by boring or honing, the bore should be properly cleaned to remove all material left from the machining operation. Refer to **Inspection** under **Cylinder** in this chapter. Improper cleaning will not remove all of the machining residue resulting in rapid wear of the new piston and rings.*

Piston Installation

1. Apply assembly oil to the needle bearing (A, **Figure 52**) and install it in the connecting rod (**Figure 50**).
2. Insert the wrist pin partway through the piston (**Figure 71**).
3. Place the piston over the connecting rod so that the arrow mark on the piston crown points toward the flywheel side of the engine. This alignment is essential so the piston ring ends will be correctly positioned and will not snag in the ports. Line up the pin with the bearing and push the pin into the piston until it is even with the wrist pin clip grooves.

NOTE
If there is no mark on the piston crown, position the piston ring locating pins, in the piston, toward the intake port.

CAUTION
If the wrist pin will not slide in the piston smoothly, heat the piston crown slightly

*with a torch (**Figure 72**), then when the piston expands slightly, install the wrist pin. Wear suitable gloves to prevent from burning your hands when performing this step.*

4. Install new wrist pin clips (**Figure 47**), making sure they are completely seated in their grooves with the open ends of the clips facing down. See **Figure 73**.

5. Check the installation by rocking the piston back and forth around the pin axis and from side-to-side along the axis. It should rotate freely back and forth but not from side-to-side.

NOTE
*Keystone piston rings should be installed with their bevel side facing toward the top of the piston. See **Figure 74**.*

6. Install the piston rings—first the bottom one, then the top—by carefully spreading the ends of the ring with your thumbs and slipping the ring over the top of the piston. Make sure manufacturer's mark on the piston rings are toward the top of the piston. If you are installing used rings, install them by referring to the identification marks made during removal.

7. Make sure the rings are seated completely in the grooves, all the way around the circumference, and that the ends are aligned with the locating pins. See **Figure 75**.

8. If new components were installed, the engine must be broken in as if it were new. Refer to *Break-In Procedure* in Chapter Three.

CRANKCASE AND CRANKSHAFT

Disassembly of the crankcase—splitting the cases—and removal of the crankshaft assembly requires engine removal from the frame. However, the cylinder head, cylinder and all other attached assemblies should be removed with the engine in the frame.

The crankcase is made in 2 halves of precision diecast aluminum alloy and is of the "thin-

walled" type. See **Figures 76-78**. To avoid damage to them, do not hammer or pry on any of the interior or exterior projected walls. These areas are easily damaged if stressed beyond what they are designed for. They are assembled without a gasket; only gasket sealer is used while dowel pins align the crankcase halves when they are bolted together. The crankcase halves are sold as a matched set only. If one crankcase half is severely damaged, both must be replaced. Crankshaft components are available as individual parts. However, crankshaft service—replacement of unsatisfactory parts or crankshaft alignment—should be entrusted to a dealer or engine specialist. Special measuring and alignment tools, a hydraulic press and experience

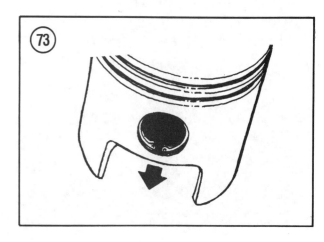

are necessary to disassemble, assemble and accurately align the crankshaft assembly, which, in the case of the average twin cylinder engine, is made up of a number of pressed-together pieces, not counting the bearings, seals and connecting rods. However, you can save considerable expense by disassembling the engine and taking the crankshaft in for service.

The procedure which follows is presented as a complete, step-by-step major lower end overhaul that should be followed if the engine is to be completely reconditioned.

Crankcase Disassembly

This procedure describes disassembly of the crankcase halves and removal of the crankshaft.

1. Remove the engine from the frame as described in this chapter.

2. *Liquid cooled models*: Remove the water pump as described in Chapter Ten.

3. Remove the pistons as described in this chapter.

4. Remove the flywheel and stator plate as described in Chapter Eight.

5. Check crankshaft runout before disassembling the crankcase halves. Refer to *Crankshaft Runout* in this chapter.

> *CAUTION*
> *Do not damage the crankcase studs when performing the following procedures.*

6. Turn the crankcase assembly so that it rests up-side-down on its crankcase studs (**Figure 79**).

7. Loosen the crankcase bolts in a crisscross pattern (**Figure 80**).

8. Tap on the large bolt bosses with a soft mallet to break the crankcase halves apart and then remove the bottom half.

> *CAUTION*
> *Make sure that you have removed all the fasteners. If the cases are hard to separate, check for any fasteners you may have missed.*

> *CAUTION*
> *Do not pry the cases apart with a screwdriver or any other sharp tool, otherwise the sealing surface will be damaged.*

9. Remove the 2 dowel pins.

10. Remove the crankshaft (**Figure 81**) from the upper case half.

11. Remove the oil pump drive shaft as described in Chapter Nine.

(75)

(74) **PISTON TOP**

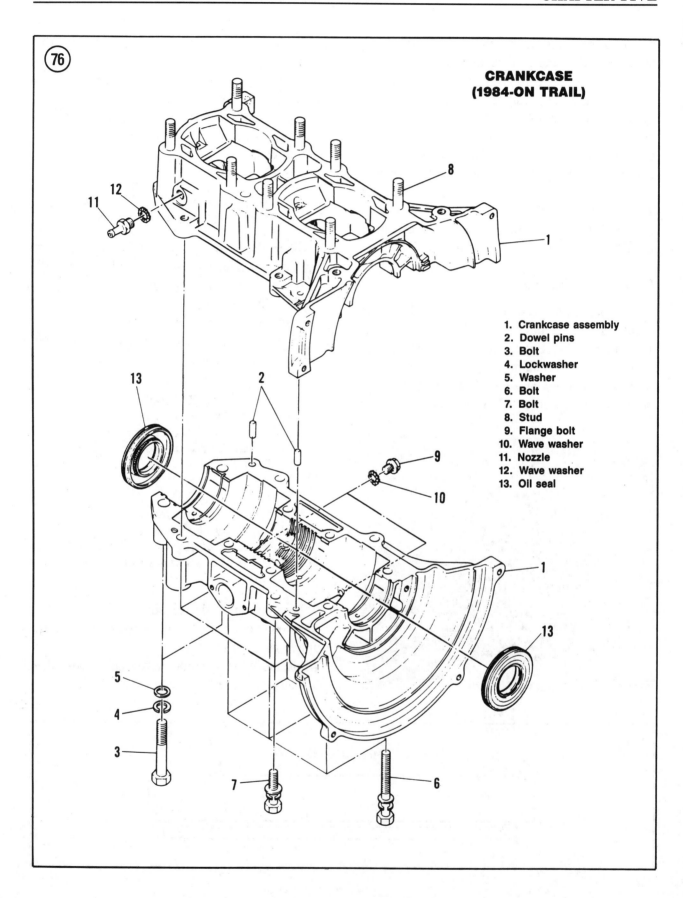

(76)

**CRANKCASE
(1984-ON TRAIL)**

8

12

11

1

13

2

9

10

1. Crankcase assembly
2. Dowel pins
3. Bolt
4. Lockwasher
5. Washer
6. Bolt
7. Bolt
8. Stud
9. Flange bolt
10. Wave washer
11. Nozzle
12. Wave washer
13. Oil seal

13

5

4

3

7

6

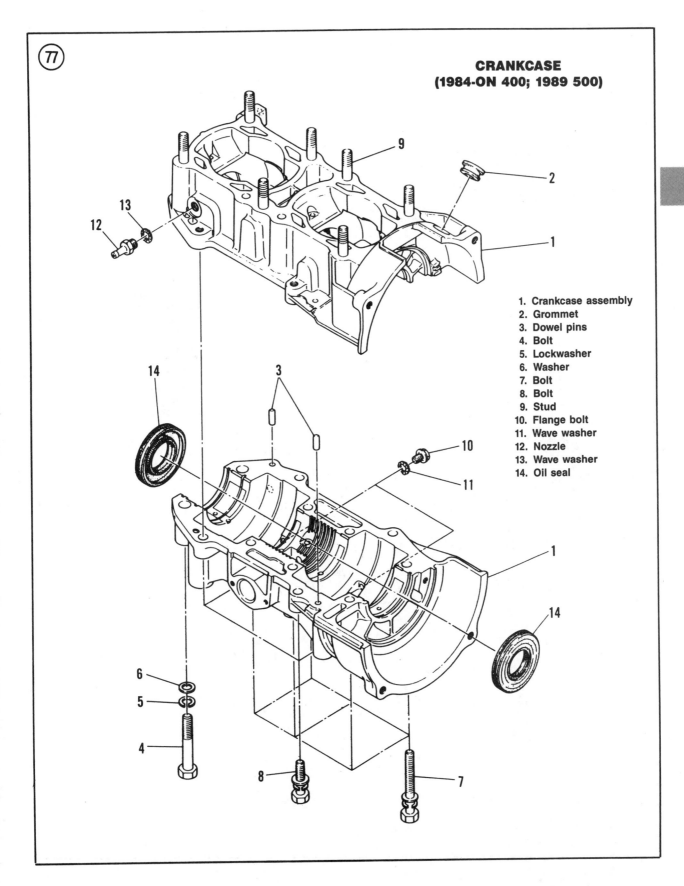

**CRANKCASE
(1984-ON 400; 1989 500)**

5

1. Crankcase assembly
2. Grommet
3. Dowel pins
4. Bolt
5. Lockwasher
6. Washer
7. Bolt
8. Bolt
9. Stud
10. Flange bolt
11. Wave washer
12. Nozzle
13. Wave washer
14. Oil seal

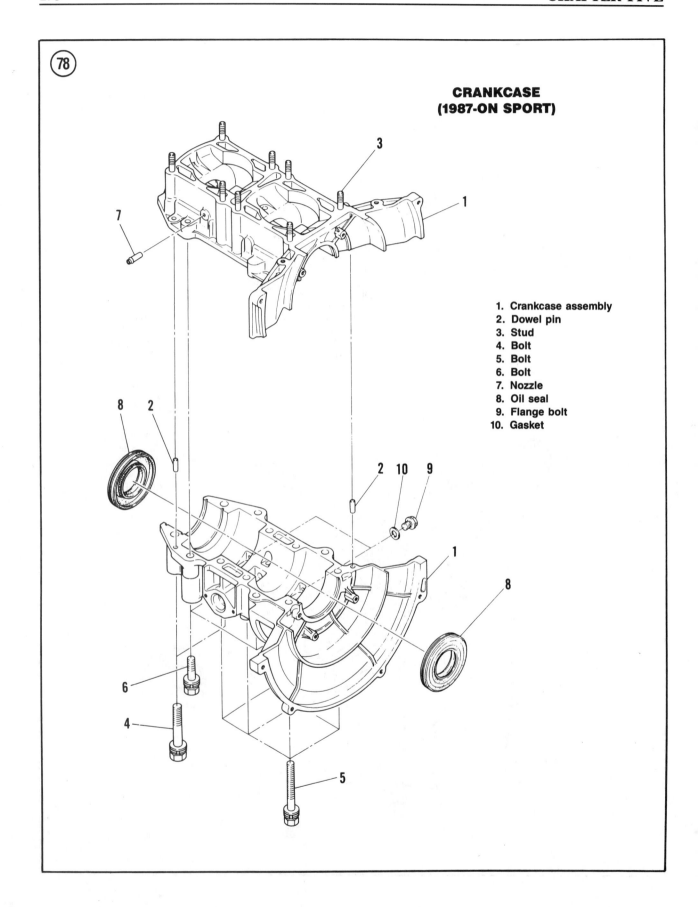

⑦⑧

**CRANKCASE
(1987-ON SPORT)**

1. Crankcase assembly
2. Dowel pin
3. Stud
4. Bolt
5. Bolt
6. Bolt
7. Nozzle
8. Oil seal
9. Flange bolt
10. Gasket

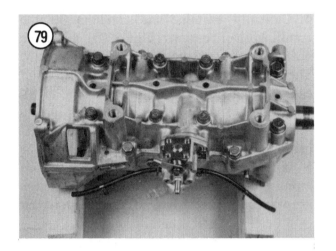

Cleaning

1. Clean both crankcase halves with cleaning solvent. Thoroughly dry with compressed air and wipe off with a clean shop cloth. Be sure to remove all traces of old gasket sealer from all mating surfaces.

2. Clean the oil passages in the transfer ports in the upper crankcase half (A, **Figure 82**). Use compressed air to ensure that they are clean.

3. Clean the crankshaft assembly in solvent and dry with compressed air. Oil bearings with engine oil after all solvent has been removed.

5

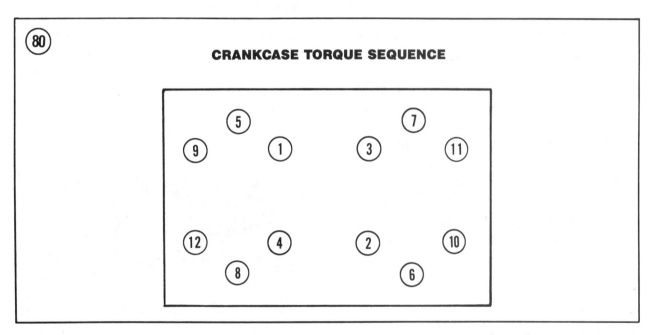

CRANKCASE TORQUE SEQUENCE

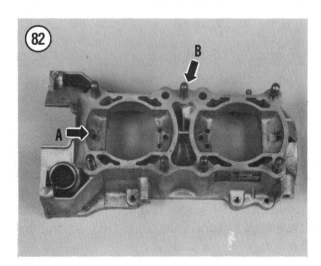

Crankcase Inspection

Refer to **Figure 76**, **Figure 77** or **Figure 78** for this procedure.

1. Carefully inspect the case halves for cracks and fractures. Also check the areas around the stiffening ribs, around bearing bosses and threaded holes. If any are found, have them repaired by a shop specializing in the repair of precision aluminum castings or replace them.
2. Check the locating pin holes in both case halves for cracks or damage.
3. Check the oil seal and bearing grooves in the upper (**Figure 83**) and lower (**Figure 84**) case halves for cracks or damage.
4. Check the bearing surface areas in the upper (**Figure 83**) and lower (**Figure 84**) case halves for roughness, cracks or other damage.
5. Check the threaded holes in both crankcase halves for thread damage, dirt or oil buildup. If necessary, clean or repair the threads with a suitable size metric tap. Coat the tap threads with kerosene or an aluminum tap fluid before use.
6. Check the cylinder studs (B, **Figure 82**) for thread damage or bending. If necessary, replace the studs as described in Chapter One.
7. If there is any doubt as to the condition of the crankcase halves, and they cannot be repaired, replace the crankcase halves as a set.

Crankshaft Inspection

Refer to **Figure 85** or **Figure 86** for this procedure.

1. Remove the left-hand and right-hand crankshaft oil seals (**Figure 87**).
2. Inspect the oil seal(s) for excessive wear, cuts or other damage; replace oil seal(s) as required.
3. Check the connecting rods (**Figure 88**) for signs of excessive heat, cracks or other damage. Check the lower rod bearings for visual signs of excessive heat or damage. With the crankshaft mounted on V-blocks (**Figure 89**), rotate the crankshaft and check the connecting rod bearings for roughness or damage.

4. Carefully examine the condition of the crankshaft ball bearings (**Figure 89**). Clean the bearings in solvent and allow to dry thoroughly. Then oil each bearing before checking it. Roll each bearing around by hand, checking that it turns quietly and smoothly and that there are no rough spots. There should be no apparent radial play. Defective bearings should be replaced.
5. If the crankshaft exceeded any of the service limits in Steps 3 or 4 or if one or more bearings are worn or damaged, have the crankshaft rebuilt by a dealer or crankshaft specialist.

Crankshaft Runout

Crankshaft runout is checked with the crankshaft installed in the crankcase with the case

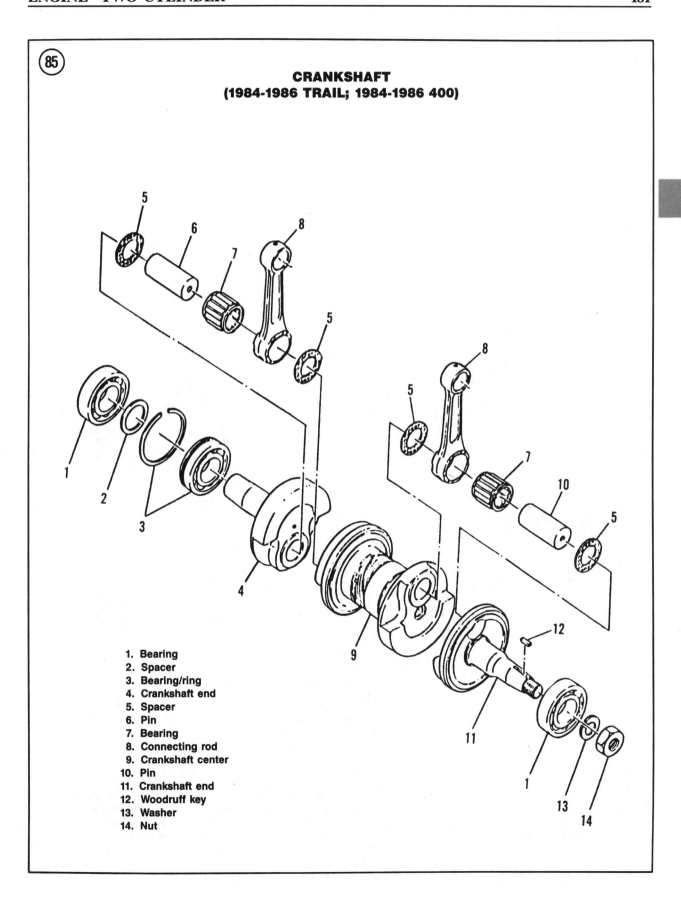

CRANKSHAFT
(1984-1986 TRAIL; 1984-1986 400)

1. Bearing
2. Spacer
3. Bearing/ring
4. Crankshaft end
5. Spacer
6. Pin
7. Bearing
8. Connecting rod
9. Crankshaft center
10. Pin
11. Crankshaft end
12. Woodruff key
13. Washer
14. Nut

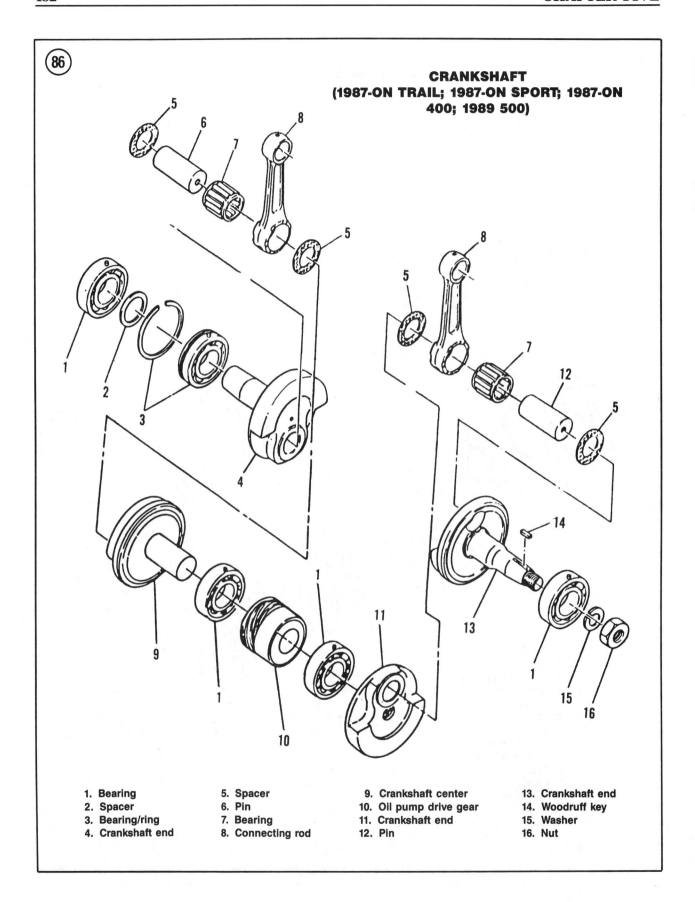

⁸⁶

**CRANKSHAFT
(1987-ON TRAIL; 1987-ON SPORT; 1987-ON
400; 1989 500)**

1. Bearing
2. Spacer
3. Bearing/ring
4. Crankshaft end
5. Spacer
6. Pin
7. Bearing
8. Connecting rod
9. Crankshaft center
10. Oil pump drive gear
11. Crankshaft end
12. Pin
13. Crankshaft end
14. Woodruff key
15. Washer
16. Nut

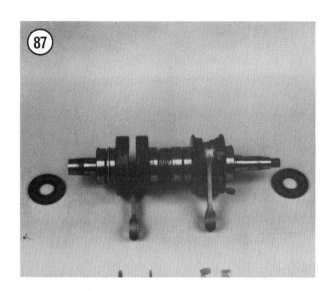

bolts tightened securely. If the engine has been disassembled, install the crankshaft and bolt the case halves together as described in this chapter.

1. Remove the primary sheave from the crankshaft.

2. Clean the end of the crankshaft with contact cleaner. If the crankshaft is nicked or lightly roughed up, clean with emery cloth.

3. Place a dial indicator pointer 1/2 in. from the end of the crankshaft as shown in **Figure 90**. Turn the crankshaft slowly and note the indicator reading. If the runout exceeds the service limit of 0.152 mm (0.006 in.), have the crankshaft serviced by a crankshaft specialist.

5

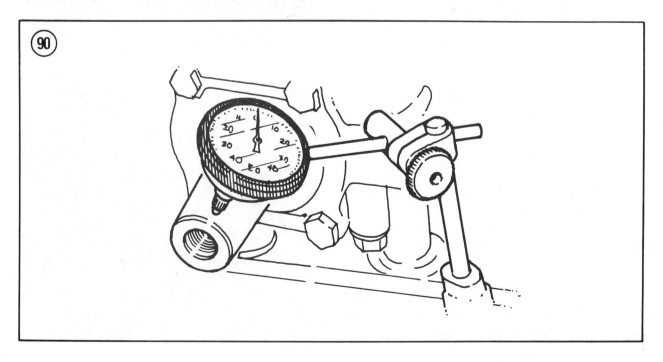

Bearing Replacement

The outer bearings (A, **Figure 91** and A, **Figure 92**) can be replaced as follows. A bearing puller will be required to remove the bearing. Read this procedure completely through before removing the bearing.

NOTE
*Before removing the bearings from the crankshaft, note the position of the dowel pin in each outer bearing race. See A, **Figure 91** and A, **Figure 92**. These pins must align with the dowel pin holes in the upper case half during crankshaft installation. Because the pins are mounted off-center in the bearings, each bearing must be installed with the pin in the same position. If a bearing is installed incorrectly, the pin will not align with the pin hole; the bearing will have to be removed and turned around.*

NOTE
Before removing the MAG side bearing with a bearing puller, thread the flywheel nut onto the crankshaft so that it is flush with the end of the crankshaft. The nut will prevent the puller screw from damaging the crankshaft threads.

NOTE
*Before removing the PTO side bearings, the bearings may have to be separated slightly to allow installation of the bearing puller. When separating the bearings, position the crankshaft so that the bearings are supported by wood blocks. Then align a bearing splitter or chisel between the 2 bearing races and drive the bearings apart enough to allow the use of a bearing puller. See **Figure 93**.*

1. Install a bearing puller and remove the bearing(s) (**Figure 94**). Remove any shims as required. See **Figure 85** or **Figure 86**.

2. Clean the crankshaft bearing area with solvent or electrical contact cleaner and dry thoroughly.

NOTE
Completely read Step 3 through before heating and installing the bearing. During bearing installation, the crankshaft should be supported securely so that the bearing can be installed quickly. If the bearing cools and tightens on the crankshaft before it is completely installed, remove the bearing with the puller and reheat.

3. Install the bearing(s) as follows:
   a. Lay the bearing(s) on a clean, lint-free surface in the order of assembly.
   b. When installing the bearing(s), make sure that the bearing dowel pin faces in the direction shown in A, **Figure 91** or A, **Figure 92**.
   c. Refer to *Shrink Fit* under *Ball Bearing Replacement* in Chapter One.

d. After referring to the information in substep c, heat and install the bearing(s) and shim (as required).

Crankshaft Installation

Refer to **Figure 85** or **Figure 86** for this procedure.

1. Install new crankshaft oil seals as follows:

5

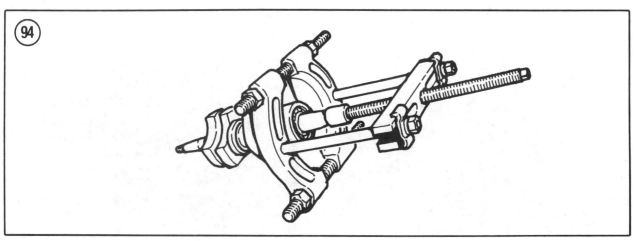

a. Fill the lips of the new seals with a low-temperature grease (**Figure 95**).

b. Install the left-hand (B, **Figure 91**) and right-hand (B, **Figure 92**) oil seals.

2. Install the oil pump drive shaft as described in Chapter Nine.

3. Install the 2 crankcase dowel pins. See **Figure 96**, typical.

4. Apply grease to the labyrinth seal grooves in the case halves (**Figure 97**).

NOTE
*Step 5 describes crankshaft (**Figure 81**) installation. However, because of the number of separate procedures required during installation, read Step 5 through first before actually installing the crankshaft.*

5. Align the crankshaft with the lower case half and install the crankshaft into the upper case half (**Figure 81**). Note the following:

a. Rotate the crankshaft ball bearings to align their locating pin with the pin recesses in the upper case half. See A, **Figure 98**, typical.

b. Fit the ring on the PTO inner bearing into the groove in the upper case half. See B, **Figure 98**, typical.

c. Fit the left- and right-hand oil seals into the case half grooves. See **Figure 99**, typical.

d. Double check to make sure that all of the bearings and seals are seated.

6. Rotate the crankshaft to make sure it turns freely.

Crankcase Assembly

1. Install the crankshaft into the upper case half as described in this chapter.

2. Install the 2 dowel pins if not previously installed. See **Figure 96**.

3. Make sure the crankcase mating surfaces are completely clean.

NOTE
A non-shimming gasket sealer, such as Loctite Gasket Eliminator 515 Sealant

(Figure 100), should be used. Always follow manufacturer's directions for surface cleaning, application and cure time.

4. Apply a gasket sealer to both case half mating surfaces.

5. Put the lower case half onto the upper half. Check the mating surfaces all the way around the case halves to make sure they are even and that a locating pin has not worked out of its pin recess.

6. Apply a light coat of oil to the crankcase bolt threads before installing them.

7. Install the crankcase bolts and run them down hand tight. See **Figure 101**.

8. Tighten the crankcase bolts in 2-steps in the numerical sequence shown in **Figure 102**. Refer to **Table 4** for torque specifications.

5

CAUTION
While tightening the crankcase fasteners, make frequent checks to ensure that the

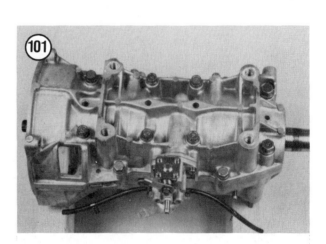

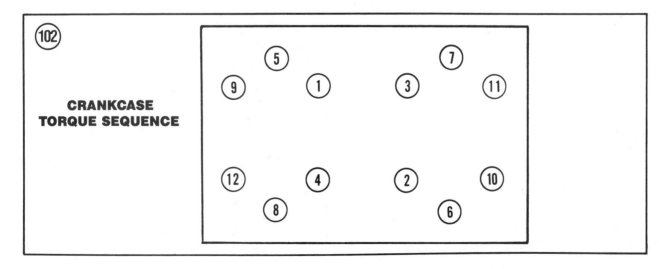

CRANKCASE TORQUE SEQUENCE

crankshaft turns freely and that the crankshaft locating pins and crankcase dowel pins fit into place in the case halves.

9. Check again that the crankshaft turns freely. If it is binding, separate the crankcase halves and determine the cause of the problem.

10. Turn the engine right side up.

11. Liberally coat the crank pins and bearings with two-stroke oil. Apply the same oil through the oil delivery holes in the transfer ports.

12. Install the engine top end as described in this chapter.

13. Install the stator plate and flywheel as described in Chapter Eight.

14. *Liquid cooled models*: Install the water pump as described in Chapter Ten.

15. Install the engine in the frame as described in this chapter.

16. If new components were installed, the engine must be broken in as if it were new. Refer to *Break-In Procedure* in Chapter Three.

Table 1 GENERAL ENGINE SPECIFICATIONS

Bore	
Sport	
1987	62 mm (2.4410 in.)
1988-on	62.3 mm (2.4546 in.)
Trail	
1984-1986	67.72 mm (2.6557 in.)
1987-on	72 mm (2.8346 in.)
400	65 mm (2.559 in.)
500	72 mm (2.8346 in.)
Stroke	
Sport	55.6 mm (2.1886 in.)
Trail	60 mm (2.362 in.)
400	60 mm (2.362 in.)
500	60 mm (2.362 in.)
Displacement	
Sport	339 cc (20.7 cu. in.)
Trail	
1984-1986	432 cc (26.4 cu. in.)
1987-on	488 cc (29.8 cu. in.)
400	398 cc (24.3 cu. in.)
500	488 cc (29.8 cu. in.)

Table 2 PISTON RING END GAP

	mm	in.
1984	*	*
1985-1987	0.15-0.40	0.006-0.016
1988-on	0.15-0.45	0.006-0.018
* Not specified		

Table 3 PISTON-TO-CYLINDER CLEARANCE*

	New mm (in.)	Wear limit mm (in.)
Sport	0.064 (0.0025)	0.114 (0.0045)
Trail		
1983-1985	0.076 (0.003)	0.127 (0.005)
1986-on	0.070 (0.0027)	0.121 (0.0047)
400	0.057-0.095 (0.0022-0.0037)	1.121 (0.0047)
500	0.053 (0.0021)	0.104 (0.0041)
* ± 0.051 mm (0.002 in.)		

Table 4 ENGINE TIGHTENING TORQUES

	N·m	ft.-lb.
Cylinder head		
Sport and Trail	23-25	17-18
All other		
8 mm	22-23	16-17
10 mm	36-40	26-29
Crankcase		
8 mm	22-23	16-17
10 mm	32-35	23-25
Cylinder studs	33-39	24-28
Flywheel	83-90	60-65
Engine mounting bolts		
3/8 bolts	46-52	34-38
7/16 bolts	75-82	55-60

5

Chapter Six

Engine—Three Cylinder

The engine is a water-cooled 2-stroke triple. The engine is equipped with ball-type main crankshaft bearings and needle bearings on both ends of the connecting rods. Crankshaft components are available as individual parts. However, other than to replace the outer seals, it is recommended that all crankshaft work be entrusted to a dealer or other competent engine specialist.

This chapter covers information to provide routine top-end service as well as crankcase disassembly and crankshaft inspection. Work on the snowmobile engine requires considerable mechanical ability. You should carefully consider your own capabilities before attempting any operation involving major disassembly of the engine.

Much of the labor charge for dealer repairs involves the removal and disassembly of other parts to reach the defective component. Even if you decide not to tackle the entire engine overhaul after studying the text and illustrations in this chapter, it can be cheaper to perform the preliminary operations yourself and then take the engine to your dealer. Since dealers have lengthy waiting lists for service (especially during the fall and winter season), this practice can reduce the time your unit is in the shop. If you have done much of the preliminary work, your repairs can be scheduled and performed much quicker. General engine specifications are listed in **Table 1**. Engine service specifications are listed in **Table 2** and **Table 3**. **Tables 1-4** are found at the end of the chapter.

ENGINE LUBRICATION

Engine lubrication is provided by the fuel/oil mixture used to power the engine.

SERVICE PRECAUTIONS

Whenever you work on your Polaris, there are several precautions that should be followed to help with disassembly, inspection and reassembly.

1. In the text there is frequent mention of the left-hand and right-hand side of the engine. This refers to the engine as it is mounted in the frame, not as it sits on your workbench. See **Figure 1**.

2. Always replace a worn or damaged fastener with one of the same size, type and torque requirements. Make sure to identify each bolt before replacing it with another. Bolt threads should be lubricated with engine oil, unless otherwise specified, before torque is applied. If a tightening torque is not listed in **Table 4** (end of this chapter), refer to the torque and fastener information in Chapter One.

3. Use special tools where noted. In some cases, it may be possible to perform the procedure with makeshift tools, but this procedure is not recommended. The use of makeshift tools can damage the components and may cause serious personal injury. Where special snowmobile tools are required, these may be purchased through any Polaris snowmobile dealer. Other tools can be purchased through your dealer, or from a motorcycle or automotive accessory store. When purchasing tools from an automotive accessory dealer or store, remember that all threaded parts must have metric threads.

4. Before removing the first bolt and to prevent frustration during assembly, get a number of boxes, plastic bags and containers and store the parts as they are removed (**Figure 2**). Also have on hand a roll of masking tape and a permanent, waterproof marking pen to label each part or assembly as required. If your snowmobile was purchased second hand and it appears that some of the wiring may have been changed or replaced, label each electrical connection before disconnecting it.

5. Use a vise with protective jaws to hold parts. If protective jaws are not available, insert wooden blocks on each side of the part(s) before clamping them in the vise.

6. Remove and install pressed-on parts with an appropriate mandrel, support and hydraulic press. *Do **not** try to pry, hammer or otherwise force them on or off.*

7. Refer to **Table 4** at the end of the chapter for torque specifications. Proper torque is essential to assure long life and satisfactory service from snowmobile components.

8. Discard all O-rings and oil seals during disassembly. Apply a small amount of heat durable grease to the inner lips of each oil seal to prevent damage when the engine is first started.

9. Keep a record of all shims and where they came from. As soon as the shims are removed,

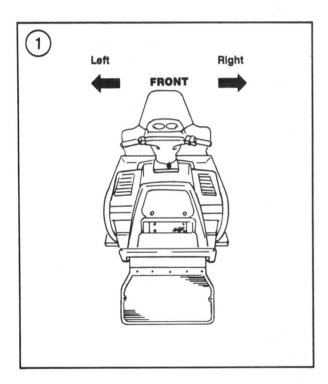

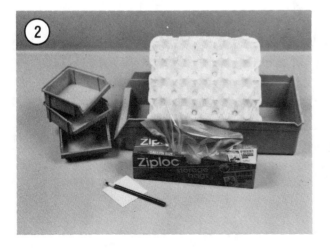

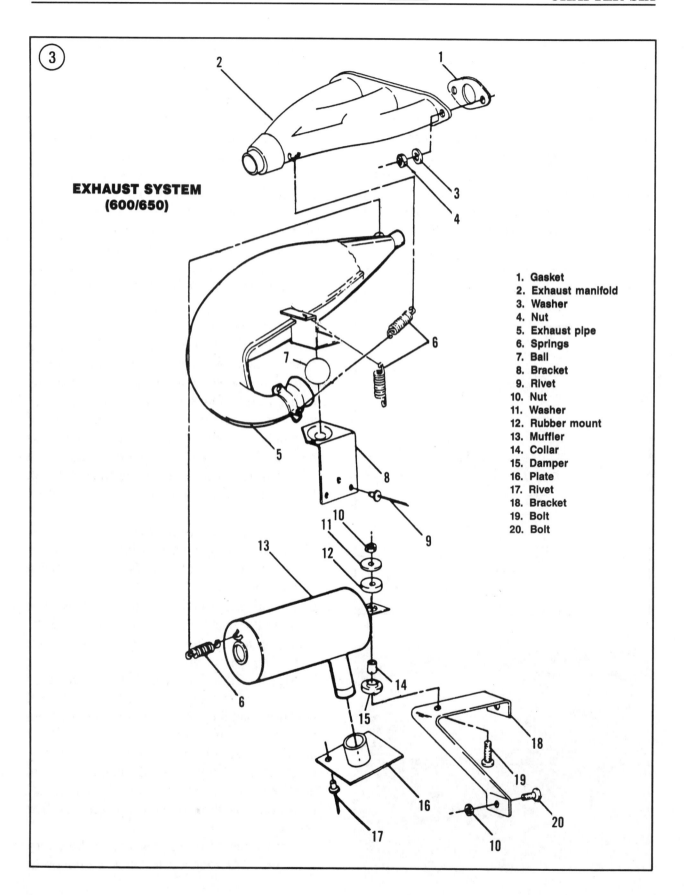

③

EXHAUST SYSTEM
(600/650)

1. Gasket
2. Exhaust manifold
3. Washer
4. Nut
5. Exhaust pipe
6. Springs
7. Ball
8. Bracket
9. Rivet
10. Nut
11. Washer
12. Rubber mount
13. Muffler
14. Collar
15. Damper
16. Plate
17. Rivet
18. Bracket
19. Bolt
20. Bolt

inspect them for damage and write down their thickness and location.

10. Work in an area where there is sufficient lighting and room for component storage.

SERIAL NUMBERS

Polaris snowmobiles are identified by frame and engine identification numbers. Refer to *Serial Numbers* in Chapter One. These numbers identify your snowmobile and should always be used when ordering replacement parts.

SERVICING ENGINE IN FRAME

Some of the components can be serviced while the engine is mounted in the frame:

   a. Cylinder heads.
   b. Cylinders.

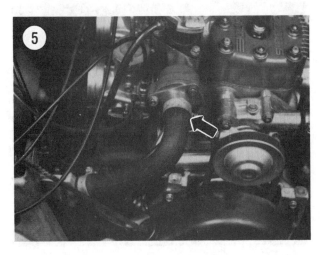

   c. Pistons.
   d. Carburetors.
   e. Magneto.
   f. Oil pump.
   g. Starter motor.
   h. Recoil starter.
   i. Drive assembly.

ENGINE REMOVAL

Engine removal and crankcase separation is required for repair of the "bottom end" (crankshaft, connecting rod and bearings).

1. Open the shroud all the way. Unplug the harness from the headlight and free the harness from the shroud. Disconnect the shroud restraining cable and with assistance, support the shroud and remove the bolts that attach the shroud hinge to the chassis. Remove the shroud and set it well out of the way to prevent it from being damaged.

2. Remove the exhaust system as follows (**Figure 3**)

   a. Disconnect the springs that secure the muffler to the exhaust pipe (**Figure 4**).
   b. Remove the springs securing the exhaust pipe to the bracket.
   c. Remove the spring securing the muffler to the exhaust pipe and remove the exhaust pipe.
   d. Remove the rubber ball from the bracket.
   e. Remove the muffler mounting nut, washer and bolt and remove the muffler. Account for the bushing and dampers when removing the muffler (**Figure 3**).

3. Remove the drive belt as described in Chapter Twelve.

4. Remove the primary sheave as described in Chapter Twelve.

5. Drain the cooling system as described under *Coolant Change* in Chapter Three.

6. Disconnect the coolant hose at the engine (**Figure 5**).

7. Disconnect the lower coolant hose at the water pump.

8. Remove the air box as described in Chapter Seven.

9. Remove the carburetors as described in Chapter Seven.

10. Label and disconnect the oil hoses at the oil pump. Plug the hoses to prevent oil from spilling into the cradle.

11. Remove the recoil starter as described in Chapter Eleven.

12. Disconnect the CDI electrical connectors.

13. Disconnect the pulse hose at the rear of the crankcase. **Figure 6** shows the pulse hose connection with the engine removed for clarity.

14. If necessary, the engine top end (cylinder heads, pistons and cylinder blocks) can be removed before removing the engine from the frame. Refer to *Cylinder* in this chapter.

NOTE
Because of the number of bushings, washers and rubber dampers used on the engine bracket assemblies, refer to Figure 7 when performing Step 15.

15. See **Figure 7**. Loosen the front (**Figure 8**) and rear engine bracket nuts. Remove the nut, washer and rubber damper from the front engine brackets. Remove the nut and washer from the rear bracket.

16. Check to make sure all of the wiring and hoses have been disconnected from the engine.

17. With at least one assistant, lift the engine up and remove it from the frame. Take it to a workbench for further disassembly.

ENGINE PLATES

Removal/Installation

The engine plates are a critical part of the snowmobile drive train. Damaged or loose engine plates will allow the engine to shift or pull out of alignment during operation. This condition will cause primary and secondary sheave misalignment that will result in drive belt wear and reduced performance. The engine plate

assembly should be inspected carefully whenever the engine is removed from the frame or if clutch misalignment becomes a problem.

Refer to **Figure 7** when performing this procedure.

1. Carefully turn the engine over to gain access to the engine plates.

2. The engine plates are different. Before removing the plates, label the left-hand plate with a "L" or the right-hand plate with a "R" to indicate correct alignment.

3. Remove the bolts and lockwashers securing the engine plates to the engine. Remove the engine plates.

4. Check the engine plates for cracks or other damage.

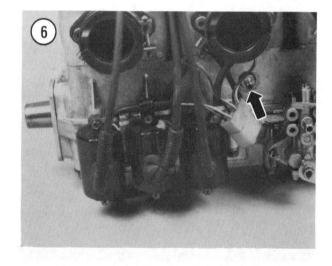

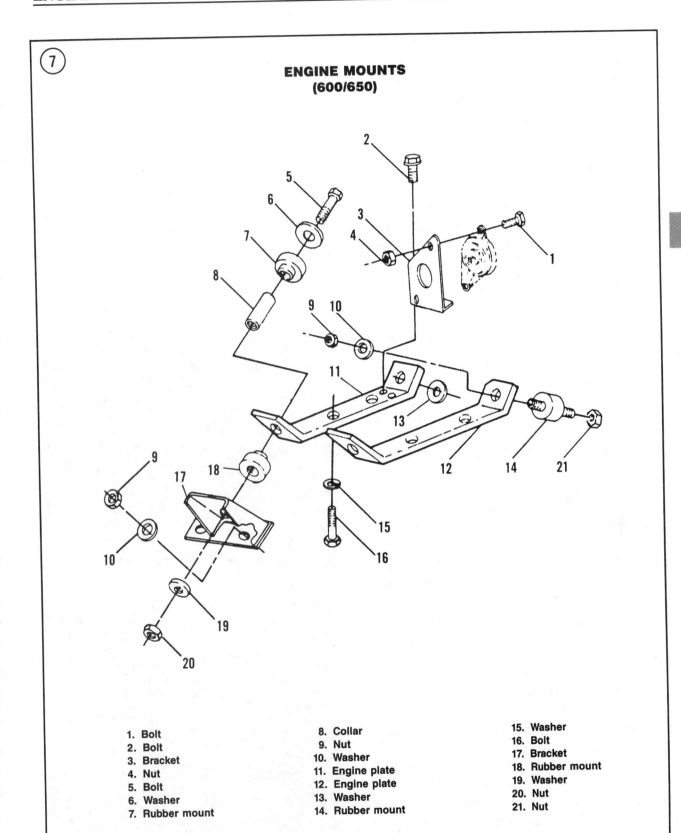

**ENGINE MOUNTS
(600/650)**

1. Bolt	8. Collar	15. Washer
2. Bolt	9. Nut	16. Bolt
3. Bracket	10. Washer	17. Bracket
4. Nut	11. Engine plate	18. Rubber mount
5. Bolt	12. Engine plate	19. Washer
6. Washer	13. Washer	20. Nut
7. Rubber mount	14. Rubber mount	21. Nut

6

5. Check all of the rubber dampers for separation or other damage. Don't forget to check the dampers still mounted in the engine compartment against the firewall.

6. Check all of the engine plate bolts for thread damage. Replace damaged bolts with the same grade bolt. Weaker bolts will loosen and allow the engine to slip.

7. Check the engine plate tapped holes in the lower crankcase for stripped threads or debris buildup. Clean threads with a suitable size metric tap. If Loctite was previously used, make sure to remove all traces of Loctite residue before reinstalling bolts.

8. Replace worn or damaged parts as required.

9. Installation is the reverse of these steps. Tighten the engine plate bolts to the torque specification listed in **Table 4**.

ENGINE INSTALLATION

1. Wash the engine compartment with clean water.

2. Spray all of the exposed electrical connectors with a spray electrical contact cleaner.

3. Before installing the engine, now is a good time to check components which are normally inaccessible for visual inspection. Note the following:

    a. Check all of the coolant hoses (A, **Figure 9**) for damage. Replace damaged hoses as required.

    b. Check all of the steering component tightening torques as described in Chapter Fifteen.

    c. Make sure the engine dampers are mounted on the firewall. See B, **Figure 9**, typical.

    d. Check that the CDI unit is secured tightly.

4. Bolt the engine plates onto the bottom of the crankcase as described in this chapter.

5. Check inside the engine compartment for tools or other objects that may interfere with engine installation. Make sure all wiring harnesses are routed and secured properly.

6. Before installing the engine in the frame, note the following:

    a. Bolt the ignition coils onto the crankcase. See Chapter Eight.

    b. Reconnect the 3 oil pump-to-check valve oil lines (**Figure 10**).

7. With an assistant, place the engine into the frame.

8. Install the engine mount bolts finger-tight at this time. Refer to **Figure 7** when installing the engine mounts, washers and dampers.

9. Reconnect the oil hose at the oil pump. Then bleed the oil pump as described in Chapter Nine.

10. Reconnect the pulse hose at the crankcase nozzle.

11. Reconnect all electrical connectors.

12. Reinstall the recoil starter housing. See Chapter Eleven.

13. Reinstall the carburetors as described in Chapter Seven. Check carburetor synchronization as described in Chapter Three.

14. Install the air box as described in Chapter Seven.

NOTE
Make sure the hoses from the carburetor to the air box are not pinched or torn. This condition would cause a lean air/fuel mixture and result in engine seizure.

15. Reconnect the coolant hoses at the engine and water pump.

16. Refill the cooling system and bleed it as described in Chapter Three.

17. Reinstall the primary sheave and drive belt as described in Chapter Twelve.

18A. Check clutch alignment as described in Chapter Twelve.

18B. Tighten engine mount bolts to the torque specification listed in **Table 4**.

19. Reinstall the exhaust pipe assembly. Secure the muffler with the springs and dampers.

20. Reinstall the engine shroud (**Figure 11**). Reconnect the headlight electrical connector.

ENGINE TOP END

The engine "top end" consists of the cylinder heads, cylinder blocks, pistons, piston rings, piston pins and the connecting rod small-end bearings. See **Figure 12** and **Figure 13**.

The engine top end can be serviced with the engine installed in the frame. However, the following service procedures are shown with the engine removed for clarity.

Cylinder Heads
Removal/Installation

Refer to **Figure 12** for this procedure.

CAUTION
To prevent warpage and damage to any component, remove the cylinder heads only when the engine is at room temperature.

NOTE
If the engine is being disassembled for inspection procedures, check the compression as described in Chapter Three before disassembly.

1. If the engine is mounted in the frame, perform the following:

a. Open the engine shroud all the way. Unplug the harness from the headlight and free the harness from the shroud. Disconnect the shroud restraining cable and with assistance, support the shroud and remove the bolts that attach the shroud hinge to the chassis. Remove the shroud (**Figure 11**) and set it well out of the way to prevent it from being damaged.

b. Drain the cooling system as described in Chapter Three.

c. Disconnect the coolant hose at the cylinder head water manifold (**Figure 14**).

d. Disconnect the spark plug caps at the spark plugs.

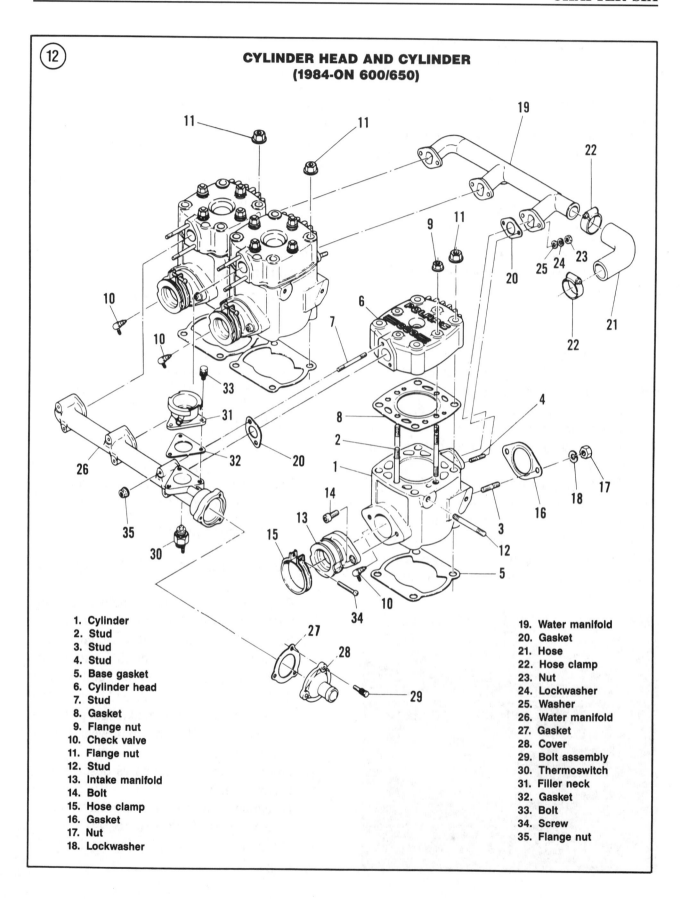

CYLINDER HEAD AND CYLINDER
(1984-ON 600/650)

1. Cylinder
2. Stud
3. Stud
4. Stud
5. Base gasket
6. Cylinder head
7. Stud
8. Gasket
9. Flange nut
10. Check valve
11. Flange nut
12. Stud
13. Intake manifold
14. Bolt
15. Hose clamp
16. Gasket
17. Nut
18. Lockwasher
19. Water manifold
20. Gasket
21. Hose
22. Hose clamp
23. Nut
24. Lockwasher
25. Washer
26. Water manifold
27. Gasket
28. Cover
29. Bolt assembly
30. Thermoswitch
31. Filler neck
32. Gasket
33. Bolt
34. Screw
35. Flange nut

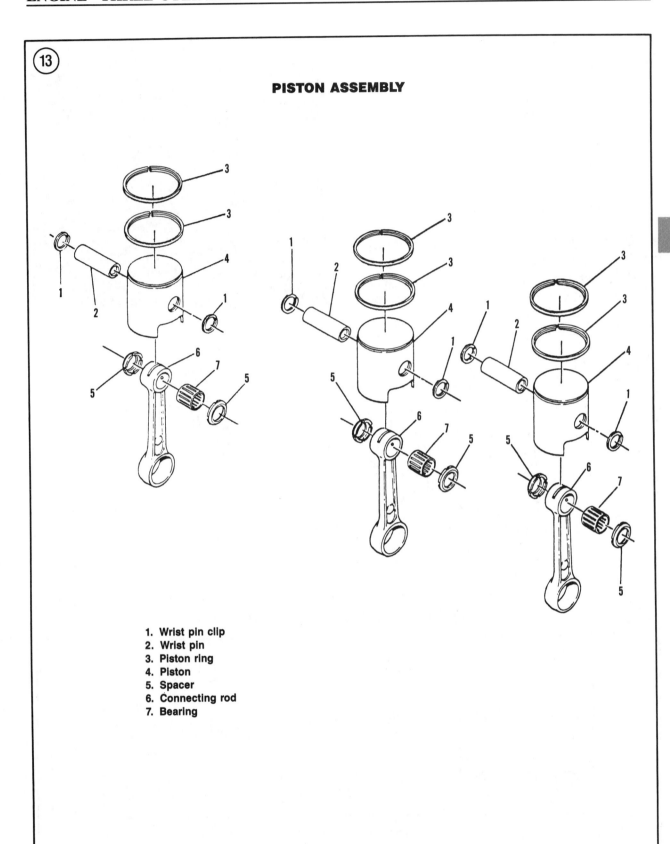

PISTON ASSEMBLY

1. Wrist pin clip
2. Wrist pin
3. Piston ring
4. Piston
5. Spacer
6. Connecting rod
7. Bearing

6

2. Loosen the spark plugs if they are going to be removed later.

NOTE
*The cylinder heads can be removed separately after the water manifold has been removed or they can be removed together with the manifold attached (**Figure 15**). If the cylinder heads are removed separately, mark them for cylinder location (left, center and right).*

3. Referring to **Figure 16**, loosen the cylinder head nuts in a crisscross pattern.

4. Remove the nuts and washers or flange nuts securing the water pipe (**Figure 17**) to the

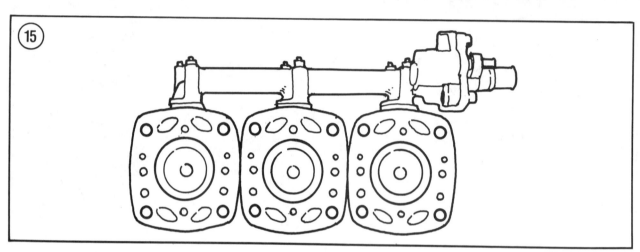

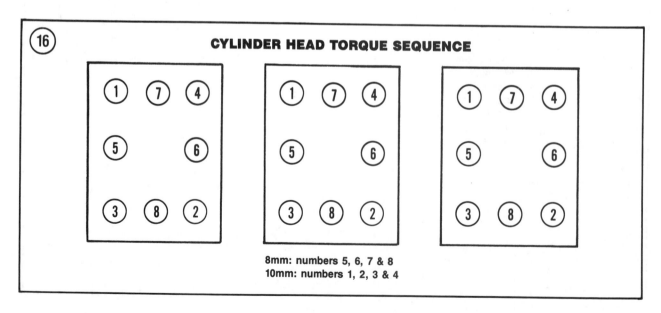

CYLINDER HEAD TORQUE SEQUENCE

8mm: numbers 5, 6, 7 & 8
10mm: numbers 1, 2, 3 & 4

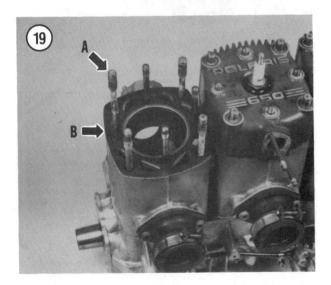

cylinder heads. Remove the water pipe and gaskets.

5. Loosen the cylinder heads by tapping around the perimeter with a rubber or plastic mallet. Remove the cylinder heads (**Figure 18**).

6. Remove and discard the cylinder head gaskets.

7. Lay a rag over the cylinder block to prevent dirt from falling into the cylinder blocks.

8. Inspect the cylinder heads as described in this chapter.

NOTE
While the cylinder heads are removed, check the crankcase studs (A, Figure 19) for thread strippage, looseness or other damage. If necessary, remove the cylinder as described in this chapter and replace damaged studs as described in Chapter One.

9. Remove all gasket residue from each cylinder head and cylinder mating surface with a gasket scraper.

10A. *600*: Spray new head gaskets with a high-temperature aluminum paint and allow them to dry before installation. Then install the gasket so that the small discharge hole in the gasket faces toward the intake side of the engine.

10B. *650*: Install the cylinder head gaskets so that all of the cylinder head studs and coolant passage holes align with the cylinder block. See B, **Figure 19**.

11. Install the cylinder heads onto their respective cylinders so that the water manifold ports face toward the intake side of the engine. See **Figure 18**.

12. Install the cylinder head washers and nuts (early models) or the cylinder head flange nuts (late models). Install all nuts fingertight.

NOTE
If the cylinder heads were installed with the water manifold (Figure 15), tighten the cylinder head nuts as described in Step 13. If the cylinder heads were installed separately, install the manifold

(Figure 17) together with new gaskets over the cylinder head manifold studs; install the manifold nuts finger-tight. Then tighten the cylinder head nuts (Step 13) and manifold nuts (Step 14) in order.

13. Tighten the cylinder head nuts in the following order:
   a. Tighten the 10 mm nuts in a crisscross pattern (**Figure 16**) to the torque specification listed in **Table 4**.
   b. Tighten the 8 mm nuts in a crisscross pattern (**Figure 16**) to the torque specification listed in **Table 4**.
14. Tighten the water manifold nuts securely.
15. Install the spark plugs and tighten securely.
16. If the engine is installed in the frame, note the following:
   a. Connect the spark plug caps.
   b. Connect the coolant hose at the water manifold (**Figure 14**). Tighten the hose clamp securely.
   c. Refill and bleed the cooling system as described in Chapter Three.
   d. Reinstall the engine shroud.

Inspection

1. Wipe away any soft deposits on the cylinder head (**Figure 20**) combustion chamber and gasket surface. Hard deposits should be removed with a wire brush mounted in a drill or drill press or with a soft-metal scraper (**Figure 21**). Be careful not to gouge the aluminum surfaces. Burrs created from improper cleaning will cause preignition and heat erosion.

NOTE
Always use an aluminum thread fluid or kerosene on the thread chaser and cylinder head threads when performing Step 2.

2. With the spark plug removed, check the spark plug threads in each cylinder head (**Figure 20**) for any sign of carbon buildup or cracking. The carbon can be removed with a 14 mm spark plug chaser.

3. Wash the cylinder head in hot soapy water and rinse thoroughly.
4. Check for cracks in the combustion chamber or along the spark plug thread hole. A cracked head must be replaced.

NOTE
Polaris does not list wear limits for cylinder head flatness. The procedure given in Step 5 should be used as a guideline when determining cylinder head flatness.

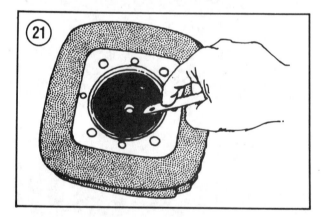

5. After the head has been thoroughly cleaned, place a straightedge across the cylinder head gasket surface at several points. Measure the warp by attempting to insert a flat feeler gauge between the straightedge and the cylinder head at each location. There should be no warpage; if a small amount is present, it can be resurfaced as follows:

a. Tape a piece of 400-600 grit wet emery sandpaper onto a piece of thick plate glass or surface plate (**Figure 22**).

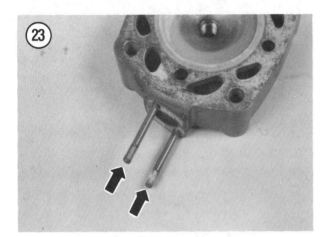

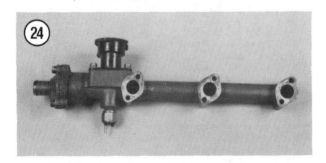

b. Slowly resurface the head by moving it in figure-eight patterns on the sandpaper.

c. Rotate the head several times to avoid removing too much material from one side. Check progress often with the straightedge and feeler gauge.

d. If there is still evidence of warpage, it will be necessary to have the head resurfaced by a machine shop familiar with snowmobile and motorcycle service. Note that removing material from the cylinder head mating surface will change the compression ratio. Consult with the machinist on how much material was removed.

6. Check the water manifold studs on the cylinder heads (**Figure 23**) for looseness or damaged threads. Replace damaged studs as described in Chapter One.

7. Check the water manifold (**Figure 24**) coolant passages for residue buildup. Flush water passages thoroughly. Remove all gasket residue from the water manifold gasket surfaces with a scraper.

CYLINDER

An aluminum cylinder block is used with a cast iron liner pressed into the block. When severe wear is experienced, the cylinder liner can be bored to 0.25 mm (0.10 in.) and 0.50 mm (0.20 in.) over and new pistons and rings installed. See **Figure 13**.

Removal

1. Remove the cylinder heads as described in this chapter.

2. If the engine is installed in the frame, perform the following:

a. Remove the carburetors as described in Chapter Seven.

b. Disconnect the oil lines at the check valves on the cylinder blocks (**Figure 10**).

c. Disconnect the coolant hose at the cylinder water manifold (A, **Figure 25**).

6

d. Remove the exhaust pipe as described under *Engine Removal* in this chapter.

3. Remove the nuts and lockwashers holding the exhaust manifold (B, **Figure 25**) to the cylinders. Remove the exhaust manifold.

4. Remove the water pump as described in Chapter Ten.

5. Remove the nuts and washers holding the water pipe to the cylinder blocks. Remove the water pipe (**Figure 26**).

NOTE
Label each cylinder for position before removal. The cylinders should be reinstalled in their original mounting position.

6. Loosen a cylinder (**Figure 27**) by tapping around the perimeter with a rubber or plastic mallet. Repeat for each cylinder.

CAUTION
When removing the cylinders in Step 7, do not twist the cylinder block so far that the piston rings could snap into the intake port. This would cause the cylinder block to bind and cause piston and ring damage. If this should happen, push the rings back into position.

7. Rotate the engine so that the piston is at the bottom of its stroke. Pull the cylinder block (**Figure 27**) straight up and off the crankcase studs.

8. Repeat for each cylinder.

9. Remove the cylinder base gasket and discard it.

10. Stuff clean rags around the connecting rods to keep dirt and loose parts from entering the crankcase.

11. If necessary, remove the bolts holding the intake manifold to the engine. Remove the manifold (A, **Figure 28**).

Inspection

Cylinder measurement requires a precision inside micrometer or bore gauge. If you don't have the right tools, have your dealer or a machine shop take the measurements.

1. Remove all gasket residue from the top and bottom (**Figure 29**) gasket surfaces. If the cylinder head surface appears warped or excessively contaminated with exhaust deposits, resurface as follows:

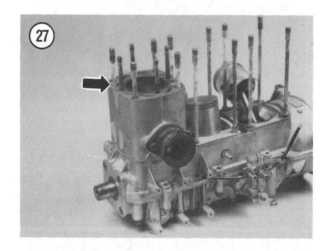

a. Remove the cylinder studs (**Figure 30**) as described in Chapter One.

b. Tape a piece of 400-600 grit wet emery sandpaper onto a piece of thick plate glass or surface plate (**Figure 22**).

c. Slowly resurface the cylinder by moving it in figure-eight patterns on the sandpaper.

d. Rotate the cylinder several times to avoid removing too much material from one side.

e. Reinstall the cylinder studs as described in Chapter One.

2. Using a soft scraper or a wire wheel mounted on a drill or hand grinder, remove all carbon deposits from the exhaust port (A, **Figure 31**).

3. Wash the cylinder in solvent to remove any oil and carbon particles. The cylinder bore must be cleaned thoroughly before attempting any measurement as incorrect readings may be obtained.

4. Measure the cylinder bore diameter as described under *Piston/Cylinder Clearance* in this chapter.

5. If the cylinder is not worn past the service limit, check the bore carefully for scratches or gouges. The bore may require reconditioning.

6. Check the threaded holes in the cylinder block for thread damage. Minor damage can be cleaned up with the correct size metric tap. Refer to Chapter One for information pertaining to threads, fasteners and repair tools. If damage is severe, a thread insert should be installed.

7. Check the cylinder studs (**Figure 30**) and (B, **Figure 31**) for thread damage. Replace studs as described in Chapter One.

8. After the cylinder has been serviced, wash the bore in hot soapy water. This is the only way to clean the cylinder wall of the fine grit material left from the boring or honing job. After washing the cylinder wall, run a clean white cloth through it. The cylinder wall should show no traces of grit or other debris. If the rag shows any sign of debris, the cylinder wall is not clean and must be rewashed. After the cylinder is thoroughly cleaned, lubricate the cylinder walls with clean engine oil to prevent the cylinder liners from rusting.

CAUTION
A combination of soap and water is the only solution that will completely clean the cylinder wall. Solvent and kerosene cannot wash fine grit out of cylinder crevices. Grit left in the cylinder will act as a grinding compound and cause premature wear to the new rings.

9. After cleaning the cylinder, check that the oil line check valve (B, **Figure 28**) in each cylinder

is clear. A plugged check valve will cause piston and cylinder damage. Clean check valves with compressed air if available.

Installation

1. Clean the cylinder bore as described under *Inspection* in this chapter.
2. Install the intake manifold (A, **Figure 28**) and its mounting bolts. Tighten the bolts securely following a crisscross pattern.
3. Check that the top surface of the crankcase and the bottom cylinder surface are clean prior to installation.
4. Install a new base gasket.
5. Make sure the end gaps of the piston rings are lined up with the locating pins in the ring grooves (**Figure 32**). Lightly oil the piston rings and the inside of the cylinder bores.
6. Place a piston holding tool under one piston and turn the crankshaft until the piston is down firmly against the tool. This will make cylinder installation easier. You can make this tool out of wood as shown in **Figure 33**.
7. Install the cylinder block (**Figure 34**) on the crankcase studs so that the intake port faces the rear of the engine. Compress each piston ring with your fingers as the cylinder starts to slide over it. After one piston is installed in its cylinder, lower the cylinder and install the remaining cylinders. If the rings are hard to compress, you can use a large hose clamp as an effective compressor. Make sure the cylinder block is fully seated on the crankcase. Remove the piston holding tool and hose clamp.
8. Make sure the cylinders are installed facing in the direction shown in **Figure 17**.
9. Install the water pipe onto the cylinder blocks, using new gaskets (**Figure 26**). Install the lockwashers and nuts and tighten securely.
10. Install the head gaskets and cylinder heads as described in this chapter.
11. Before installing the exhaust manifold, check the gaskets at the mouth of the manifold. Replace the gaskets if they appear worn, damaged or if there are signs of exhaust leakage.
12. Install the water pump as described in Chapter Ten.
13. Install the exhaust manifold. Secure it with the lockwashers and nuts. Tighten the nuts securely.
14. If the engine is installed in the frame, note the following:
   a. Reconnect the oil lines at the check valves on the cylinder blocks (**Figure 10**).
   b. Install the carburetors as described in Chapter Seven.

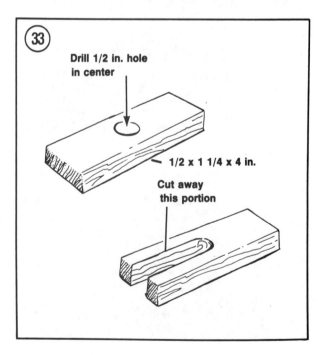

Drill 1/2 in. hole in center

1/2 x 1 1/4 x 4 in.

Cut away this portion

c. Reconnect the coolant hose at the cylinder water manifold (A, **Figure 25**).

d. Install the exhaust pipe assembly. Secure the muffler with the springs and dampers.

e. Refill and bleed the cooling system as described in Chapter Three.

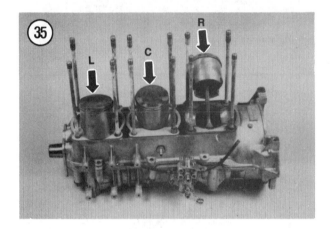

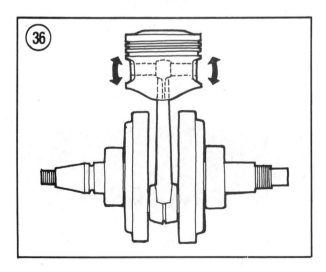

15. If new components were installed or if the cylinders were bored or honed, the engine must be broken in as if it were new. Refer to *Break-In Procedure* in Chapter Three.

PISTON, WRIST PIN AND PISTON RINGS

The piston is made of an aluminum alloy. The wrist pin is a precision fit and is held in place by a clip at each end. A caged needle bearing is used on the small end of the connecting rod. See **Figure 13**.

Piston and Piston Ring Removal

1. Remove the cylinders as described in this chapter.

2. Identify the pistons by scratching an "L," "C" and "R" on the piston crown (**Figure 35**). In addition, keep each piston together with its own pin, bearing and piston rings to avoid confusion during reassembly.

3. Before removing the piston, hold the rod tightly and rock the piston as shown in **Figure 36**. Any rocking motion (do not confuse with the normal sliding motion) indicates wear on the wrist pin, needle bearing, wrist pin bore, or more likely a combination of all three.

NOTE
Wrap a clean shop cloth under the piston so that the clip will not fall into the crankcase.

WARNING
Safety glasses should be worn when performing Step 4.

4. Remove the wrist pin clips from the outside of the piston with an awl, scribe or other sharp pointed tool. Hold your thumb over one edge of the clip when removing it to prevent it from springing out.

5. Use a proper size wooden dowel or socket extension and push out the wrist pin.

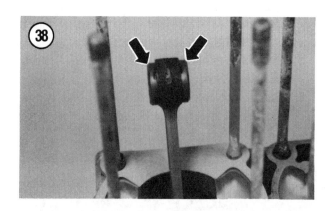

> *CAUTION*
> *If the engine ran hot or seized, the wrist pin may be difficult to remove. However, do not drive the wrist pin out of the piston. This will damage the piston, needle bearing and connecting rod. If the wrist pin will not push out by hand, remove it as described in Step 6.*

6. If the wrist pin is tight, fabricate the tool shown in **Figure 37**. Assemble the tool onto the piston and pull the wrist pin out of the piston. Make sure to install a pad between the piston and piece of pipe to prevent from scoring the side of the piston.

7. Lift the piston (**Figure 35**) off the connecting rod.

8. Remove the 2 spacers (**Figure 38**) from the connecting rod and remove the needle bearing (**Figure 39**).

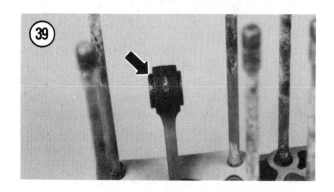

9. Repeat for each piston.

10. If the pistons are going to be left off for some time, place a piece of foam insulation tube, or shop cloth, over the end of each rod to protect it.

> *NOTE*
> *On pistons which use 2 piston rings, always remove the top piston ring first.*

11. The top and bottom piston rings are identical. However, if they are going to be reused, mark each ring after removal to ensure correct installation. Remove the upper ring by spreading

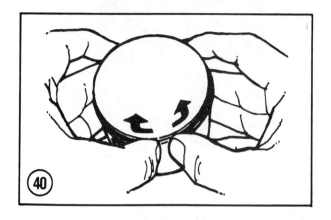

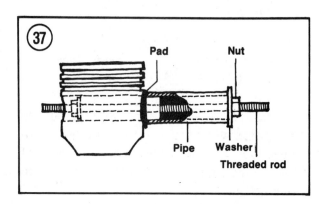

Pad   Nut

Pipe   Washer
Threaded rod

A

B

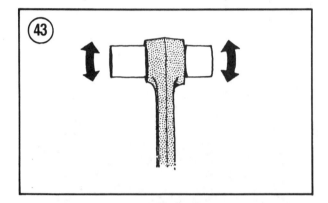

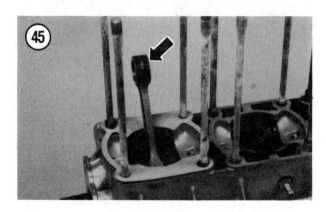

the ends with your thumbs just enough to slide it up over the piston (**Figure 40**). Repeat for the lower ring.

Wrist Pin and Needle Bearing Inspection

1. Clean the needle bearing (A, **Figure 41**) in solvent and dry it thoroughly. Use a magnifying glass and inspect the bearing cage for cracks at the corners of the needle slots (**Figure 42**) and inspect the needles themselves for cracking. If any cracks are found, the bearing must be replaced.

2. Check the wrist pin (B, **Figure 41**) for severe wear, scoring or chrome flaking. Also check the wrist pin for cracks along the top and side. Replace the wrist pin if necessary.

3. Oil the needle bearing and pin and install them into the connecting rod. Slowly rotate the pin and check for radial and axial play (**Figure 43**). If any play exists, the pin and bearing should be replaced, providing the rod bore is in good condition. If the condition of the rod bore is in question, the old pin and bearing can be checked with a new connecting rod.

4. Oil the wrist pin and install it in the wrist pin hole (**Figure 44**). Check for up and down play between the pin and piston. There should be no noticeable play. If play is noticeable, replace the wrist pin and/or piston.

CAUTION
If there are signs of piston seizure or overheating, replace the wrist pins and bearings. These parts have been weakened from excessive heat and may fail later.

5. Check the needle bearing spacers for cracks or other damage; replace if necessary.

Connecting Rod Inspection

1. Wipe the wrist pin bore (**Figure 45**) in the connecting rod with a clean rag and check it for

galling, scratches, or any other signs of wear or damage. If any of these conditions exist, replace the connecting rods as described in this chapter.

2. Check the connecting rod big end bearing play. You can make a quick check by simply rocking the connecting rod back and forth (**Figure 46**). If there is more than a very slight rocking motion (some side-to-side sliding is normal), the connecting rods may require replacement.

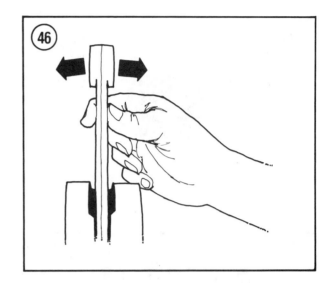

Piston and Ring Inspection

1. Carefully check the piston for cracks at the top edge of the transfer cutaways (**Figure 47**) and replace if found. Check the piston skirt (**Figure 48**) for brown varnish buildup. More than a slight amount is an indication of worn or sticking rings which should be replaced.

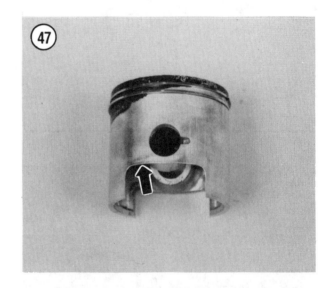

2. Check the piston skirt for galling and abrasion which may have resulted from piston seizure. If light galling is present, smooth the affected area with No. 400 emery paper and oil or a fine oilstone. However, if galling is severe or if the piston is deeply scored, replace it.

3. Check the condition of the piston crown (**Figure 49**). Normal carbon buildup can be removed with a wire wheel mounted on a drill press. If the piston is damaged, it is important to pinpoint the cause so that the failure will not repeat after engine assembly. Note the following when checking damaged pistons:

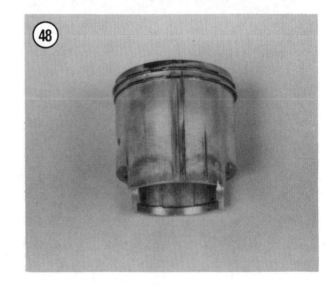

  a. If the piston damage is contained to the area above the wrist pin bore, the engine is probably overheating. Seizure or galling conditions contained to the area below the wrist pin bore is usually caused by a lack of lubrication, rather than overheating.

  b. If the piston has seized and appears very dry (apparent lack of oil or lubrication on the piston), a lean fuel mixture probably caused the overheating. Overheating can result from incorrect jetting, air leaks or over advanced ignition timing.

c. Preignition will cause a sand-blasted appearance on the piston crown. This condition is discussed in Chapter Two.

d. If the piston damage is confined to the exhaust port area on the front of the piston, look for incorrect jetting (too lean) or over advanced ignition timing.

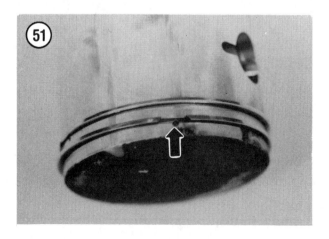

e. If the piston has a melted pocket starting in the crown or if there is a hole in the piston crown, the engine is running too lean. This can be due to incorrect jetting, an air leak or over advanced ignition timing. A spark plug that is too hot will also cause this type of piston damage.

f. If the piston is seized around the skirt but the dome color indicates proper lubrication (no signs of dryness or excessive heat), the damage may result from a condition referred to as cold seizure. This condition typically results from running a water-cooled engine too hard without first properly warming it up. A lean fuel mixture can also cause skirt seizure.

4. Check the piston ring locating pins in the piston (**Figure 50**). The pins should be tight and the piston should show no signs of cracking around the pins. If a locating pin is loose, replace the piston. A loose pin will fall out and cause severe engine damage. Do not attempt to repair a loose locating pin.

5. Check the wrist pin clip grooves (**Figure 51**) in the piston for cracks or other damage that could allow a clip to fall out. This would cause severe engine damage. Replace the piston if any one groove shows signs of wear or damage.

NOTE
Maintaining proper piston ring end gap helps to ensure peak engine performance. Excessive ring end gap reduces engine performance and can cause overheating. Insufficient ring end gap will cause the ring ends to butt together and cause the ring to break, resulting in severe engine damage. So that you don't have to wait for parts, always order extra cylinder head and base gaskets to have on hand for routine top end inspection and maintenance.

6. Measure piston ring end gap. Place a ring into the cylinder approximately 9.5 mm (3/8 in.) from the top; square the ring in the cylinder by pushing

it with the base of the piston. This ensures that the ring is square in the cylinder bore. Measure the gap with a flat feeler gauge (**Figure 52**) and compare to the wear limit in **Table 2**. If the gap is greater than specified, the rings should be replaced as a set.

> *NOTE*
> *When installing new rings, measure the end gap in the same manner as for old ones. If the gap is less than specified, make sure you have the correct piston rings. If the replacement rings are correct but the end gap is too small, carefully file the ends with a fine cut file until the gap is correct (**Figure 53**). Insufficient ring end gap will allow the rings to butt together. This can result in ring breakage, ring seizure or rapid cylinder wear. These conditions will reduce engine performance and may cause seizure or severe damage to the cylinder wall.*

7. Carefully remove all carbon buildup from the ring grooves with a broken ring (**Figure 54**). Inspect the grooves carefully for burrs, nicks, or broken and cracked lands. Recondition or replace the piston if necessary.

8. Check piston ring side clearance by rolling each ring around its piston groove as shown in **Figure 55**. Check the ring for ease of movement and squareness. If you are checking side clearance with old rings, and severe wear is indicated, recheck with new rings. If the new rings still show wear, replace the piston.

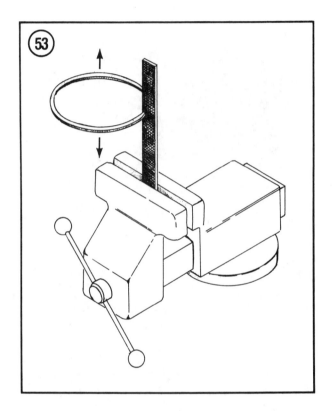

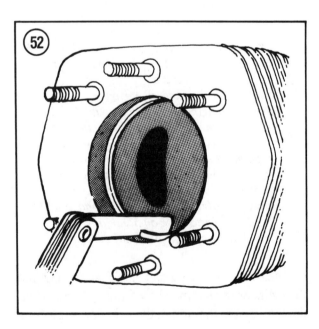

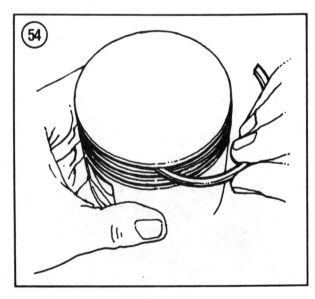

9. If the piston checked out okay after performing these inspection procedures, measure the piston outside diameter as described under *Piston/Cylinder Clearance* in this chapter.

10. If new piston rings are required, the cylinders should be honed before assembling the engine. Refer to *Cylinder Honing* in this chapter.

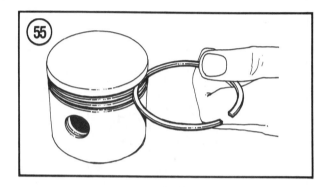

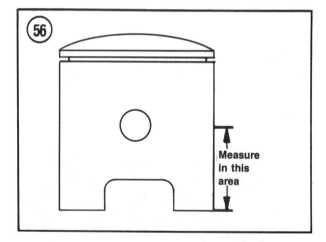

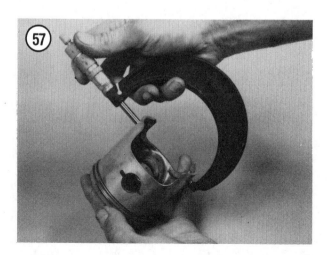

Piston/Cylinder Clearance

The following procedure requires the use of highly specialized and expensive measuring tools. If such equipment is not readily available, have the measurements performed by a dealer or machine shop. Always replace pistons as a set.

1. Measure the outside diameter of the piston with a micrometer at the point indicated in **Figure 56** above the bottom of the piston skirt and at a 90° angle to the piston pin (**Figure 57**). Record the measurement.

> *NOTE*
> *Always install new rings when installing a new piston.*

2. Wash the cylinder block in solvent to remove any oil and carbon particles. The cylinder bore must be cleaned thoroughly before attempting any measurement as incorrect readings may be obtained.

3. Measure the cylinder bore with a bore gauge or telescoping gauge (**Figure 58**). Then measure the bore gauge or telescoping gauge with a micrometer to determine the bore diameter.

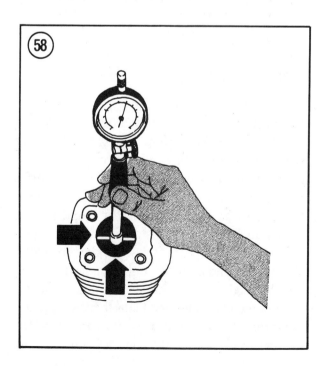

Measure the cylinder bore at the points shown in **Figure 59**. Measure in 3 axes—in line with the piston pin and at 90° to the pin. Record the measurement.

4. Piston clearance is the difference between the maximum piston diameter and the minimum cylinder diameter. For a run-in (used) piston and cylinder, subtract the dimension of the piston from the cylinder dimension. If the clearance exceeds the dimension in **Table 3**, the piston should be replaced.

NOTE
Polaris does not list specifications for the piston outside diameter or the cylinder bore inside diameter. Thus piston and cylinder wear is monitored by checking piston-to-cylinder clearance; when clearance is excessive, the piston is replaced to renew the clearance specification. If you find that the clearance is still excessive with a new piston, the cylinder bore is worn. Oversize pistons are required.

NOTE
The new pistons should be obtained first before the cylinders are bored so that the pistons can be measured. The cylinders must be bored to match the pistons. Piston-to-cylinder clearance is specified in Table 3.

Cylinder Honing

The surface condition of a worn cylinder bore is normally very shiny and smooth. When installing new piston rings for use in a cylinder with minimum wear, they would not seat properly and engine performance would suffer from compression losses. Cylinder honing, often referred to as bead breaking or deglazing, is required whenever new piston rings are installed. When a cylinder bore is honed, the surface is slightly roughed up to provide a textured or crosshatched surface. This surface finish controls wear of the new rings and helps them to seat and seal properly. *Whenever* new rings are installed, the cylinder surface should be honed. This

service can be performed by a Polaris dealer or independent repair shop. The cost of having the cylinder honed by a dealer is usually minimal compared to the cost of purchasing a hone and doing the job yourself. If you choose to hone the cylinder yourself, follow the hone manufacturer's directions closely.

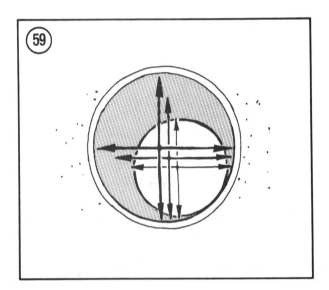

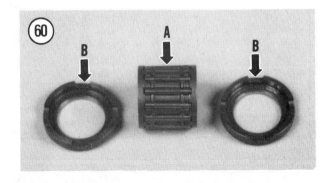

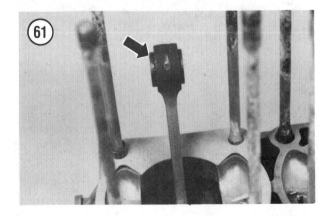

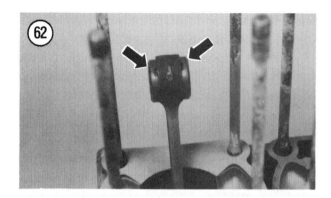

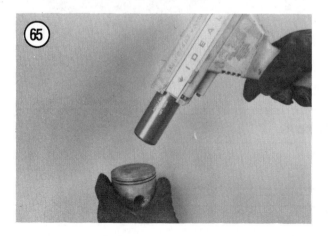

6

CAUTION
After a cylinder has been reconditioned by boring or honing, the bore should be properly cleaned to remove all material left from the machining operation. Refer to **Inspection** *under* **Cylinder** *in this chapter. Improper cleaning will not remove all of the machining residue resulting in rapid wear of the new piston and rings.*

Piston Installation

1. Apply assembly oil to the needle bearing (A, **Figure 60**) and install it in the connecting rod (**Figure 61**).

2. Install the 2 spacers (B, **Figure 60**) onto the connecting rod as shown in **Figure 62**.

NOTE
*If the spacers installed in Step 2 fall off of the connecting rod as the wrist pin is pushed into the piston, first secure the piston onto the connecting rod with an aluminum pilot plug (**Figure 63**). The pilot plug will hold the piston and spacer assembly intact as the wrist pin is installed. A pilot plug can be fabricated by a machine shop.*

3. Place the piston over the connecting rod so that the arrow mark (**Figure 64**) on the piston crown points toward the flywheel side if the engine. This alignment is essential so the piston ring ends will be correctly positioned and will not snag in the ports. Line up the pin with the bearing and push the pin into the piston until it is even with the wrist pin clip grooves.

NOTE
If there is no mark on the piston crown, position the piston ring locating pins, on the piston, toward the intake port.

CAUTION
*If the wrist pin will not slide in the piston smoothly, heat the piston crown slightly with a torch (**Figure 65**), then when the piston expands slightly, install the wrist pin. Wear suitable gloves to prevent from burning your hands when performing this step.*

4. Install new wrist pin clips, making sure they are completely seated in their grooves with the open ends of the clips facing down. See **Figure 66**.

5. Check the installation by rocking the piston back and forth around the pin axis and from side-to-side along the axis. It should rotate freely back and forth but not from side-to-side.

NOTE
Keystone piston rings should be installed with their bevel side facing toward the top of the piston. See Figure 67.

6. Install the piston rings—first the bottom one, then the top—by carefully spreading the ends of the ring with your thumbs and slipping the ring over the top of the piston. Make sure manufacturer's mark on the piston rings are toward the top of the piston. If you are installing used rings, install them by referring to the identification marks made during removal.

7. Make sure the rings are seated completely in the grooves, all the way around the

circumference, and that the ends are aligned with the locating pins. See **Figure 68**.

8. If new components were installed, the engine must be broken in as if it were new. Refer to *Break-In Procedure* in Chapter Three.

CRANKCASE AND CRANKSHAFT

Disassembly of the crankcase—splitting the cases—and removal of the crankshaft assembly requires engine removal from the frame. However, the cylinder head, cylinder and all other attached assemblies should be removed with the engine in the frame.

The crankcase is made in 2 halves of precision diecast aluminum alloy and is of the "thin-walled" type (**Figure 69**). To avoid damaging them, do not hammer or pry on any of the interior or exterior projected walls. These areas are easily damaged if stressed beyond what they are designed for. They are assembled without a gasket; only gasket sealer is used while dowel

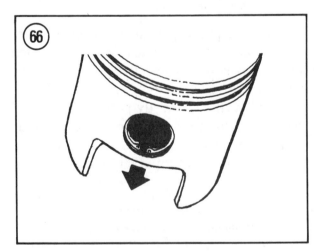

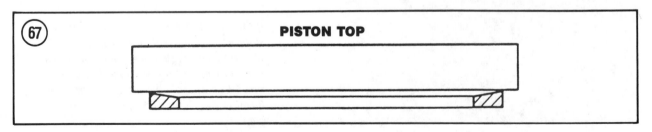
PISTON TOP

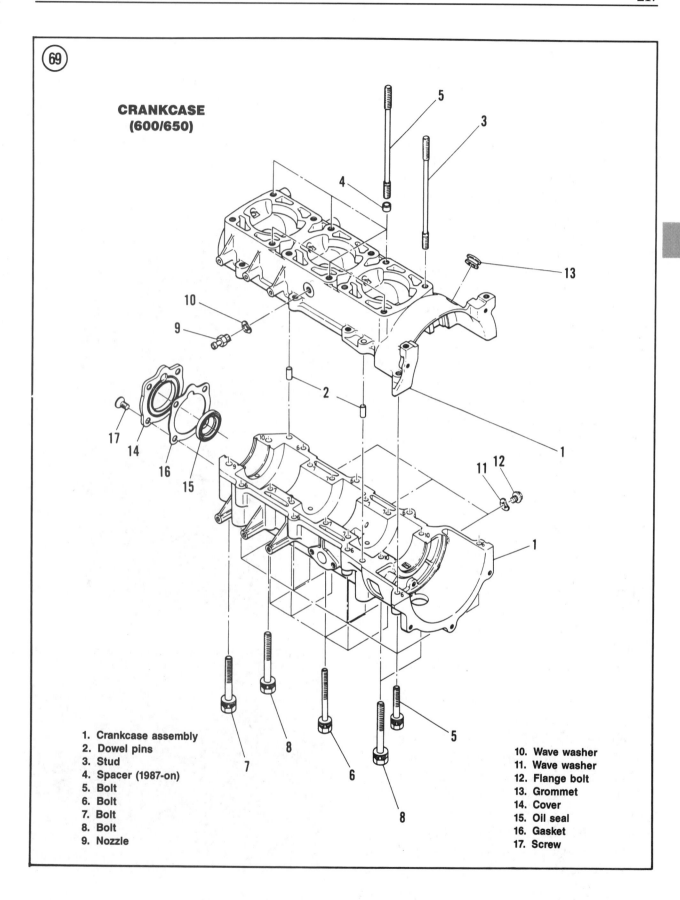

69

**CRANKCASE
(600/650)**

6

1. Crankcase assembly
2. Dowel pins
3. Stud
4. Spacer (1987-on)
5. Bolt
6. Bolt
7. Bolt
8. Bolt
9. Nozzle

10. Wave washer
11. Wave washer
12. Flange bolt
13. Grommet
14. Cover
15. Oil seal
16. Gasket
17. Screw

pins align the crankcase halves when they are bolted together. The crankcase halves are sold as a matched set only (**Figure 70**). If one crankcase half is severely damaged, both must be replaced.

Crankshaft components are available as individual parts. However, crankshaft service—replacement of unsatisfactory parts or crankshaft alignment—should be entrusted to a dealer or engine specialist. Special measuring and alignment tools, a hydraulic press and experience are necessary to disassemble, assembly and accurately align the crankshaft assembly, which, in the case of the average three cylinder engine, is made up of a number of pressed-together pieces, not counting the bearings, seals and connecting rods. However, you can save considerable expense by disassembling the engine and taking the crankshaft in for service.

The procedure which follows is presented as a complete, step-by-step major lower end overhaul that should be followed if the engine is to be completely reconditioned.

Crankcase Disassembly

This procedure describes disassembly of the crankcase halves and removal of the crankshaft.
1. Remove the engine from the frame as described in this chapter.
2. Remove the water pump as described in Chapter Ten.
3. Remove the pistons as described in this chapter.
4. Remove the flywheel and stator plate as described in Chapter Eight.
5. Remove the Phillips screws holding the crankcase cover/oil seal assembly onto the crankshaft. Remove the cover and oil seal (**Figure 71**). Remove and discard the cover gasket.
6. Check crankshaft runout before disassembling the crankcase halves. Refer to *Crankshaft Runout* in this chapter.

CAUTION
Do not damage the crankcase studs when performing the following procedures.

7. Turn the crankcase assembly so that it rests up-side-down on its crankcase studs (**Figure 72**).
8. Loosen the crankcase bolts in a crisscross pattern (**Figure 73**).

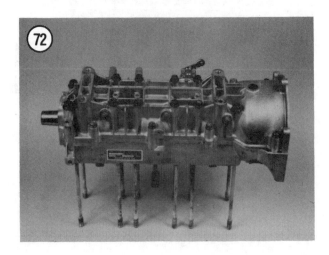

9. Tap on the large bolt bosses with a soft mallet to break the crankcase halves apart and then remove the bottom half.

> *CAUTION*
> *Make sure that you have removed all the fasteners. If the cases are hard to separate, check for any fasteners you may have missed.*

> *CAUTION*
> *Do not pry the cases apart with a screwdriver or any other sharp tool, otherwise, the sealing surface will be damaged.*

10. Remove the 2 dowel pins (**Figure 74**).
11. Remove the crankshaft from the upper case half.

12. Remove the oil pump drive shaft as described in Chapter Nine.

Cleaning

Refer to **Figure 69** for this procedure.
1. Clean both crankcase halves (**Figure 70**) with cleaning solvent. Thoroughly dry with compressed air and wipe off with a clean shop cloth. Be sure to remove all traces of old gasket sealer from all mating surfaces.
2. Clean the oil passages in the transfer ports in the upper crankcase half (A, **Figure 75**). Use compressed air to ensure that they are clean.
3. Clean the crankshaft assembly in solvent and dry with compressed air. Oil bearings with engine oil after all solvent has been removed.

6

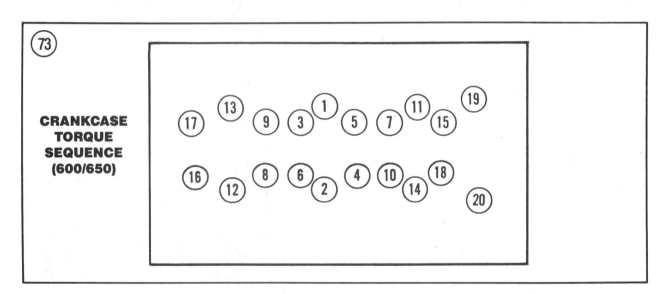

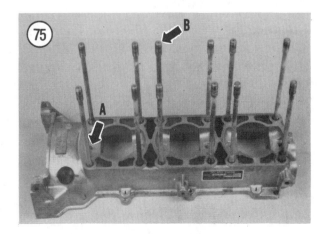

Crankcase Inspection

Refer to **Figure 69** for this procedure.

1. Carefully inspect the case halves (**Figure 70**) for cracks and fractures. Also check the areas around the stiffening ribs, around bearing bosses and threaded holes. If any are found, have them repaired by a shop specializing in the repair of precision aluminum castings or replace them.

2. Check the locating pin holes (A, **Figure 76**) in the lower case half for cracks or damage.

3. Check the oil seal and bearing grooves (B, **Figure 76**) in the upper and lower case halves for cracks or damage.

4. Check the bearing surface areas in the upper (**Figure 77**) and lower (**Figure 78**) case halves for roughness, cracks or other damage.

5. Check the threaded holes in both crankcase halves for thread damage or dirt or oil buildup. If necessary, clean or repair the threads with the correct size metric tap. Coat the tap threads with kerosene or an aluminum tap fluid before use.

6. Check the cylinder studs (B, **Figure 75**) for thread damage or bending. If necessary, replace damaged studs as described in Chapter One.

7. Check the oil seal in the cover (**Figure 79**) for excessive wear or damage; replace oil seal if necessary. Install oil seal so that open end of seal faces toward back of cover.

8. If there is any doubt as to the condition of the crankcase halves, and they cannot be repaired, replace the crankcase halves as a set.

Crankshaft Inspection

Refer to **Figure 80** (1984-1986) or **Figure 81** (1987-on) for this procedure.

1. Remove the right-hand oil seal (A, **Figure 82**).

2. Inspect the oil seal(s) for excessive wear, cuts or other damage; replace oil seal(s) as required.

3. Check the connecting rods (**Figure 83**) for signs of excessive heat, cracks or other damage. Check the lower rod bearings for visual signs of excessive heat or damage. With the crankshaft mounted on V-blocks (**Figure 84**), rotate the

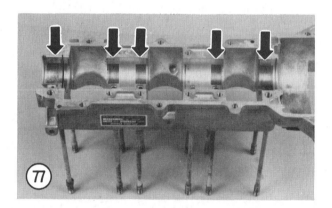

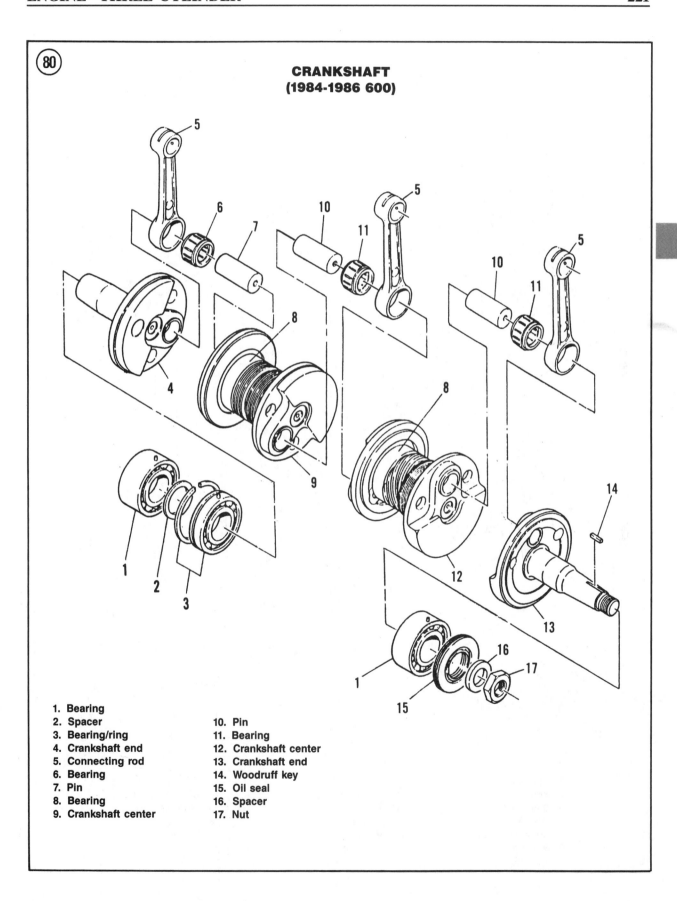

**CRANKSHAFT
(1984-1986 600)**

1. Bearing
2. Spacer
3. Bearing/ring
4. Crankshaft end
5. Connecting rod
6. Bearing
7. Pin
8. Bearing
9. Crankshaft center
10. Pin
11. Bearing
12. Crankshaft center
13. Crankshaft end
14. Woodruff key
15. Oil seal
16. Spacer
17. Nut

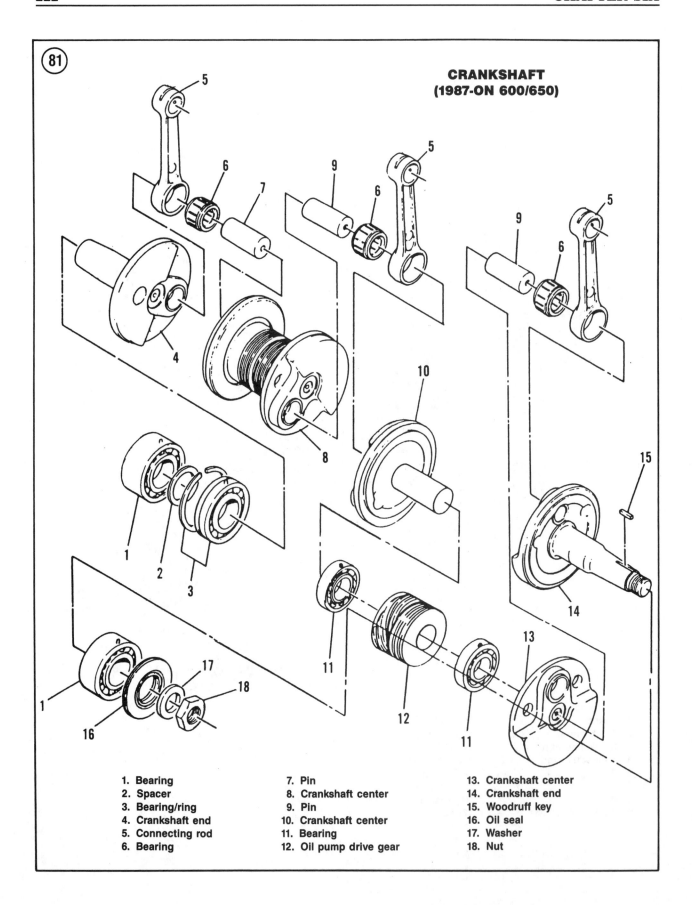

CRANKSHAFT
(1987-ON 600/650)

1. Bearing
2. Spacer
3. Bearing/ring
4. Crankshaft end
5. Connecting rod
6. Bearing
7. Pin
8. Crankshaft center
9. Pin
10. Crankshaft center
11. Bearing
12. Oil pump drive gear
13. Crankshaft center
14. Crankshaft end
15. Woodruff key
16. Oil seal
17. Washer
18. Nut

(82)

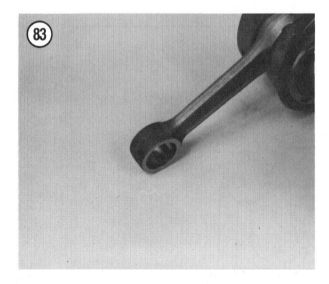

(83)

(84)

crankshaft and check the connecting rod bearings for roughness or damage.

4. Carefully examine the condition of the crankshaft ball bearings (**Figure 84**). Clean the bearings in solvent and allow to dry thoroughly. Then oil each bearing before checking it. Roll each bearing around by hand, checking that it turns quietly and smoothly and that there are no rough spots. There should be no apparent radial play. Defective bearings should be replaced.

5. Check the keyseat in the crankshaft (B, **Figure 82**) for cracks or other damage.

6. If the crankshaft exceeded any of the service limits in Steps 3-5 or if one or more bearings are worn or damaged, have the crankshaft rebuilt by a dealer or crankshaft specialist.

Crankshaft Runout

Crankshaft runout is checked with the crankshaft installed in the crankcase with the case bolts tightened securely. If the engine has been disassembled, install the crankshaft and bolt the case halves together as described in this chapter.

1. Remove the primary sheave from the crankshaft.

2. Clean the end of the crankshaft with contact cleaner. If the crankshaft is nicked or lightly roughed up, clean with emery cloth.

3. Place a dial indicator pointer 1/2 in. from the end of the crankshaft as shown in **Figure 85**. Turn the crankshaft slowly and note the indicator reading. If the runout exceeds the service limit of 0.152 mm (0.006 in.), have the crankshaft serviced by a crankshaft specialist.

Bearing Replacement

The outer bearings (**Figure 84**) can be replaced as follows. A bearing puller will be required to remove the bearings. Read this procedure completely through before removing the bearings.

NOTE
Before removing the bearings from the crankshaft, note the position of the dowel

6

pin in each outer bearing race. See A, *Figure 86* and A, *Figure 87*. These pins must align with the dowel pin holes in the upper crankcase during crankshaft installation. Because the pins are mounted off-center in the bearings, each bearing must be installed with the pin in the same position. If a bearing is installed incorrectly, the pin will not align with the pin hole; the bearing will have to be removed and turned around.

NOTE
Before removing the MAG side bearing with a bearing puller, thread the flywheel nut onto the crankshaft so that it is flush

with the end of the crankshaft. The nut will prevent the puller screw from damaging the crankshaft threads.

NOTE
Before removing the PTO side bearings, the bearings may have to be separated slightly to allow installation of the bearing puller. When separating the bearings, position the crankshaft so that the bearings are supported by wood blocks. Then align a bearing splitter or chisel between the 2 bearing races and drive the bearings apart enough to allow the use of a bearing puller. See **Figure 88**.

1. Install a bearing puller and remove the bearing(s) (**Figure 89**). Remove any shims as required. See **Figure 80** or **Figure 81**.

2. Clean the crankshaft bearing area with solvent or electrical contact cleaner and dry thoroughly.

NOTE
Completely read Step 3 through before heating and installing the bearing.

During bearing installation, the crankshaft should be supported securely so that the bearing can be installed quickly. If the bearing cools and tightens on the crankshaft before it is completely installed, remove the bearing with the puller and reheat.

3. Install the bearing as follows:

   a. Lay the bearing on a clean, lint-free surface in the order of assembly.

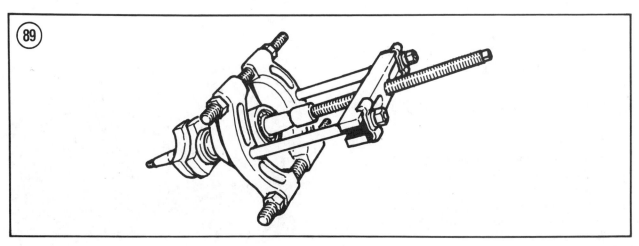

b. When installing the bearing, make sure that the bearing dowel pin faces in the direction shown in A, **Figure 86** or A, **Figure 87**.

c. Refer to *Shrink Fit* under *Ball Bearing Replacement* in Chapter One.

d. After referring to the information in sub-step c, heat and install the bearing(s) and shim (as required).

Crankshaft Installation

Proper crankshaft installation is the most critical aspect of crankcase assembly. The basic procedure is to install the crankshaft into one case half and then install the opposite half. When assembling the engine, it us easier to install the crankshaft first into the lower case half. Then after the crankshaft oil seal(s), bearing locating pins and ring have been located in the lower case half recesses and holes, the upper case half can be installed. If the crankshaft is first installed into the upper case half, it is more difficult to properly align the bearing locating pins with their mating pin recesses in the lower case half.

Refer to **Figure 80** (1984-1986) or **Figure 81** (1987-on) for this procedure.

1. Install new crankshaft oil seals as follows:
   a. Fill the lips of the new seals with a low-temperature grease (**Figure 90**).
   b. Install the right-hand oil seal (A, **Figure 82**).

2. Install the oil pump drive shaft as described in this chapter.

> *NOTE*
> *It is easier to install the crankshaft in the lower case half (**Figure 78**).*

3. Install the 2 crankcase dowel pins (**Figure 91**).

> *NOTE*
> *Step 4 describes crankshaft (**Figure 92**) installation. However, because of the number of separate procedures required during installation, read Step 4 through first before actually installing the crankshaft.*

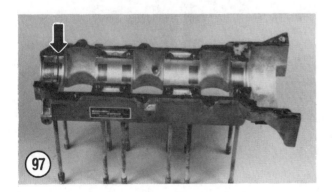

4. Align the crankshaft with the lower case half and install the crankshaft (**Figure 92**). Note the following:

    a. Rotate the crankshaft ball bearings races to align their locating pin with the pin recess in the lower crankcase. See A, **Figure 86** and A, **Figure 76**. When aligning the pins with the pin recess, it will be easier to start on one side of the crankshaft and work toward the opposite end, fitting the locating pins in order.

    b. Fit the ring on the PTO inner bearing (B, **Figure 86**) into the groove in the lower case half (B, **Figure 76**). See **Figure 93**.

    c. Fit the oil seal(s) into the case half groove(s) as shown in **Figure 94**.

    d. Double check to make sure that all of the bearings and seals are seated.

5. Rotate the crankshaft to make sure it turns freely. Then apply a low-temperature grease to the labyrinth seals (**Figure 95**).

6

Crankcase Assembly

1. Install the crankshaft in the lower crankcase half as described in this chapter.

2. Install the 2 dowel pins into the lower crankcase half (**Figure 74**), if not previously installed.

3. Make sure the crankcase mating surfaces are completely clean.

NOTE
*A non-shimming gasket sealer, such as Loctite Gasket Eliminator 515 Sealant (**Figure 96**), should be used. Always follow manufacturer's directions for surface cleaning, application and cure time.*

4. Apply a gasket sealer to both case half mating surfaces (**Figure 97**).

5. Put the upper case half onto the lower half. Check the mating surfaces all the way around the case halves to make sure they are even and

that a locating pin has not worked out of its pin recess (A, **Figure 98**).

6. Apply a light coat of oil to the crankcase bolt threads before installing them.

7. Install the crankcase bolts (B, **Figure 98**) and run them down hand tight.

8. Tighten the crankcase bolts in 2-steps in the numerical sequence shown in **Figure 99**. Refer to **Table 4** for torque specifications.

CAUTION
While tightening the crankcase fasteners, make frequent checks to ensure that the crankshaft turns freely and that the

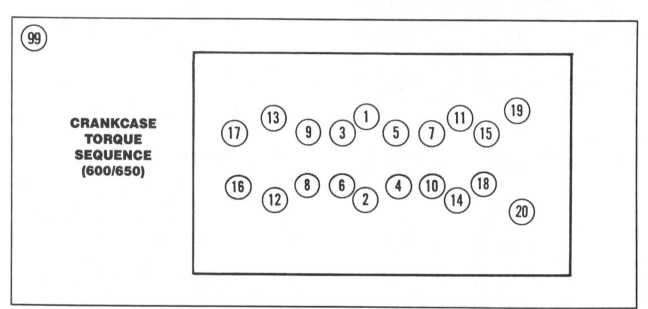

CRANKCASE TORQUE SEQUENCE (600/650)

crankshaft locating pins and crankcase dowel pins fit into place in the case halves.

9. Check again that the crankshaft turns freely. If it is binding, separate the crankcase halves and determine the cause of the problem.

10. Turn the engine right side up.

11. Install the crankcase cover/oil seal assembly as follows:

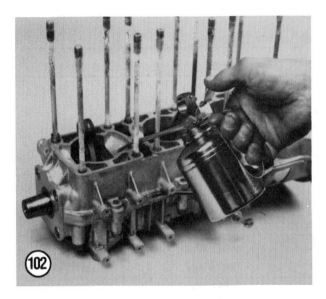

a. Fill the oil seal lip cavity with a low-temperature grease.

b. Install a new gasket (**Figure 100**), aligning the tab on the gasket with the outline on the crankcase halves.

c. Slide the crankcase cover/oil seal assembly (**Figure 101**) over the end of the crankshaft and fit it against the case halves.

d. Apply Loctite 242 (blue) to the Phillips screws and install them; tighten screws securely.

12. Liberally coat the crank pins and bearings with two-stroke oil. Apply the same oil though the oil delivery holes in the transfer ports (**Figure 102**).

13. Install the engine top end as described in this chapter.

14. Install the stator plate and flywheel as described in Chapter Eight.

15. Install the water pump as described in Chapter Ten.

16. Install the engine in the frame as described in this chapter.

17. If new components were installed, the engine must be broken in as if it were new. Refer to *Break-In Procedure* in Chapter Three.

Tables are on the following page.

Table 1 GENERAL ENGINE SPECIFICATIONS

Bore	
600	65 mm (2.559 in.)
650	67.72 mm (2.668 in.)
Stroke	
600	60 mm (2.362 in.)
650	60 mm (2.362 in.)
Displacement	
600	597 cc (36.4 cu. in.)
650	648 cc (39.5 cu. in.)

Table 2 PISTON RING END GAP

	mm	in.
1984	*	*
1985-1987	0.15-0.40	0.006-0.016
1988-on	0.15-0.45	0.006-0.016
*Not specified.		

Table 3 PISTON-TO-CYLINDER CLEARANCE*

	New mm (in.)	Wear limit mm (in.)
600	0.140 (0.0055)	0.241 (0.0095)
650	0.140 (0.0055)	0.241 (0.0095)
*+ 0.051 mm (0.002 in.); – 0.025 mm (0.001 in.)		

Table 4 ENGINE TIGHTENING TORQUES

	N·m	ft.-lb.
Cylinder head		
8 mm	22-23	16-17
10 mm	36-40	26-29
Crankcase		
8 mm	22-23	16-17
10 mm	32-35	23-25
Cylinder studs	33-39	24-28
Flywheel	83-90	60-65
Engine mounting bolts		
3/8 bolts	46-52	34-38
7/16 bolts	75-82	55-60

Chapter Seven

Fuel and Exhaust Systems

The fuel system consists of the fuel tank, fuel pump, carburetors and air silencer. There are slight differences among the various models and they are noted in the various procedures. The exhaust system consists of an exhaust pipe assembly and a muffler.

This chapter includes service procedures for all parts of the fuel and exhaust systems.

Carburetor specifications are listed in **Tables 1-6** at the end of the chapter.

AIR BOX

The air box (**Figure 1**), sometimes referred to as the air silencer box, should be periodically

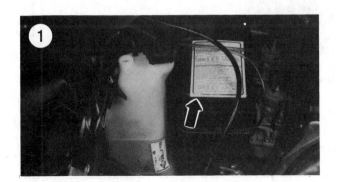

inspected for cleanliness, cracks or debris buildup.

> *CAUTION*
> *Never run the engine without the air box installed. Running the engine without the air box will cause a lean fuel mixture and engine seizure. The air box must be installed during carburetor adjustment.*

Removal/Installation

Refer to **Figure 1** for this procedure.
1. Open the shroud.
2. Loosen the intake boot clamps at the carburetor.
3. Disconnect the intake hoses at the air box.
4. Remove the air box (**Figure 1**).
5. Cover the carburetors to prevent dirt or moisture from entering.
6. Installation is the reverse of these steps.

Inspection

Check the air box for cracks or other damage. Repair cracks before reinstalling the air box onto the engine.

CARBURETOR

All models are equipped with slide type Mikuni carburetors. Refer to **Tables 1-6** for carburetor identification and specifications.

A hand-operated choke lever located on the cowl is used for cold starting.

Fuel is supplied by a remote pulse type fuel pump.

CARBURETOR OPERATION

For proper operation, a gasoline engine must be supplied with fuel and air mixed in proper proportions by weight. A mixture in which there is an excess of fuel is said to be rich. A lean mixture is one which contains insufficient fuel. A properly adjusted carburetor supplies the proper mixture to the engine under all operating conditions.

The carburetors installed on all models consist of several major systems. A float and float valve mechanism maintain a constant fuel level in the float bowl. The pilot system supplies fuel at low speeds. The main fuel system supplies fuel at medium and high speeds. Finally, a choke system

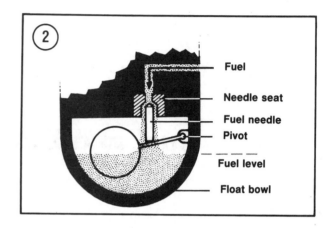

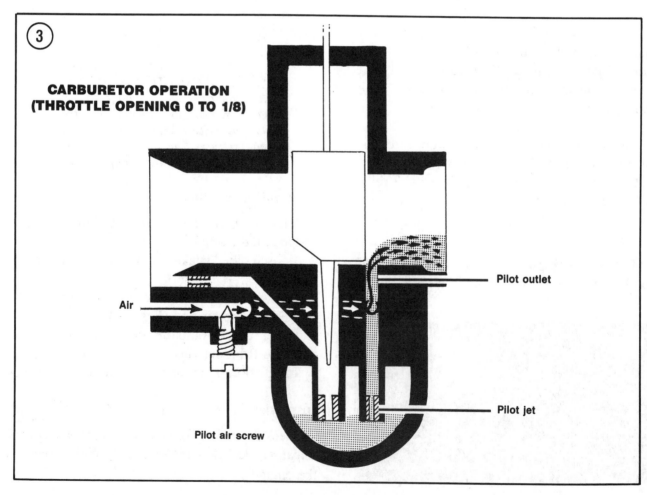

supplies the rich mixture needed to start a cold engine.

Float Mechanism

To assure a steady supply of fuel, the carburetor is equipped with a float valve through which fuel flows by the pulse operated fuel pump into the float bowl (**Figure 2**). Inside the bowl is a combined float assembly that moves up and down with the fuel level. Resting on the float arm is a float needle, which rides inside the float valve. The float valve regulates fuel flow into the float bowl. The float needle and float valve contact surfaces which are accurately machined to insure correct fuel flow calibration. As the float rises, the float needle rises inside the float valve and blocks it, so that when the fuel has

reached the required level in the float bowl, no more fuel can enter.

Pilot and Main Fuel Systems

The carburetor's purpose is to supply and atomize fuel and mix it in correct proportions with air that is drawn in through the air intake. At primary throttle openings (from idle to 1/8 throttle), a small amount of fuel is siphoned through the pilot jet by suction from the incoming air (**Figure 3**). As the throttle is opened further, the air stream begins to siphon fuel through the main jet and needle jet. The tapered needle increases the effective flow capacity of the needle jet as it rises with the throttle slide, in that it occupies decreasingly less of the area of the needle jet (**Figure 4**). In addition, the amount

7

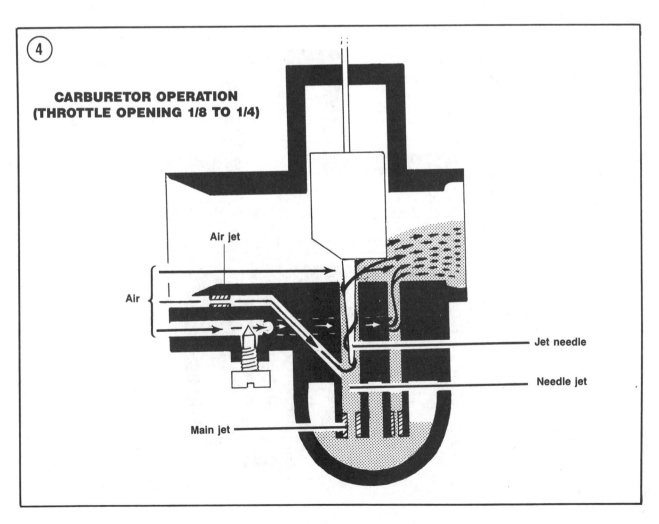

④ **CARBURETOR OPERATION (THROTTLE OPENING 1/8 TO 1/4)**

Air jet

Air

Jet needle

Needle jet

Main jet

of cutaway in the leading edge of the throttle slide aids in controlling the fuel/air mixture during partial throttle openings.

At full throttle, the carburetor venturi is fully open and the needle is lifted far enough to permit the main jet to flow at full capacity. See **Figure 5** and **Figure 6**.

Removal/Installation

1. Open the shroud.
2. Remove the air box (**Figure 1**) as described in this chapter.
3. Label the hoses at the carburetors and disconnect them. Plug the hoses with golf tees to prevent fuel leakage and contamination.

4. If necessary, loosen the carburetor caps and remove the throttle valve assembly from the carburetor body.

CAUTION
Handle the throttle valve carefully to prevent from scratching or damaging the valve and needle jet.

5. Loosen the hose clamps at the intake manifolds.
6. Remove the carburetors (**Figure 7**).
7. Installation is the reverse of these steps. Note the following.
8. Adjust the carburetors as described in Chapter Three.

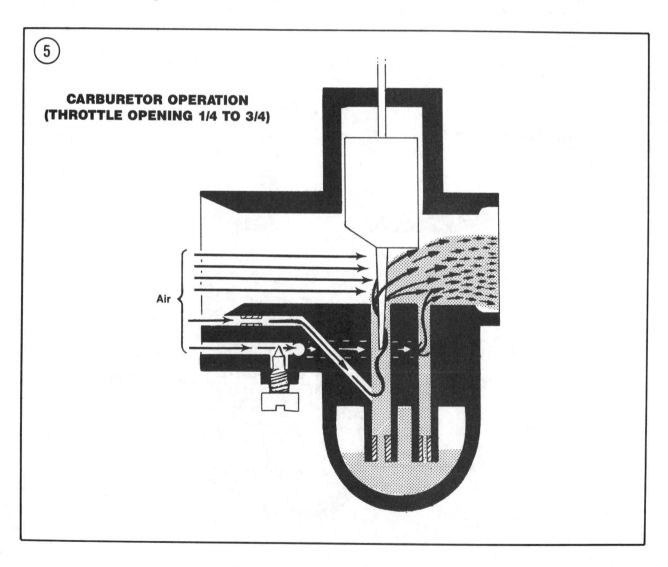

(5)

**CARBURETOR OPERATION
(THROTTLE OPENING 1/4 TO 3/4)**

Air

9. Adjust the oil pump cable as described in Chapter Three.

10. Make sure the fuel hoses are properly connected to prevent a fuel leak.

> *WARNING*
> *Do not start the engine if the fuel hoses are leaking.*

11. Make sure the air hoses are properly positioned and the hose clamps tightened securely to prevent an air leak.

Intake Manifolds

The intake manifolds (**Figure 8**) should be inspected frequently for damage that could cause a lean fuel mixture.

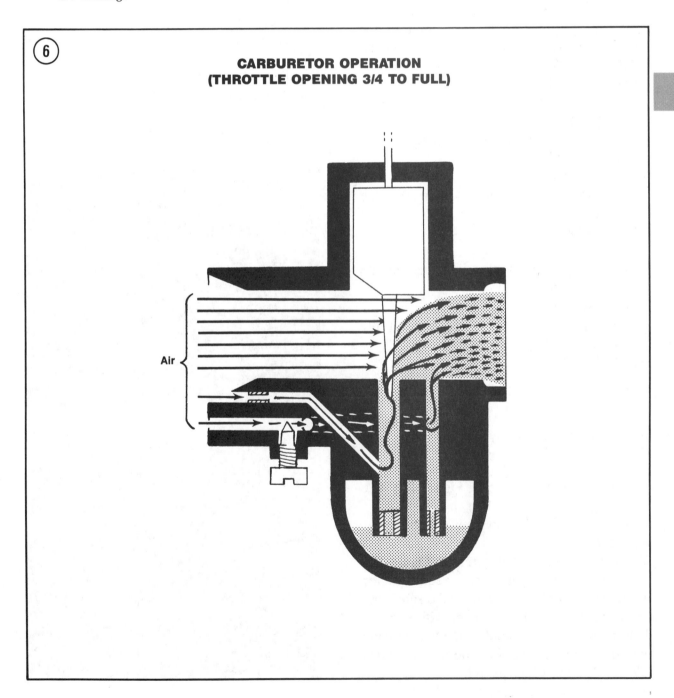

⑥

**CARBURETOR OPERATION
(THROTTLE OPENING 3/4 TO FULL)**

Air

7

Carburetor Identification

Due to the number of models covered in this manual and the slight differences in the specific carburetors, the following will help identify your specific carburetor.

The Mikuni carburetor used on all of these models is basically the same and to help clarify this, the carburetors are split into 5 types. Slight differences do occur between models, so it is important to pay particular attention to the location and order of parts during disassembly.

The Type I carburetor is found on the following models:

   a. 1986-1989 Trail.
   b. 1984-1985 400.

The Type II carburetor is found on the following model:

   1986-1989 400.

The Type III carburetor is found on the following models:

   a. 1984-1985 Trail.
   b. 1987-1989 Sport.

The Type IV carburetor is found on the following model:

   1984-1985 600.

The Type V carburetor is found on the following models:

   a. 1989 500.
   b. 1986-1989 600/650.

Disassembly/Reassembly
(Type I, Type II and Type III)

Refer to **Figure 9** (Type I), **Figure 10** (Type II) or **Figure 11** (Type III).

1. Remove the throttle stop screw and spring (A, **Figure 12**).

2. Lightly seat the pilot air screw (B, **Figure 12**), counting number of turns required for reassembly reference, then back the screw out and remove from the carburetor with spring.

3. Remove the screws holding the float bowl to the carburetor housing. Remove the float bowl (**Figure 13**) and paper gasket.

4. Remove the pilot jet (**Figure 14**) with a flat-tipped screwdriver.

5A. *Type I and Type II*: Remove the main jet (**Figure 15**) and washer (**Figure 16**).

5B. *Type III*: Remove the main jet from the Banjo bolt installed in the float bowl (**Figure 11**).

6. Remove the needle jet through the top of the carburetor (**Figure 17**).

7. Remove the needle jet setter assembly on Type III carburetors. See **Figure 11**.

8. Remove the float pin and float arm (**Figure 18**).

9. Remove the needle valve assembly as follows:

   a. Remove the needle (**Figure 19**).
   b. Remove the seat (**Figure 20**).
   c. Remove the washer (**Figure 21**).

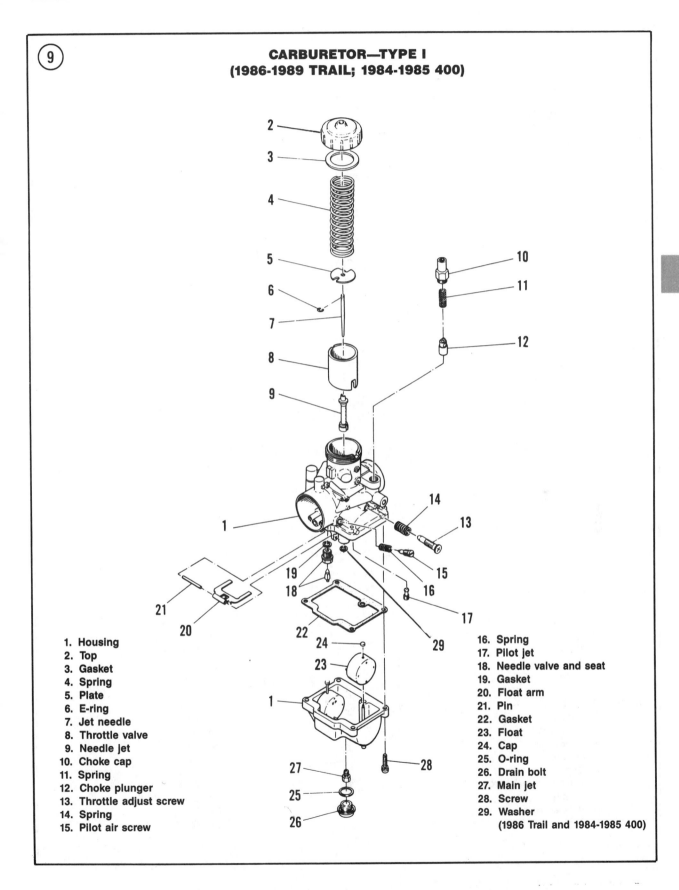

⑨

CARBURETOR—TYPE I
(1986-1989 TRAIL; 1984-1985 400)

1. Housing
2. Top
3. Gasket
4. Spring
5. Plate
6. E-ring
7. Jet needle
8. Throttle valve
9. Needle jet
10. Choke cap
11. Spring
12. Choke plunger
13. Throttle adjust screw
14. Spring
15. Pilot air screw
16. Spring
17. Pilot jet
18. Needle valve and seat
19. Gasket
20. Float arm
21. Pin
22. Gasket
23. Float
24. Cap
25. O-ring
26. Drain bolt
27. Main jet
28. Screw
29. Washer
    (1986 Trail and 1984-1985 400)

7

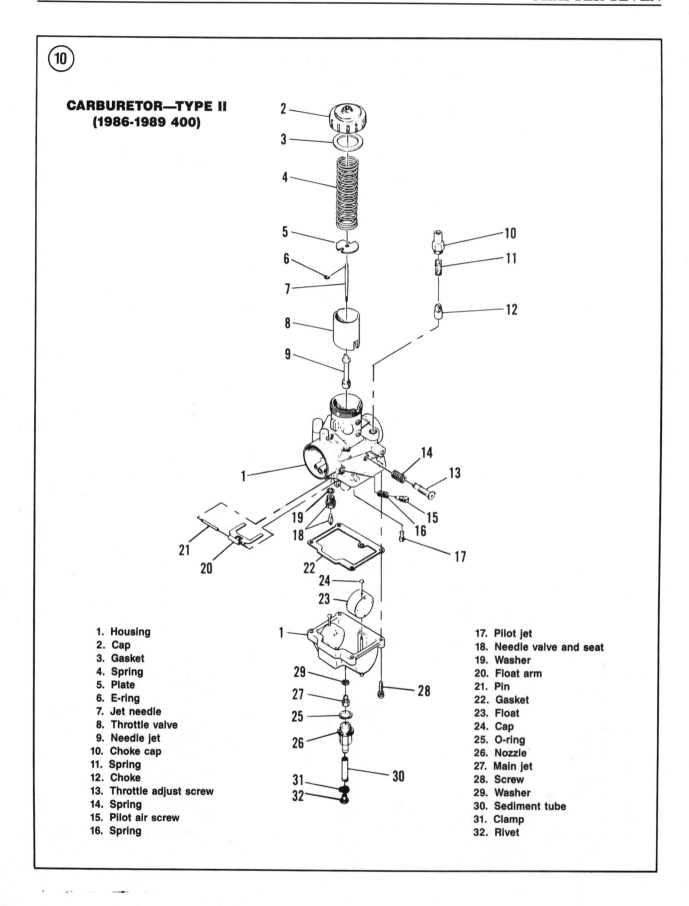

⑩

CARBURETOR—TYPE II
(1986-1989 400)

1. Housing
2. Cap
3. Gasket
4. Spring
5. Plate
6. E-ring
7. Jet needle
8. Throttle valve
9. Needle jet
10. Choke cap
11. Spring
12. Choke
13. Throttle adjust screw
14. Spring
15. Pilot air screw
16. Spring

17. Pilot jet
18. Needle valve and seat
19. Washer
20. Float arm
21. Pin
22. Gasket
23. Float
24. Cap
25. O-ring
26. Nozzle
27. Main jet
28. Screw
29. Washer
30. Sediment tube
31. Clamp
32. Rivet

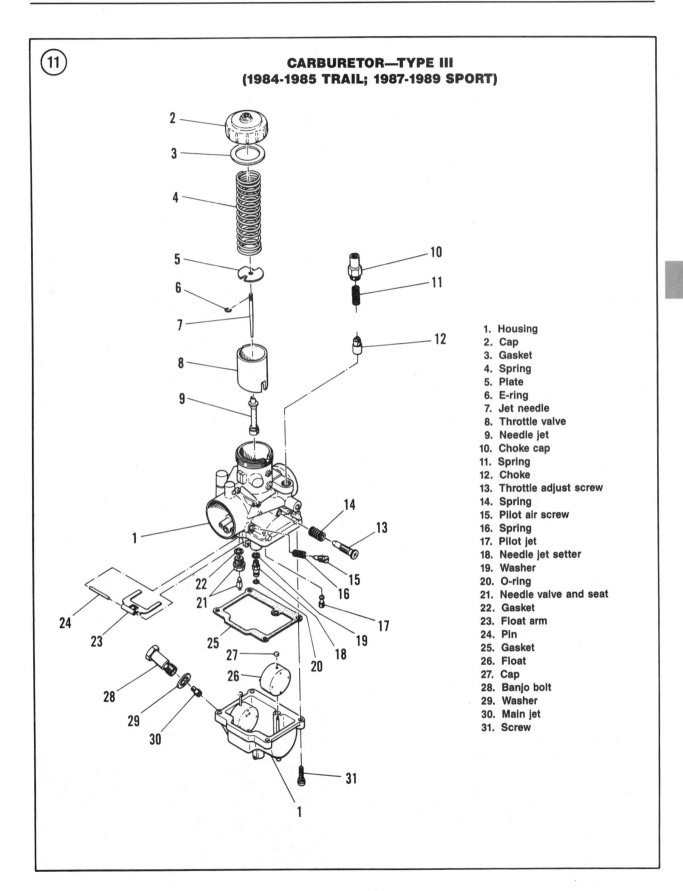

**CARBURETOR—TYPE III
(1984-1985 TRAIL; 1987-1989 SPORT)**

1. Housing
2. Cap
3. Gasket
4. Spring
5. Plate
6. E-ring
7. Jet needle
8. Throttle valve
9. Needle jet
10. Choke cap
11. Spring
12. Choke
13. Throttle adjust screw
14. Spring
15. Pilot air screw
16. Spring
17. Pilot jet
18. Needle jet setter
19. Washer
20. O-ring
21. Needle valve and seat
22. Gasket
23. Float arm
24. Pin
25. Gasket
26. Float
27. Cap
28. Banjo bolt
29. Washer
30. Main jet
31. Screw

7

10. Clean and inspect the carburetor assembly as described in this chapter.

Assembly
(Type I, Type II and Type III)

Refer to **Figure 9** (Type I), **Figure 10** (Type II) or **Figure 11** (Type III).
1. Install the needle valve assembly as follows:
   a. Install the washer (**Figure 21**).
   b. Install the seat (**Figure 20**) and tighten securely.
   c. Install the needle (**Figure 19**) so that the tapered portion faces down.
2. Install the float arm (**Figure 18**) and secure it with the float pin (**Figure 22**). Push the pin all the way in.

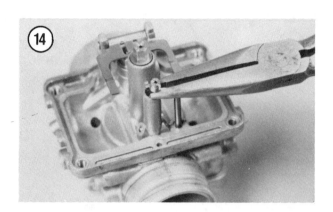

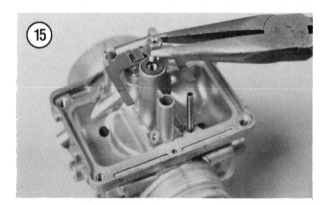

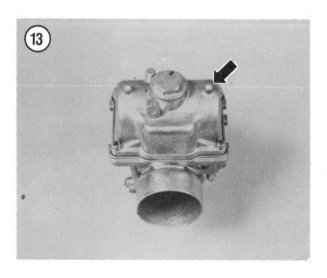

3. Install the needle jet setter assembly on Type III carburetors (**Figure 11**).

4. Install the needle jet (**Figure 17**) so that the notch in the bottom of the needle jet aligns with the pin in the needle jet bore. See **Figure 23**.

5A. *Type I and Type II*: Install the washer onto the top of the needle jet (**Figure 16**) and install the main jet (**Figure 15**).

5B. *Type III*: Install the main jet into the Banjo bolt and install the Banjo bolt into the float bowl. See **Figure 11**.

6. Install the pilot jet (**Figure 14**) with a suitable screwdriver. Tighten the pilot jet securely.

7. Install the float bowl gasket. Make sure it fits around the carburetor flange correctly.

7

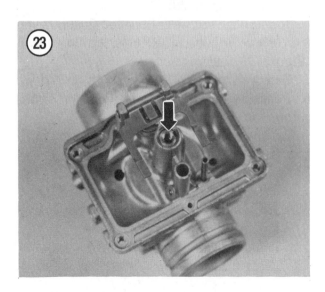

NOTE
Before installing the float bowl in Step 8, make sure the float caps are installed securely onto the float pins in the float bowl. See **Figure 24.**

8. Install the float bowl (**Figure 13**) and screws. Tighten the screws securely.

9. Install the throttle stop screw and spring (A, **Figure 12**).

10. Slide the spring onto the pilot air screw. Install the screw (B, **Figure 12**) into the carburetor body until it lightly seats, then back out the number of turns noted during disassembly.

11. After installing the carburetor, perform the carburetor adjustments as described in Chapter Three.

Disassembly
(Type IV and Type V)

Refer to **Figure 25** (Type IV) or **Figure 26** (Type V).

1. Remove the throttle stop screw and spring (A, **Figure 27**).

2. Lightly seat the pilot air screw (B, **Figure 27**), counting number of turns required for reassembly reference, then back screw out and remove from carburetor with spring.

3. *Type V:* Disconnect the sediment tube at the float bowl (**Figure 28**).

4. Remove the screws holding the float bowl to the carburetor housing. Remove the float bowl (**Figure 29**) and paper gasket.

5. Remove the main jet (**Figure 30**) and ring.

6. Remove the needle jet through the top of the carburetor (**Figure 31**).

7. Remove the float pin (A, **Figure 32**) and float arm (B, **Figure 32**).

8. Remove the needle valve assembly as follows:
   a. Remove the needle valve and seat (**Figure 33**). Do not loose the needle clip.
   b. Remove the washer (**Figure 34**).
   c. Remove the plate (**Figure 35**).
   d. Remove the washer (**Figure 36**).

9. Remove the pilot jet (**Figure 37**) with a flat-tipped screwdriver.

10. Clean and inspect the carburetor assembly as described in this chapter.

Assembly
(Type IV and Type V)

Refer to **Figure 25** (Type IV) or **Figure 26** (Type V).

1. Install the pilot jet (**Figure 37**).

2. Install the needle valve assembly as follows:
   a. Install the washer (**Figure 36**).
   b. Install the plate (**Figure 35**).
   c. Install the washer (**Figure 34**).
   d. Install the needle seat and tighten securely (**Figure 33**).
   e. Install the needle valve and secure it with the clip.

3. Install the needle jet (**Figure 31**) so that the notch in the bottom of the needle jet aligns with the pin in the needle jet bore.

4. Install the ring and main jet (**Figure 30**).

5. Install the float bowl gasket. Make sure it fits around the carburetor flange correctly.

NOTE
Before installing the float bowl in Step 6, make sure the float caps are installed securely onto the float pins in the float bowl. See **Figure 38.**

6. Install the float bowl (**Figure 29**) and screws. Tighten the screws securely.

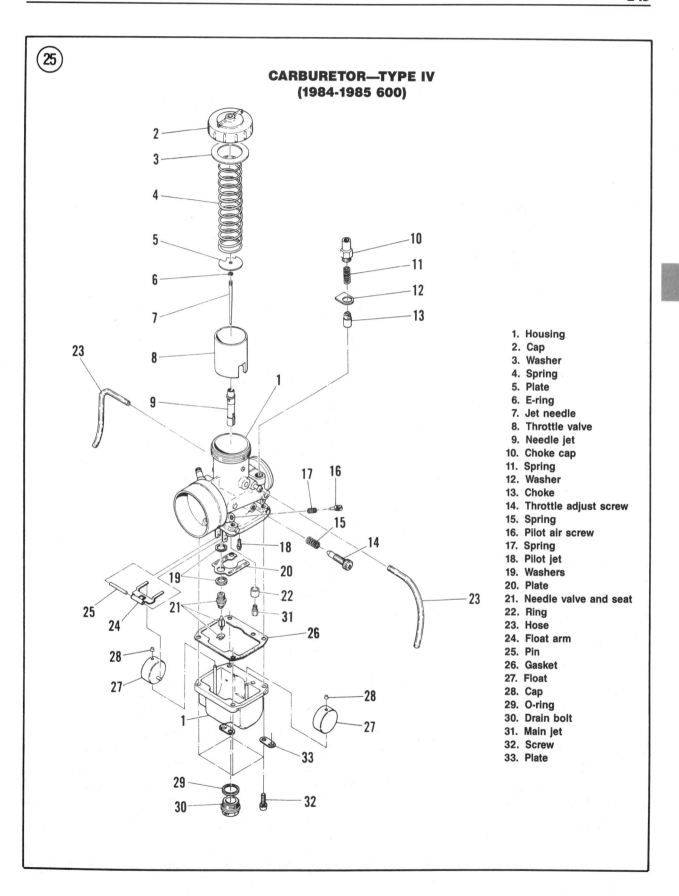

CARBURETOR—TYPE IV
(1984-1985 600)

1. Housing
2. Cap
3. Washer
4. Spring
5. Plate
6. E-ring
7. Jet needle
8. Throttle valve
9. Needle jet
10. Choke cap
11. Spring
12. Washer
13. Choke
14. Throttle adjust screw
15. Spring
16. Pilot air screw
17. Spring
18. Pilot jet
19. Washers
20. Plate
21. Needle valve and seat
22. Ring
23. Hose
24. Float arm
25. Pin
26. Gasket
27. Float
28. Cap
29. O-ring
30. Drain bolt
31. Main jet
32. Screw
33. Plate

7

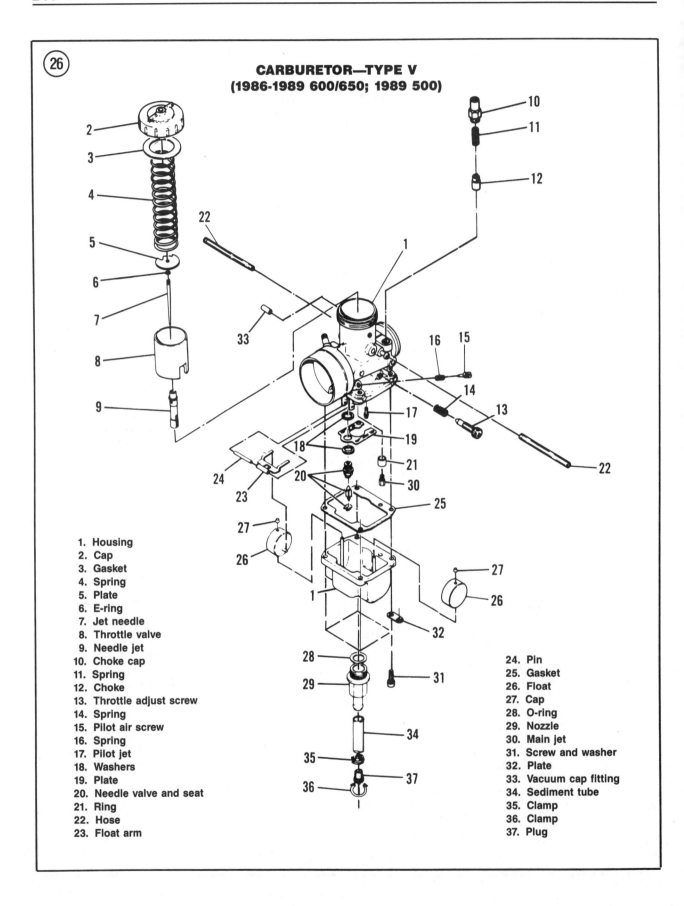

㉖ **CARBURETOR—TYPE V**
(1986-1989 600/650; 1989 500)

1. Housing
2. Cap
3. Gasket
4. Spring
5. Plate
6. E-ring
7. Jet needle
8. Throttle valve
9. Needle jet
10. Choke cap
11. Spring
12. Choke
13. Throttle adjust screw
14. Spring
15. Pilot air screw
16. Spring
17. Pilot jet
18. Washers
19. Plate
20. Needle valve and seat
21. Ring
22. Hose
23. Float arm
24. Pin
25. Gasket
26. Float
27. Cap
28. O-ring
29. Nozzle
30. Main jet
31. Screw and washer
32. Plate
33. Vacuum cap fitting
34. Sediment tube
35. Clamp
36. Clamp
37. Plug

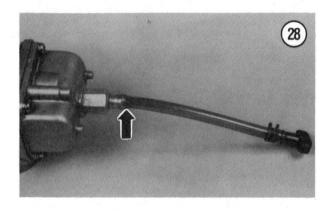

7

7. Insert the sediment hose onto the float bowl (**Figure 28**), if used.

8. Install the throttle stop screw and spring (A, **Figure 27**).

9. Slide the spring onto the pilot air screw. Install the screw (B, **Figure 27**) into the carburetor body until it lightly seats, then back out the number of turns noted during disassembly.

10. After installing the carburetor, perform the carburetor adjustments as described in Chapter Three.

Cleaning/Inspection
(All Models)

Submersion of the carburetor housing assembly in a carburetor cleaner will damage rubber O-rings, seals and plastic parts.

1. Clean the carburetor castings and metal parts with aerosol solvent and a brush. Spray the aerosol solvent on the casting and scrub off any gum or varnish with a small bristle brush.

2. After cleaning the castings and metal parts, wash them thoroughly in hot water and soap. Then rinse with clean water and dry thoroughly.

3. Inspect the carburetor body and float bowl for fine cracks or evidence of fuel leaks. Minor damage can be repaired with an epoxy or liquid aluminum type filler.

4. Blow out the jets with compressed air.

CAUTION
Do not use wire or a drill bit to clean carburetor passages or jets. This can enlarge the passages and change the carburetor calibration. If a passage or jet is severely clogged, use a piece of broom straw to clean it.

5. Inspect the tip of the float valve for wear or damage (**Figure 39**). Replace the valve and seat as a set (**Figure 40**) if they are less than perfect.

NOTE
A damaged float valve will result in flooding of the carburetor float chamber

and impair performance. In addition, accumulation of raw gasoline in the engine compartment presents a severe fire hazard.

6. Inspect the pilot air screw taper for scoring and replace it if less than perfect.

7. Inspect the jets for internal damage and damaged threads. Replace any jet that is less than perfect. Make certain replacement jets are the same size as the originals.

CAUTION
The jets must be scrupulously clean and shiny. Any burring, roughness or abrasion could cause a lean mixture that could result in major engine damage.

8. Check the movement of the float arm on the pivot pin. It must move freely without binding.

9. O-ring seals tend to become hardened after prolonged use and heat and therefore lose their ability to seal properly. Inspect all O-rings and replace if necessary.

10. Check the floats (**Figure 38**) for fuel saturation, deterioration or excessive wear where it contacts the float arm. Replace as required. If the float is in good condition, check it for leakage as follows. Fill the float bowl with water and push the floats down. There should be no sign of bubbles. Replace the floats if necessary.

Float Height
Check and Adjustment

1. Remove the float bowl as described in this chapter.

2. Remove the float bowl gasket.

3. Invert the carburetor. Allow the float arm to contact the fuel valve, but don't compress the spring-loaded plunger on the needle.

4. The float arm should be parallel with the float body as shown in **Figure 41**.

5. If the float height is incorrect, remove the float pin (A, **Figure 32**) and float arm (B, **Figure 32**). Bend tang on end of float arm to adjust.

6. Reinstall float arm and pin. Recheck adjustment.

7. Install float bowl as described in this chapter.

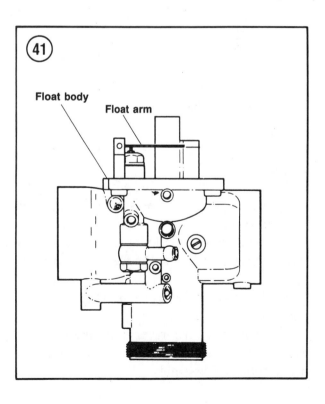

Float body

Float arm

7

Jet Needle/Throttle Valve
Removal/Installation

Refer to illustration for your model when performing the following:

   a. Type I: **Figure 9**.
   b. Type II: **Figure 10**.
   c. Type III: **Figure 11**.
   d. Type IV: **Figure 25**.
   e. Type V: **Figure 26**.

1. Unscrew the carburetor cap and pull the throttle valve out of the carburetor.
2. At the end of the throttle cable, push up on the throttle spring. Then disconnect the cable from the retainer.
3. Remove the plate and jet needle.
4. Installation is the reverse of these steps. Note the following:

   a. Carburetor tuning is described in Chapter Four.
   b. When installing the throttle valve into the carburetor, align the groove in the throttle valve with the pin in the carburetor bore while at the same time aligning the jet needle with the needle jet opening.

FUEL PUMP

Removal/Installation

1. Open the shroud.
2. Remove the air box as described in this chapter.
3. Label the hoses at the fuel pump (**Figure 42**). Plug the hoses to prevent fuel leakage or contamination.
4. Remove the bolts holding the fuel pump to the bulkhead and remove the fuel pump (**Figure 42**).
5. Replace the wire hose clamps when they have lost tension.
6. Installation is the reverse of these steps. Reconnect the hoses according to the ID marks made before disassembly.

Fuel Pump Identification

Due to the number of models covered in this manual and the slight differences in the specific fuel pump the following will help identify the fuel pump installed on your snowmobile. Slight differences do occur between models, so it is important to pay particular attention to the location and order of parts during disassembly.

The Type I fuel pump is found on the following model:

1987-1989 Sport.

The Type II fuel pump is found on the following models:

   a. 1984-1989 Trail.
   b. 1984-1989 400.
   c. 1989 500.

The Type III fuel pump is found on the following model:

1984-1987 600.

The Type IV fuel pump is found on the following model:

1988-1989 650.

Disassembly/Assembly
(Type I)

Refer to **Figure 43** for this procedure.

1. Plug the 4 pump body fuel fittings and clean the pump in solvent. Thoroughly dry.
2. Remove the Phillips screws holding the covers on each side of the pump body. Separate the covers from the body.

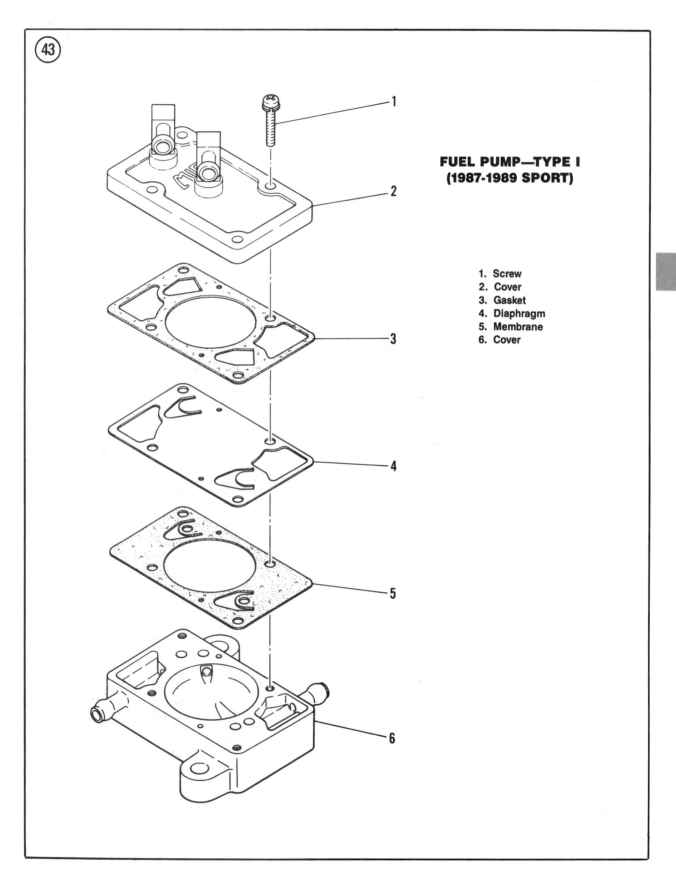

**FUEL PUMP—TYPE I
(1987-1989 SPORT)**

1. Screw
2. Cover
3. Gasket
4. Diaphragm
5. Membrane
6. Cover

7

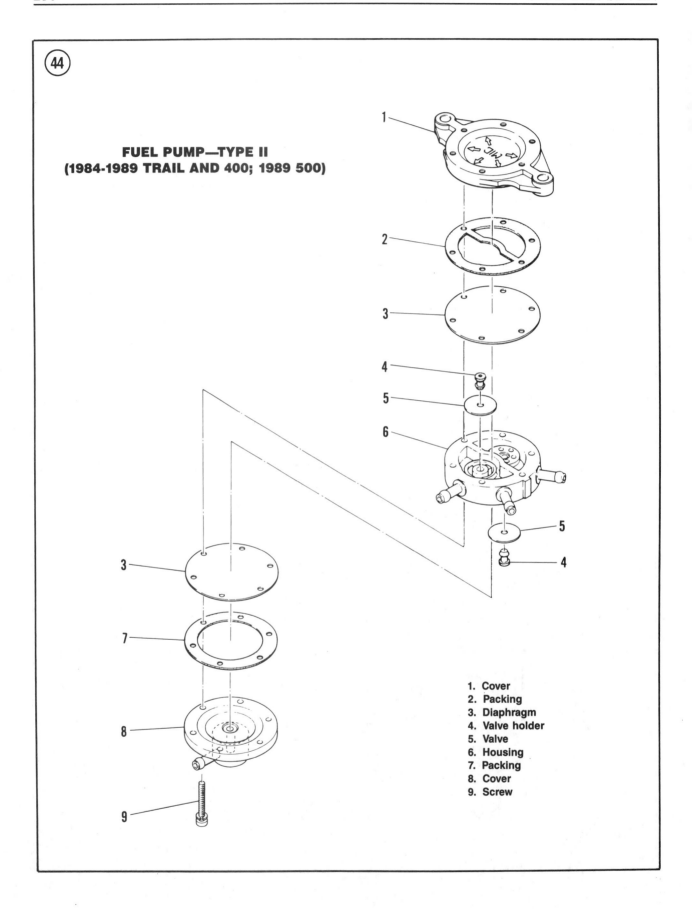

44

FUEL PUMP—TYPE II
(1984-1989 TRAIL AND 400; 1989 500)

1. Cover
2. Packing
3. Diaphragm
4. Valve holder
5. Valve
6. Housing
7. Packing
8. Cover
9. Screw

3. Remove the gasket, diaphragm and membrane.

4. Clean and inspect the pump components as described in this chapter.

5. Install the membrane, diaphragm and gasket in the order shown in **Figure 43**. Secure the pump with the Phillips screws.

**Disassembly/Assembly
(Type II, Type III and Type IV)**

Refer to **Figure 44** (Type II), **Figure 45** (Type III) or **Figure 46** (Type IV).

NOTE
The following procedures show the disassembly and assembly of a Type II

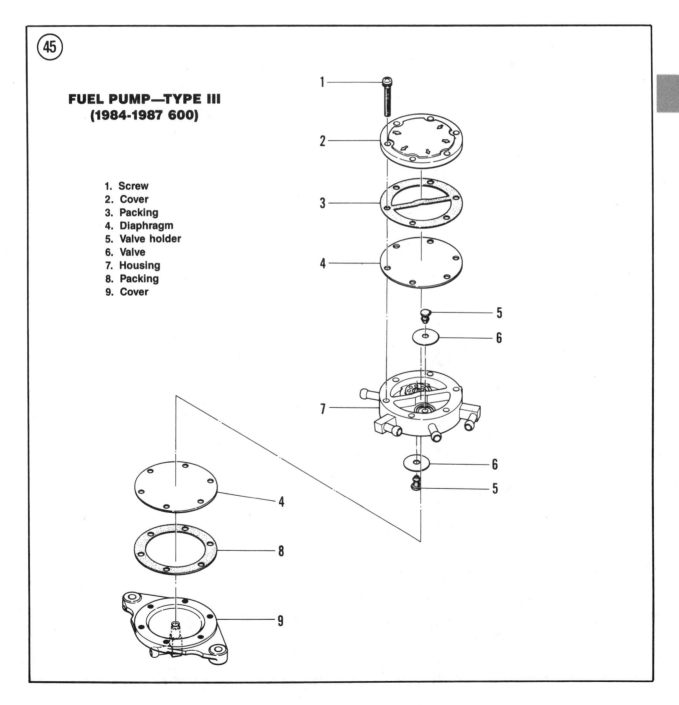

(45)

**FUEL PUMP—TYPE III
(1984-1987 600)**

1. Screw
2. Cover
3. Packing
4. Diaphragm
5. Valve holder
6. Valve
7. Housing
8. Packing
9. Cover

7

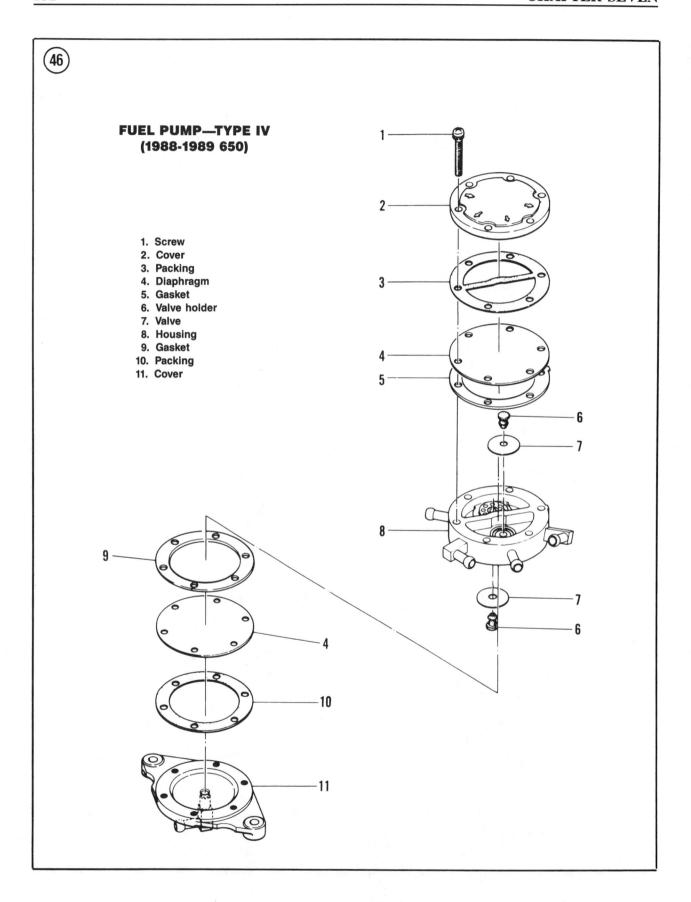

46

FUEL PUMP—TYPE IV
(1988-1989 650)

1. Screw
2. Cover
3. Packing
4. Diaphragm
5. Gasket
6. Valve holder
7. Valve
8. Housing
9. Gasket
10. Packing
11. Cover

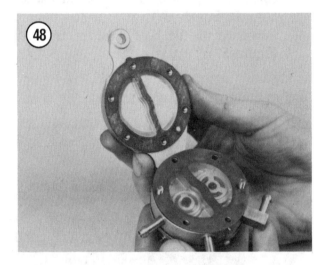

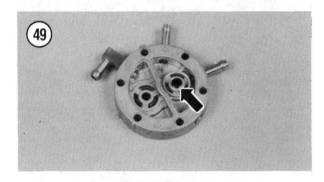

fuel pump. Procedures for the Type III and Type IV are similar.

1. Plug the 4 pump body fuel fittings and clean the pump in solvent. Thoroughly dry.

2. Scribe an alignment mark across the side of the pump to provide an alignment reference for reassembly.

3. Remove the Phillips screws (**Figure 47**) holding the covers on each side of the pump body. Separate the covers from the body (**Figure 48**).

4. Remove the diaphragms and packing. Separate the diaphragms from the packing. Discard the packing.

> *NOTE*
> *Do not remove the valve holders and valves unless replacement is required. Do not reinstall valves which have been previously used.*

5. Remove the valve holders and valves. Discard the valves. See **Figure 49** and **Figure 50**.

6. Clean and inspect the pump components as described in this chapter.

7. Install new valves as follows:
   a. Align the new valve with its seat. Make sure the valve is flat.
   b. Lubricate the tip of the valve holder with a drop of oil. Then push the holder through the valve with a pin punch or rod as shown in **Figure 51**.
   c. Repeat for both valves.

8. Install the diaphragms and gaskets in the order shown in **Figure 44**, **Figure 45** or **Figure 46**. Secure the pump with the 6 Phillips screws.

Cleaning and Inspection

1. Clean pump body and covers in solvent. Dry housing and covers with compressed air.

2. Check body condition. Make sure valve seats provide a flat contact area for the valve disc. Replace the body if cracks or rough gasket mating surfaces are found.

7

3. Check diaphragms (**Figure 52**) for holes or tearing. Replace diaphragms as required.

4. Check membrane (Type I) or packing (**Figure 53**) for tears or other damage. Replace as required.

FUEL TANK

Removal/Installation

Refer to **Figure 54** (typical) for this procedure.

1. Disconnect the taillight connector. Then remove the seat bolts and remove the seat.

2. Remove the fuel tank cover (**Figure 55**).

3. Remove the bracket at the rear of the fuel tank (**Figure 56**).

4. Label and disconnect the fuel hose at the front of the tank (A, **Figure 57**). Plug the hoses to prevent leakage.

5. Disconnect the hold-down spring (B, **Figure 57**) at the front of the fuel tank and remove the fuel tank.

6. Installation is the reverse of these steps. Note the following:

   a. Check all hose connections for leaks.

   b. Align the wire harness with the guide on the bottom side of fuel tank (**Figure 58**) when positioning tank on frame.

   c. When connecting the hold-down spring, it is easier to first connect the spring to the right-hand bracket hole (**Figure 59**). Then place the fuel tank against the bracket and connect the spring onto the left-hand bracket hole (B, **Figure 57**).

   d. Reconnect the fuel hose at the fuel tank (A, **Figure 57**).

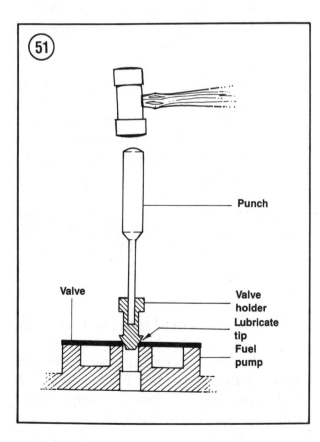

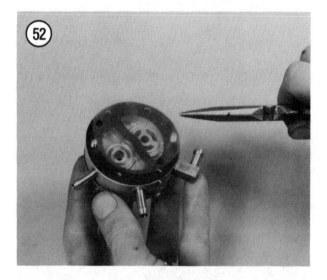

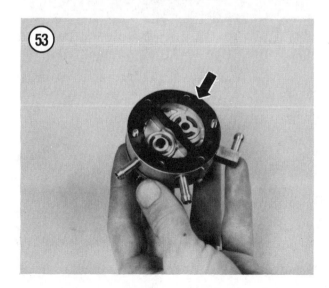

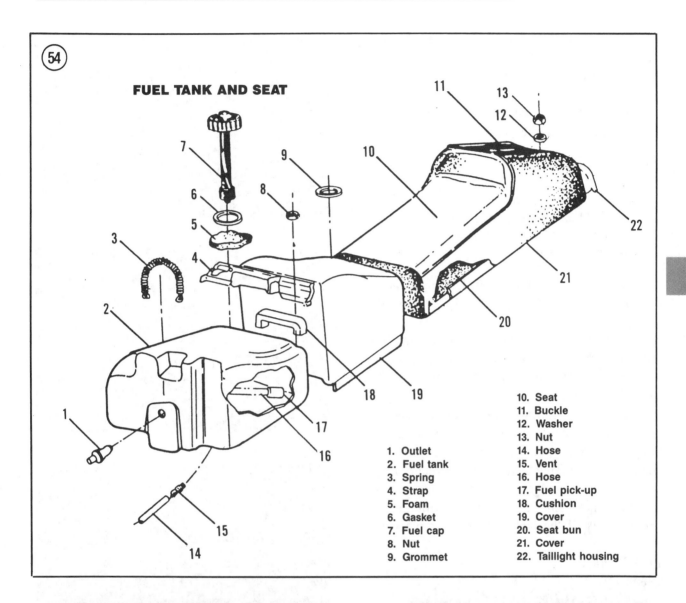

FUEL TANK AND SEAT

1. Outlet
2. Fuel tank
3. Spring
4. Strap
5. Foam
6. Gasket
7. Fuel cap
8. Nut
9. Grommet
10. Seat
11. Buckle
12. Washer
13. Nut
14. Hose
15. Vent
16. Hose
17. Fuel pick-up
18. Cushion
19. Cover
20. Seat bun
21. Cover
22. Taillight housing

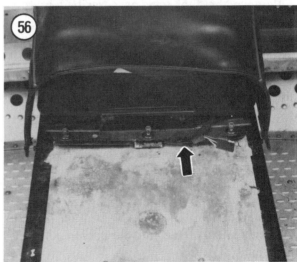

7

e. When installing the rear bracket (**Figure 56**), make sure the wiring harness connection is positioned over the top of the bracket.

f. Reconnect the taillight connector when installing the seat.

Cleaning/Inspection

> *WARNING*
> *The fuel tank should be cleaned in an open area well away from all sources of flames or sparks.*

1. Pour old gasoline from the tank into a sealable container manufactured specifically for gasoline storage.

2. Pour about 1 quart of fresh gasoline into the tank and slosh it around for several minutes to loosen sediment. Then pour the contents into a sealable container.

3. Examine the tank for cracks and abrasions, particularly at points where the tank contacts the body. Abraded areas can be protected and cushioned by coating them with a non-hardening silicone sealer and allowing it to dry before installing the tank. However, if abrading is extensive, or if the tank is leaking, replace it.

Fuel Pickup
Cleaning

1. Remove the fuel tank as described in this chapter. Pour old gasoline from the tank into a sealable container manufactured specifically for gasoline storage.

2. Unscrew the fuel valve (**Figure 60**) and remove the fuel hose/pickup assembly (**Figure 61**).

3. Flush the pickup assembly.

4. Reverse to install. Check the fuel tank for leaks before starting the engine.

THROTTLE CABLE REPLACEMENT

When replacing the throttle cable, note the following:

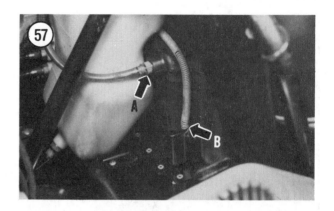

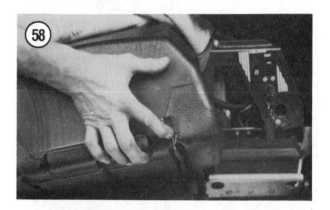

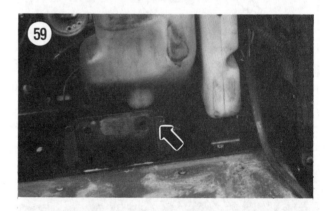

a. Adjust the carburetors as described in Chapter Three. See *Synchronization*.

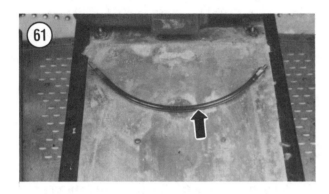

b. Adjust the oil injection cable as described in Chapter Three.
c. Operate the throttle lever to make sure the carburetor throttle valves operate correctly.
d. Check cable routing.

EXHAUST SYSTEM

Removal/Installation
(Trail, Sport, 400 and 500)

Refer to **Figure 62** or **Figure 63** for this procedure.

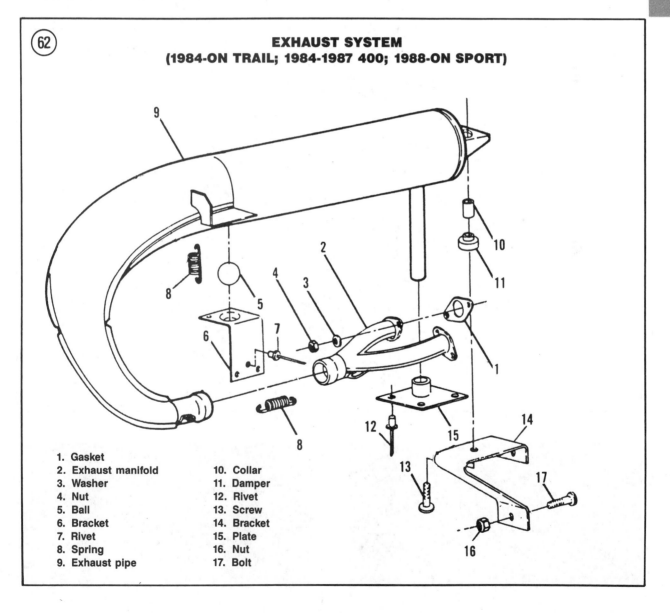

62

EXHAUST SYSTEM
(1984-ON TRAIL; 1984-1987 400; 1988-ON SPORT)

1. Gasket
2. Exhaust manifold
3. Washer
4. Nut
5. Ball
6. Bracket
7. Rivet
8. Spring
9. Exhaust pipe
10. Collar
11. Damper
12. Rivet
13. Screw
14. Bracket
15. Plate
16. Nut
17. Bolt

7

1. Open the shroud.

WARNING
If the exhaust system is hot, wait until it cools down before removing it.

2. Disconnect the springs that secure the muffler to the exhaust pipe.

3. Remove the nut and washers or spring securing the exhaust pipe to the bracket. Account for the spacers and grommet, if used.

4. Remove the spring securing the muffler to the exhaust pipe and remove the exhaust pipe.

5. Remove the rubber ball (**Figure 64**) from the bracket.

6. Remove the nut and washer securing the exhaust manifold to the cylinder blocks and remove the manifold. See **Figure 65**, typical.

7. Installation is the reverse of these steps.

8. Install new exhaust manifold gaskets.

9. Check the exhaust system for leaks after installation.

Removal/Installation (600 and 650)

Refer to **Figure 66** for this procedure.

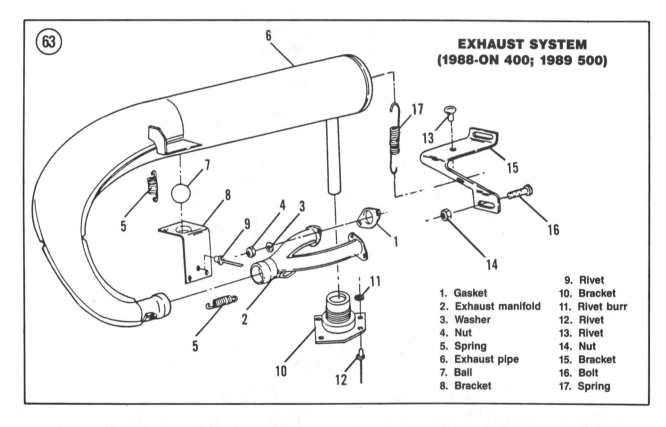

EXHAUST SYSTEM (1988-ON 400; 1989 500)

1. Gasket
2. Exhaust manifold
3. Washer
4. Nut
5. Spring
6. Exhaust pipe
7. Ball
8. Bracket
9. Rivet
10. Bracket
11. Rivet burr
12. Rivet
13. Rivet
14. Nut
15. Bracket
16. Bolt
17. Spring

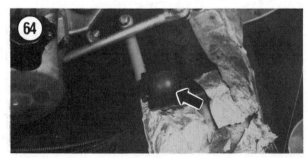

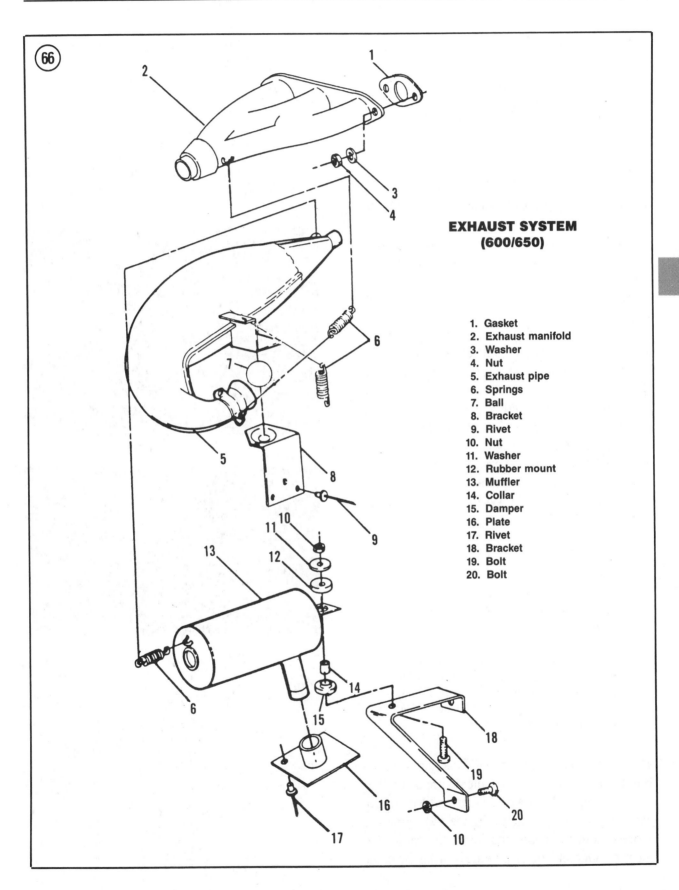

**EXHAUST SYSTEM
(600/650)**

1. Gasket
2. Exhaust manifold
3. Washer
4. Nut
5. Exhaust pipe
6. Springs
7. Ball
8. Bracket
9. Rivet
10. Nut
11. Washer
12. Rubber mount
13. Muffler
14. Collar
15. Damper
16. Plate
17. Rivet
18. Bracket
19. Bolt
20. Bolt

7

1. Open the shroud.

WARNING
If the exhaust system is hot, wait until it cools down before removing it.

2. Disconnect the springs that secure the muffler to the exhaust pipe (A, **Figure 67**).
3. Remove the springs securing the exhaust pipe to the bracket.
4. Remove the spring (A, **Figure 68**) securing the muffler to the exhaust pipe and remove the exhaust pipe.
5. Remove the rubber ball from the bracket (**Figure 64**).
6. Remove the muffler mounting nut, washer and bolt and remove the muffler. Account for the bushing and dampers when removing the muffler (B, **Figure 67**).
7. Remove the nut and washers securing the muffler to the bracket and remove the muffler (B, **Figure 68**).
8. Installation is the reverse of these steps.
9. Install new exhaust manifold gaskets.
10. Check the exhaust system for leaks after installation.

Cleaning

WARNING
When performing Step 1, do not start the drill until the cable is inserted into the pipe. At no time should the drill be running when the cable is free of the pipe. The whipping action of the cable could cause serious personal injury. Wear heavy shop gloves and a face shield when using this equipment.

1. Clean all accessible exhaust passages with a blunt-roundnose tool. To clean areas further down the pipe, chuck a piece of discarded control cable in an electric drill (**Figure 69**). Fray the loose end of the cable and insert the cable into the pipe. Operate the drill while moving the cable

back and forth inside the pipe. Take your time and do a thorough job.
2. Shake the large pieces out into a trash container.

EXHAUST SYSTEM REPAIR

A dent in the exhaust pipe will alter the system's flow characteristics and degrade

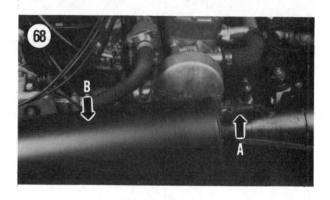

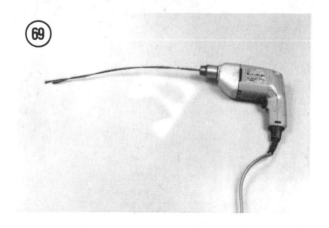

performance. Minor damage can be easily repaired if you have welding equipment, some simple body tools, and a bodyman's slide hammer.

Small Dents

1. Drill a small hole in the center of the dent. Screw the end of the slide hammer into the hole.

2. Heat the area around the dent evenly with a torch.

3. When the dent is heated to a uniform orange-red color, operate the slide hammer to raise the dent.

4. When the dent is removed, unscrew the slide hammer and weld the drilled hole closed.

Large Dents

Large dents that are not crimped can be removed with heat and a slide hammer as previously described. However, several holes must be drilled along the center of the dent so that it can be pulled out evenly.

If the dent is sharply crimped along the edges, the affected section should be cut out with a hacksaw, straightened with a body dolly and hammer and welded back into place.

Before cutting the exhaust pipe apart, scribe alignment marks over the area where the cuts will be made to aid correct alignment when the pipe is rewelded.

After the welding is completed, wire brush and clean up all welds. Paint the entire pipe with a high-temperature paint to prevent rusting.

7

Tables are on the following pages.

Table 1 CARBURETOR SPECIFICATIONS—TRAIL

Type	Mikuni
Model	VM34SS
Main jet	
1984-1985	230
1986	210
1987-1988	220
1989	230
Pilot jet	
1984-1985	35
1986-1987	30
1988-on	35
Jet needle/clip position	
1984-1986	6DH7/2
1987	6DH7/3
1988-on	6DH7/2
Needle jet	
1984-1985	P-6
1986-on	P-8
Throttle valve cutaway	
1984-on	3.0

Table 2 CARBURETOR SPECIFICATIONS—SPORT

Type	Mikuni
Size	VM30SS
Main jet	180
Pilot jet	35
Jet needle/clip position	5DP7/3
Needle jet	0-8
Throttle valve cutaway	3.0

Table 3 CARBURETOR SPECIFICATIONS—400

Type	Mikuni
Size	VM345SS
Main jet	
1984	*
1985-1988	220
1989	230
Pilot jet	
1984	*
1985	35
1986-on	30
Jet needle/clip position	
1984	*
1985	6DH7/2
1986	6DP17/2
1987-on	6DP17/3

(continued)

Table 3 CARBURETOR SPECIFICATIONS—400 (continued)

Needle jet	
1984	*
1985	P-8
1986	Q-2
Throttle valve cutaway	
1984	*
1985	3.0
1986-on	2.0
*Not specified.	

Table 4 CARBURETOR SPECIFICATIONS—500

Type	Mikuni
Size	VM38SS
Main jet	280
Pilot jet	40
Jet needle/clip position	6F9/3
Needle jet	Q-0
Throttle valve cutaway	3.0

Table 5 CARBURETOR SPECIFICATIONS—600

Type	Mikuni
Size	VM38SS
Main jet	
1984	250
1985-1987	260
Pilot jet	
1984	35
1985-1986	40
1987	35
Jet needle/clip position	
1984-1985	6F4/3
1986-1987	6F9/3
Needle jet	
1984	Q-2
1985-1987	P-8
Throttle valve cutaway	
1984	3.5
1985-1987	3.0

Table 6 CARBURETOR SPECIFICATIONS—650

Type	Mikuni
Size	VM38SS
Main jet	260
Pilot jet	50
Jet needle/clip position	
1988	6F9-3
1989	6DH4/3
Needle jet	P-8
Throttle valve cutaway	3.0

7

Chapter Eight

Electrical System

This chapter provides service procedures for the ignition system and lights. Electrical troubleshooting procedures are described in Chapter Two. Wiring diagrams are at the end of the book. **Tables 1** and **2** are at the end of the chapter.

BATTERY

A 12-volt battery (**Figure 1**) is used on all electric start models. Refer to **Table 1** for battery capacity.

With proper care, the battery should last for 2 to 3 years. Incorrect care or neglect (described below) can cut battery service life short and lead to chronic starting problems.

  a. *Electrolyte level*: Correct electrolyte level must be maintained at all times for the battery to be able to accept and maintain a full charge.

  b. *Charge level*: The battery must be maintained at full charge all the time to prevent premature sulfation which will result in internal shorts that destroy the battery. This is a point often overlooked during summer storage. The battery should be removed from the machine and cleaned and serviced and fully charged. It should be periodically tested and recharged to prevent it from sulfating.

  c. *Overcharging*: Overcharging or charging the battery at too high a rate creates excessive heat that will destroy the battery.

  d. *Freezing*: If the machine is left outdoors for long periods in freezing temperatures, freezing of the electrolyte may occur if the battery is discharged. Guard against this by keeping the battery at full charge when used in sub-zero weather. If the machine is going to be stored in cold weather for long periods, remove the battery and store it in a warm place.

CAUTION
A discharged battery will freeze, causing battery damage. The battery should be kept fully charged when used in sub-zero weather.

Care and Inspection

The battery is the heart of the electrical system. Most electrical system troubles can be attributed to neglect of this vital component.

In order to correctly service the electrolyte level, it is necessary to remove the battery from the frame. The electrolyte level should be maintained between the two marks on the battery case (**Figure 1**). If the electrolyte level is low, it's a good idea to completely remove the battery so that it can be thoroughly cleaned, serviced and checked.

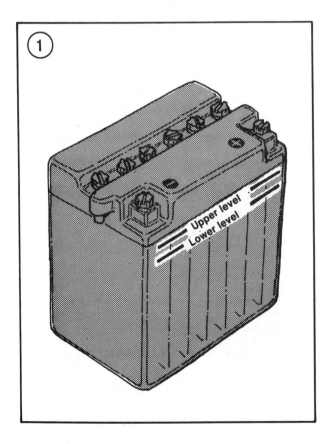

On all models, the negative side is grounded. When removing the battery, disconnect the negative (−) ground cable first, then the positive (+) cable. This minimizes the chance of a tool shorting to ground when disconnecting the "hot" positive cable.

1. Open the shroud.
2. Disconnect the negative battery cable at the battery.
3. Disconnect the positive battery cable at the battery.
4. Unlatch the battery hold-down straps and lift the battery (**Figure 2**) out of the frame.

WARNING
Protect your eyes, skin and clothing. If electrolyte gets into your eyes, flush your eyes thoroughly with clean water and get prompt medical attention.

CAUTION
Be careful not to spill battery electrolyte on painted or polished surfaces. The liquid is highly corrosive and will damage the finish. If it is spilled, wash it off immediately with soapy water and thoroughly rinse with clean water.

5. Check the entire battery case for cracks. Replace the battery if cracked.
6. Inspect the battery tray (**Figure 2**) for corrosion and clean if necessary with a solution of baking soda and water.

NOTE
Keep cleaning solution out of the battery cells in Step 7 or the electrolyte will be seriously weakened.

7. Clean the top of the battery with a stiff bristle brush using a baking soda and water solution. Rinse the battery case with clean water and wipe dry with a clean cloth or paper towel.
8. Check the battery cable clamps for corrosion and damage. If corrosion is minor, clean the battery cable clamps with a stiff wire brush. Replace severely worn or damaged cables.

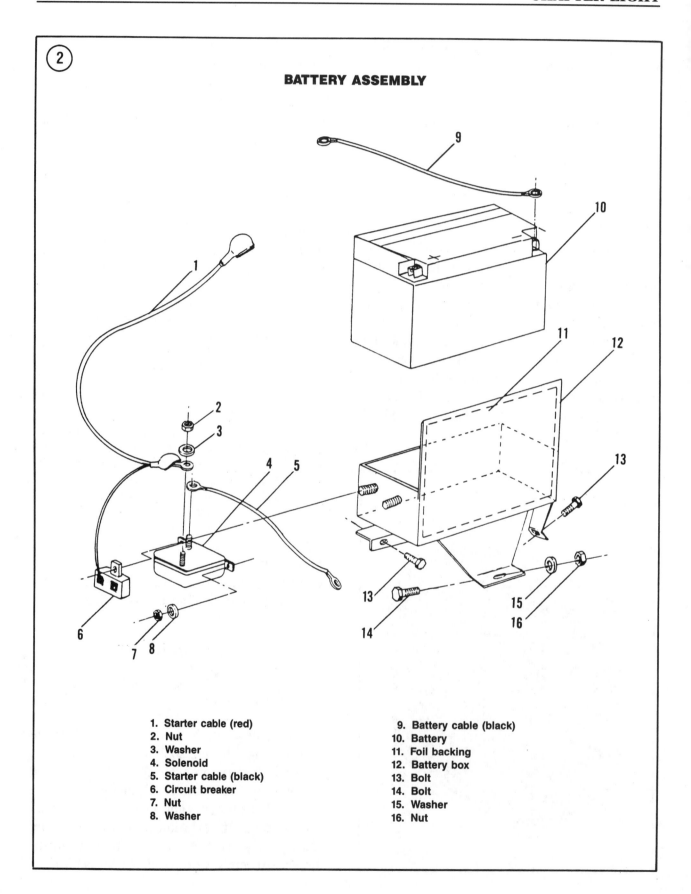

BATTERY ASSEMBLY

1. Starter cable (red)
2. Nut
3. Washer
4. Solenoid
5. Starter cable (black)
6. Circuit breaker
7. Nut
8. Washer

9. Battery cable (black)
10. Battery
11. Foil backing
12. Battery box
13. Bolt
14. Bolt
15. Washer
16. Nut

NOTE
Do not overfill the battery cells in Step 9. The electrolyte expands due to heat from charging and will overflow if the level is above the upper level line.

9. Remove the caps from the battery cells and check the electrolyte level. Add distilled water,

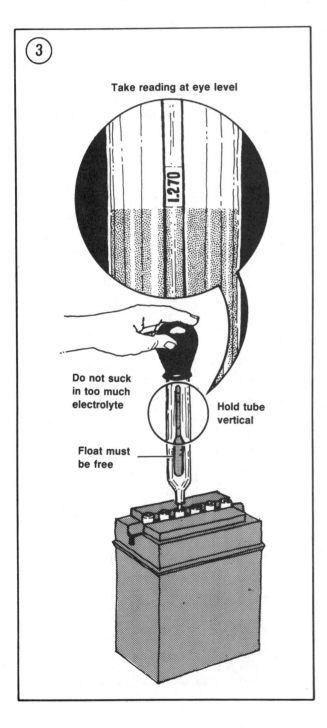

Take reading at eye level

1.270

Do not suck in too much electrolyte

Hold tube vertical

Float must be free

if necessary, to bring the level within the upper and lower level lines on the battery case (**Figure 1**).

10. Position the battery in the battery tray as shown in **Figure 2**.

11. Reconnect the positive battery cable, then the negative cable.

CAUTION
Be sure the battery cables are connected to their proper terminals. Connecting the battery backwards will reverse the polarity and damage the rectifier.

WARNING
After installing the battery, make sure the vent tube is not pinched. A pinched or kinked tube would allow high pressure to accumulate in the battery and cause the battery to explode. If the tube is pinched or otherwise damaged, install a new tube.

12. Tighten the battery connections and coat with a petroleum jelly such as Vaseline or a light mineral grease.

13. Reconnect the battery hold down straps.

Testing

Hydrometer testing is the best way to check battery condition. Use a hydrometer with numbered graduations from 1.100 to 1.300 rather than one with just color-coded bands. To use the hydrometer, squeeze the rubber ball, insert the tip into the cell and release the ball (**Figure 3**).

NOTE
Do not attempt to test a battery with a hydrometer immediately after adding water to the cells. Charge the battery for 15-20 minutes at a high rate to cause vigorous gassing and allow the water and electrolyte to mix thoroughly.

Draw enough electrolyte to float the weighted float inside the hydrometer. When using a temperature-compensated hydrometer, release the electrolyte and repeat this process several times to make sure the thermometer has adjusted to the electrolyte temperature before taking the reading.

Hold the hydrometer vertically and note the number in line with the surface of the electrolyte (**Figure 4**). This is the specific gravity for this cell. Return the electrolyte to the cell from which it came.

The specific gravity of the electrolyte in each battery cell is an excellent indication of that cell's condition (**Table 2**). A fully charged cell will read 1.260-1.280 while a cell in good condition reads from 1.230-1.250 and anything below 1.140 is discharged. Charging is also necessary if the specific gravity varies more than 0.050 from cell-to-cell.

NOTE
If a temperature-compensated hydrometer is not used. add 0.004 to the specific gravity for every 10° above 80° F (25° C). For every 10° below 80° F (25° C), subtract 0.004.

Charging

A good state of charge should be maintained in batteries used for starting. When charging the battery, note the following:

a. During charging, the cells will show signs of gas bubbling. If one cell has no gas bubbles or if its specific gravity is low, the cell is probably shorted.

b. If a battery not in use loses its charge within a week after charging or if the specific gravity drops quickly, the battery is defective. A good battery should only self-discharge approximately 1% each day.

CAUTION
Always remove the battery from the frame before connecting charging equipment.

WARNING
During charging, highly explosive hydrogen gas is released from the battery.

The battery should be charged only in a well-ventilated area, and open flames and cigarettes should be kept away. Never check the charge of the battery by arcing across the terminals; the resulting spark can ignite the hydrogen gas.

1. Remove the battery from the frame as described in this chapter.

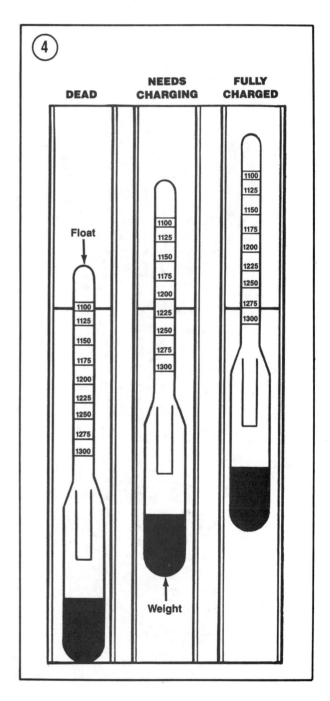

2. Connect the positive (+) charger lead to the positive battery terminal and the negative (−) charger lead to the negative battery terminal. See **Figure 5**.

3. Remove all vent caps from the battery, set the charger at 12 volts, and switch it on. Normally, a battery should be charged at a slow charge rate of 1/10 its given capacity. See **Table 1** for battery capacity.

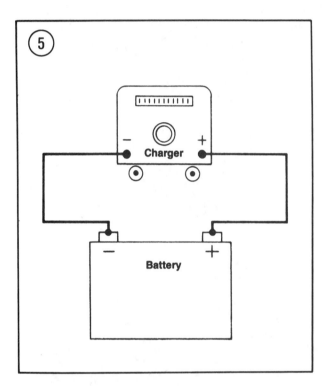

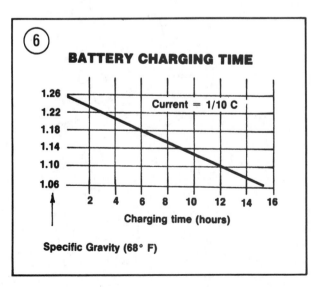

BATTERY CHARGING TIME

Current = 1/10 C

Charging time (hours)

Specific Gravity (68° F)

CAUTION
The electrolyte level must be maintained at the upper level during the charging cycle; check and refill with distilled water as necessary.

4. The charging time depends on the discharged condition of the battery. The chart in **Figure 6** can be used to determine approximate charging times at different specific gravity readings. For example, if the specific gravity of your battery is 1.180, the approximate charging time would be 6 hours.

5. After the battery has been charged for about 6-8 hours, turn the charger off, disconnect the leads and check the specific gravity. It should be within the limits specified in **Table 2**. If it is, and remains stable for one hour, the battery is charged.

New Battery Installation

When replacing the old battery with a new one, be sure to charge it completely (specific gravity, 1.260-1.280) before installing it in the frame. Failure to do so, or using the battery with a low electrolyte level will permanently damage the battery.

NOTE
When purchasing a new battery from a dealer or parts accessory store, the retailer usually services the battery by adding electrolyte. However, before purchasing the battery, make sure the battery has been properly charged.

CAPACITOR DISCHARGE IGNITION (CDI)

All models use a capacitor discharge ignition system. This section describes removal and installation procedures for the flywheel, stator plate, primary coils and lighting coil.

Refer to **Figure 7**, **Figure 8** or **Figure 9** when performing procedures in this section.

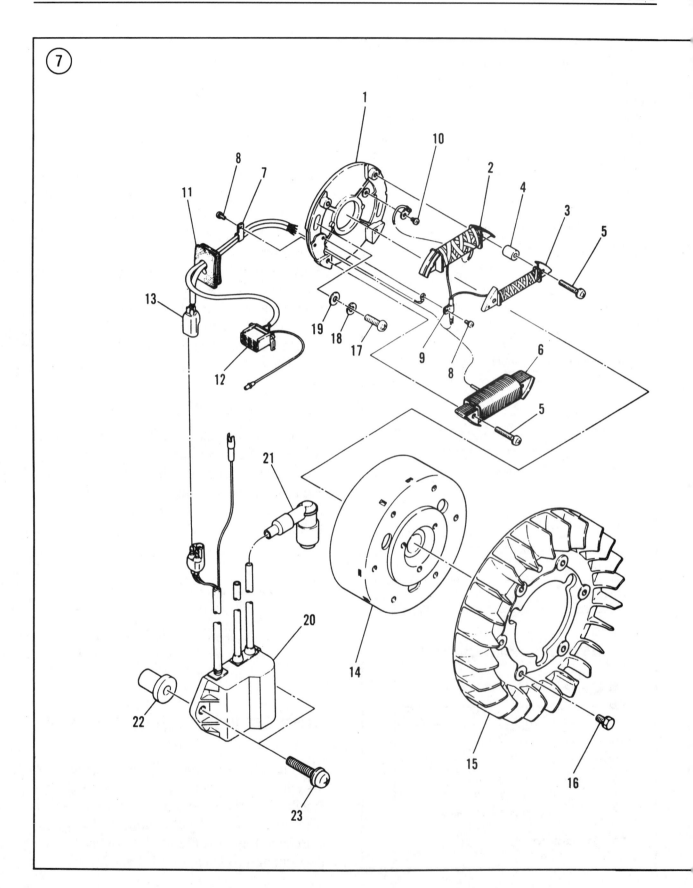

8

MAGNETO ASSEMBLY
(1984-1989 TRAIL; 1984-1989 400; 1989 500)

1. Stator plate
2. Exciter coil
3. Pulser coil
4. Collar
5. Screw
6. Lighting coil
7. Clamp
8. Screw
9. Clamp
10. Screw
11. Grommet
12. Connector
13. Coupler
14. Flywheel
15. Fan
16. Bolt
17. Screw
18. Lockwasher
19. Washer
20. CDI control unit
21. Spark plug cap
22. Rubber nut
23. Screw

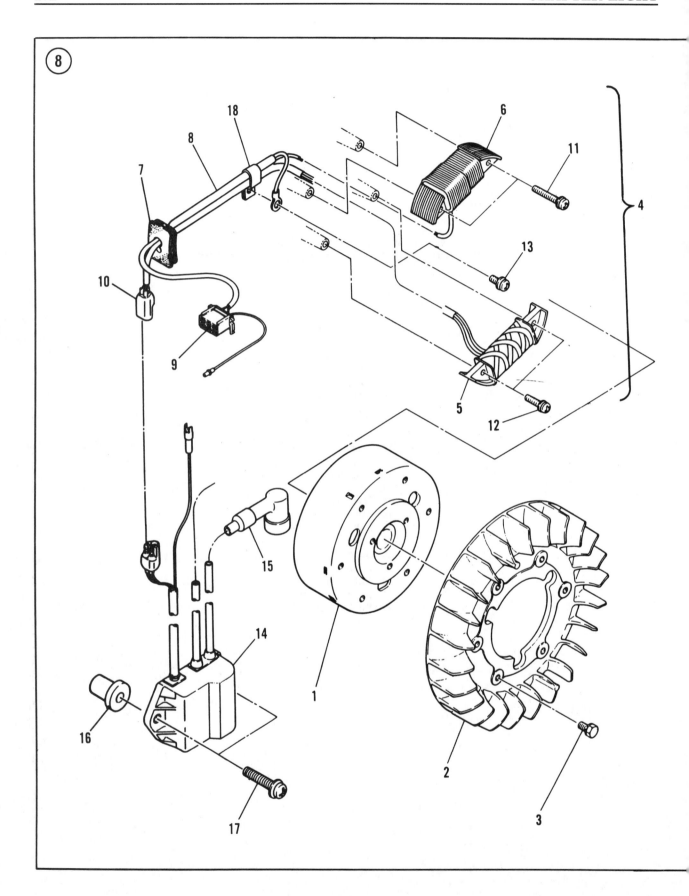

MAGNETO ASSEMBLY
(1987-1989 SPORT)

1. Flywheel
2. Fan
3. Bolt
4. Coil assembly
5. Exciter coil
6. Lighting coil
7. Grommet
8. Wire harness
9. Coupler
10. Coupler
11. Screw assembly
12. Screw assembly
13. Screw assembly
14. CDI control unit
15. Spark plug cap
16. Rubber nut
17. Screw assembly
18. Clamp

8

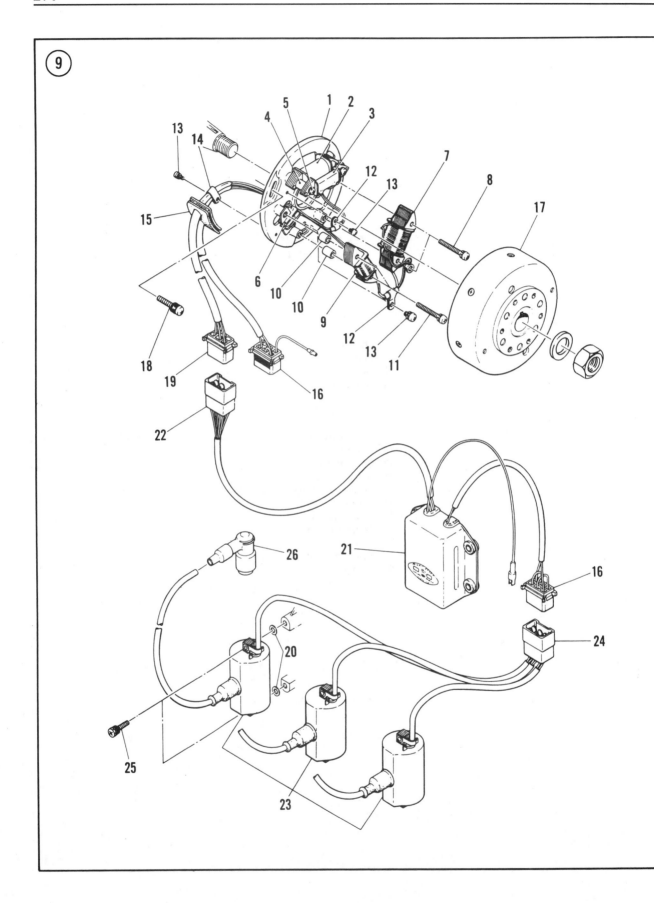

MAGNETO ASSEMBLY
(1984-1989 600/650)

1. Stator plate
2. Exciter coil
3. Pulser coil
4. Collar
5. Screw
6. Control coil
7. Lighting coil
8. Screw
9. Lighting coil
10. Collar
11. Screw
12. Clamp
13. Screw
14. Clamp
15. Grommet
16. Coupler
17. Flywheel
18. Screw assembly
19. Coupler
20. Spacers
21. CDI
22. Coupler
23. Ignition coils
24. Coupler
25. Screw assembly
26. Spark plug cap

Troubleshooting

Refer to Chapter Two.

Flywheel
Removal/Installation

The flywheel must be removed to service the stator coils. Flywheel replacement is usually necessary only if the magnets have been damaged by mechanical heat or shock.

NOTE
The following procedures are shown with the various engine assemblies removed for clarity. It is not necessary to remove the engine from the frame to remove the flywheel.

1. Remove the recoil starter housing as described in Chapter Eleven.
2. *Liquid cooled models*: Loosen the water pump mounting bolts and pivot the water pump to loosen the drive belt. See **Figure 10** (2-cylinder) or **Figure 11** (3-cylinder). Then remove the drive belt.

8

3. Secure the starter pulley with a spanner wrench (**Figure 12**) and loosen the flywheel nut. Remove the flywheel nut and washer assembly.

4A. *Liquid cooled models*: Remove the bolts holding the starter pulley onto the flywheel. Remove the starter pulley (**Figure 13**) and belt pulley (**Figure 14**).

4B. *Fan cooled models*: Remove the bolts holding the starter pulley onto the flywheel. Remove the starter pulley.

> *CAUTION*
> *If the puller used in Step 5 does not have a centering adaptor on the end of the pressure screw, a protective cap should be used together with the puller to prevent crankshaft thread damage. A cap can be made by welding a small diameter plate onto the end of a stock flywheel nut.*

> *NOTE*
> *See a Polaris dealer for the flywheel puller for your model.*

5. Mount a flywheel puller onto the face of the flywheel, using the 3 or 4 threaded holes in the flywheel (**Figure 15**).

> *NOTE*
> *On liquid cooled models, hold the flywheel with a spanner wrench when removing the flywheel. On fan cooled models, hold the fan with a strap wrench (**Figure 16**).*

> *CAUTION*
> *Do not use heat or a hammer on the flywheel to remove it in Step 6. Heat may cause the flywheel to seize on the*

crankshaft, while hammering can damage the flywheel or bearings.

6. Tighten the puller bolt and break the flywheel free of the crankshaft taper. You may have to alternate tapping on the pressure screw sharply with a hammer and tightening the bolt some more, but don't hit the flywheel.

7. Remove the flywheel (**Figure 17**) and the puller assembly. Remove the puller from the flywheel.

8. Remove the Woodruff key from the flywheel (**Figure 18**).

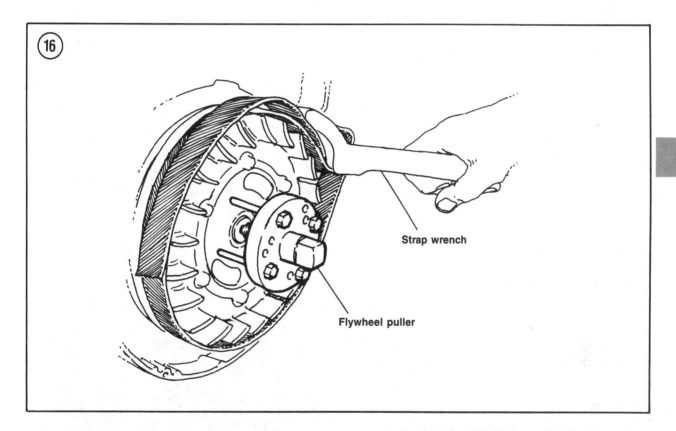

Strap wrench

Flywheel puller

NOTE
*Before installing the flywheel, check the magnets (**Figure 19**) for metal trash they may have picked up. Debris stuck to the magnets can cause coil damage.*

9. Spray the flywheel and crankshaft tapers with a rust inhibitor, such as WD-40.

10. Place the Woodruff key in the crankshaft key slot (**Figure 18**). Position flywheel over crankshaft with key slot in flywheel hub aligned with key in crankshaft (**Figure 17**). Push flywheel onto crankshaft firmly.

11. Install flywheel lockwasher and nut finger tight (**Figure 20**).

12A. *Liquid cooled models*: Install the belt pulley (**Figure 14**) and starter pulley (**Figure 13**). Install and tighten the mounting bolts securely.

12B. *Fan cooled models*: Install the starter pulley onto the face of the flywheel. Install and tighten the mounting bolts securely.

13. Hold the flywheel with the same tool used during removal and tighten the flywheel nut to 83-90 N·m (60-65 ft.-lbs.). See **Figure 12**.

14. Install the water pump drive belt as described in Chapter Ten.

15. Install the recoil starter housing as described in Chapter Eleven.

Inspection

1. Check the flywheel (**Figure 19**) carefully for cracks or breaks. On fan cooled models, check the fan blades (**Figure 21**) for cracks or other damage.

WARNING
A cracked or chipped flywheel must be replaced. A damaged flywheel may fly apart at high rpm, causing severe engine damage. Do not attempt to repair a damaged flywheel.

2. Check tapered bore of flywheel and crankshaft taper for signs of scoring, cracks or other damage.

3. Check key slot in flywheel for cracks or other damage. Check keyseat in crankshaft (**Figure 22**) for cracks or other damage.

4. Check the Woodruff key (**Figure 22**) for cracks or damage.

5. Check flywheel nut for thread damage or rounding off of the nut shoulders.

6. Check the flywheel lockwasher and the flat washer for fatigue, cracks or other damage.

7. Replace worn or damaged parts as required.

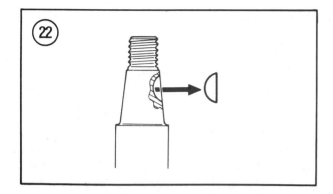

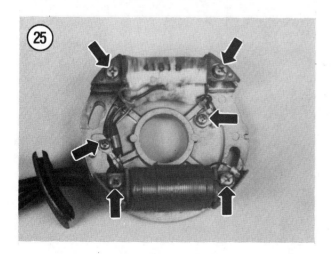

Stator Plate
Removal/Installation

NOTE
Refer to Chapter Two for troubleshooting and test procedures.

1. Remove the flywheel as described in this chapter.

2. If the engine is installed in the frame, disconnect the stator plate electrical connectors.

3. Before removing the stator plate, check the static timing mark on the stator plate. It should be aligned with the crankcase mating surfaces as shown in **Figure 23**.

4. Remove the screws (A, **Figure 24**) securing the stator plate (B, **Figure 24**, typical) to the crankcase. Remove the stator plate by removing the grommet and pulling the wires through the case opening.

5. Installation is the reverse of these steps. Note the following.

   a. Check the coil wires for chafing or other damage. Replace coil harness, if necessary.

   b. When reinstalling the stator plate, align the static timing mark on the stator plate with the crankcase mating surfaces. See **Figure 23**.

   c. Check and adjust ignition timing as described in Chapter Three.

   d. Make sure all electrical connections are tight and free from corrosion. This is absolutely necessary with electronic ignition systems.

Primary Coil
Replacement

Refer to **Figure 7**, **Figure 8** or **Figure 10** when replacing primary coils.

1. Remove the stator plate as described in this chapter.

2. Remove the Phillips screws (**Figure 25**) holding the primary coils to the stator plate.

Remove the screw from the back of the plate (**Figure 26**).

3. Some of the soldered joints (**Figure 27**) are protected with epoxy. Before separating these soldered joints, break the epoxy bond with pliers.

4. Unsolder the wires from the coils.

5. Reverse to install the new coil. Note the following:
   a. Install new coils so that their number (if marked) faces up.
   b. Resolder the wires to the new coil terminals with rosin core solder.
   c. Position the new primary coil assembly onto the stator plate.
   d. Coat the new screws with Loctite 242 (blue) and install the screws and lockwashers. Tighten securely.

CDI CONTROL UNIT (TWIN CYLINDER)

The CDI unit and ignition coils are combined into the same housing. See **Figure 7** or **Figure 8**.

Troubleshooting

Refer to Chapter Two.

Removal/Installation

1. Open the shroud.

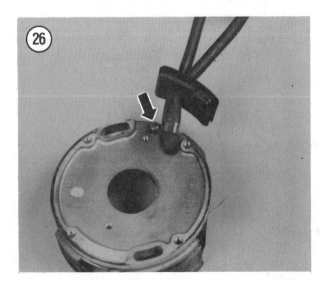

2. Disconnect the connectors at the CDI control unit.

3. Disconnect the 2 spark plug caps.

4. Remove the screws holding the CDI control unit to the recoil starter housing and remove the control unit. See **Figure 28**, typical. Remove the rubber nuts from behind the control unit.

5. Install by reversing these removal steps. Before connecting the electrical wire connectors at the unit, make sure the connectors are clean of all dirt and moisture residue. Use electrical contact cleaner to clean the connectors.

CDI BOX (600/650)

Troubleshooting

Refer to Chapter Two.

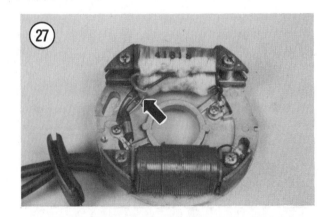

Removal/Installation

1. Open the shroud.

2. Remove the air box as described in Chapter Seven.

3. Remove the left-hand carburetor as described in Chapter Seven.

4. Disconnect the connectors at the CDI box. See **Figure 9**.

5. Remove the nuts holding the CDI to the bulkhead and remove the CDI unit. See **Figure 29**.

6. Install by reversing these removal steps. Before connecting the electrical wire connectors at the unit, make sure the connectors are clean of all

dirt and moisture residue. Use electrical contact cleaner to clean the connectors.

IGNITION COILS
(600/650)

Troubleshooting

Refer to Chapter Two.

Removal/Installation

The ignition coils are mounted on the back of the engine (**Figure 9**).

1. Open the shroud.

2. Remove the air box and carburetors as described in Chapter Seven.

3. Label and disconnect the spark plug caps at the spark plugs.

4. Disconnect the CDI-to-ignition coil electrical connector (**Figure 9**).

5. Remove the Phillips screws holding the ignition coils to the crankcase. Remove the ignition coils (**Figure 30**).

NOTE
*Spacers are used on the left-hand ignition coil between the coil and crankcase (**Figure 31**). Remove the spacers after removing the Phillips screw.*

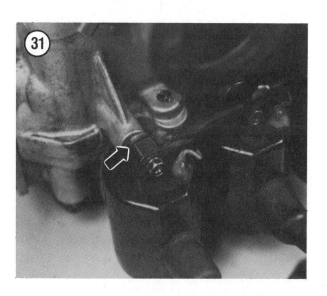

6. Installation is the reverse of these steps. Note the following:
   a. Make sure to install the 2 left-hand coil spacers (**Figure 31**).
   b. Before connecting the connector at the coil, make sure both connector halves are clean of all dirt and moisture residue. Use electrical contact cleaner to clean the connectors.

Spark Plug Caps

The spark plug caps (**Figure 32**) can be replaced by pulling the old caps off of the coil's high tension wire. Reverse to install. Make sure the cap is pushed all the way on the high tension wire.

VOLTAGE REGULATOR

The voltage regulator (**Figure 33**) is mounted at the bottom of the console.
1. Disconnect the voltage regulator wires.
2. Remove the rivets or screws securing the voltage regulator to the console.
3. Installation is the reverse of these steps. Before connecting the electrical wire connectors at the unit, make sure the connectors are clean of all dirt and moisture residue. Use electrical contact cleaner to clean the connectors.

Testing

Refer to Chapter Two.

ELECTRIC STARTING SYSTEM

An electric starter motor is available on some models. The starter motor is mounted horizontally to the front of the engine (**Figure 34**). When battery current is supplied to the starter motor, its pinion gear is thrust forward to engage the teeth on the engine ring gear. Once the engine starts, the pinion gear disengages from the flywheel.

The electric starting system requires a fully charged battery to provide the large amount of current required to operate the starter motor.

Starting system troubleshooting is described in Chapter Two.

Starter Motor

The starter motor produces a very high torque but only for a brief period of time, due to heat buildup. Never operate the starter motor continuously for more than 5 seconds. Let the motor cool for at least 15 seconds before operating it again.

If the starter motor does not turn over, check the battery and all connecting wiring for loose

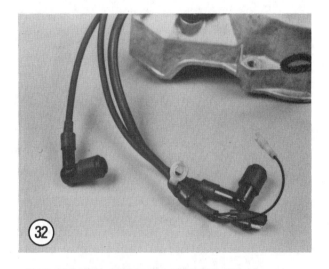

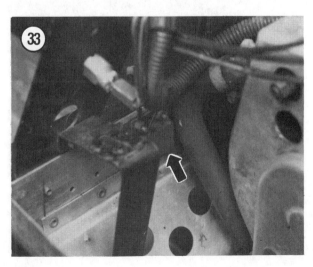

or corroded connections. If this does not solve the problem refer to Chapter Two.

Starter Motor
Removal/Installation

1. Open the shroud.
2. Disconnect the negative battery lead.
3. Remove the exhaust pipe as described in Chapter Seven.
4. Remove the nut, lockwasher and flat washer and disconnect the positive cable lead at starter motor.
5. Remove the mounting bolts securing the starter to engine and mounting bracket to engine. Remove the starter and mounting bracket.
6. Remove mounting bracket from starter.
7. Installation is the reverse of these steps. Note the following:
   a. Before installing the starter motor, check the tightness of the starter bracket bolts. Also check bracket for cracks or other damage, especially around the front bracket hole.
   b. Position the starter motor into the starter bracket and secure the starter with the 2 nuts.
   c. Attach the positive cable lead at the starter motor. Then install the flat washer,

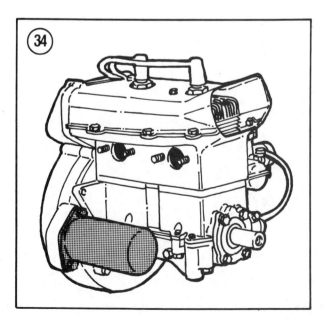

lockwasher and nut. Position the cable so that it is parallel with the starter and tighten the nut securely. Check cable routing to make sure it does not contact any moving parts or where it can become cut or damaged.
   d. Reconnect the negative battery lead.

Starter Motor
Disassembly/Reassembly

Refer to **Figure 35** for this procedure.
1. Remove the starter motor as described in this chapter.
2. Mark the front cover, center case and rear cover to aid during reassembly.
3. Remove the nuts and washers holding the starter bracket to the rear cover. Remove the starter bracket.
4. Loosen and remove the 2 case through bolts.
5. Remove the rear cover.
6. Lightly tap on the front cover with a rubber mallet until it breaks free of the starter housing.

NOTE
Step 7 describes removal of the pinion gear assembly. However, the pinion gear does not require removal unless it is damaged or requires replacement. The clip (Figure 36) securing the pinion gear assembly is difficult to remove and may tax your patience a bit. It is best to have an assistant on hand to help with its removal. Because the clip will be damaged during removal, purchase a new clip from a Polaris dealer before removing the pinion gear.

7. If necessary, remove the pinion gear as follows:
   a. Support the front cover in a vise so that the armature is positioned vertically.
   b. See **Figure 36**. Place a wrench over the pinion stopper and lightly tap the wrench downward to uncover the clip.

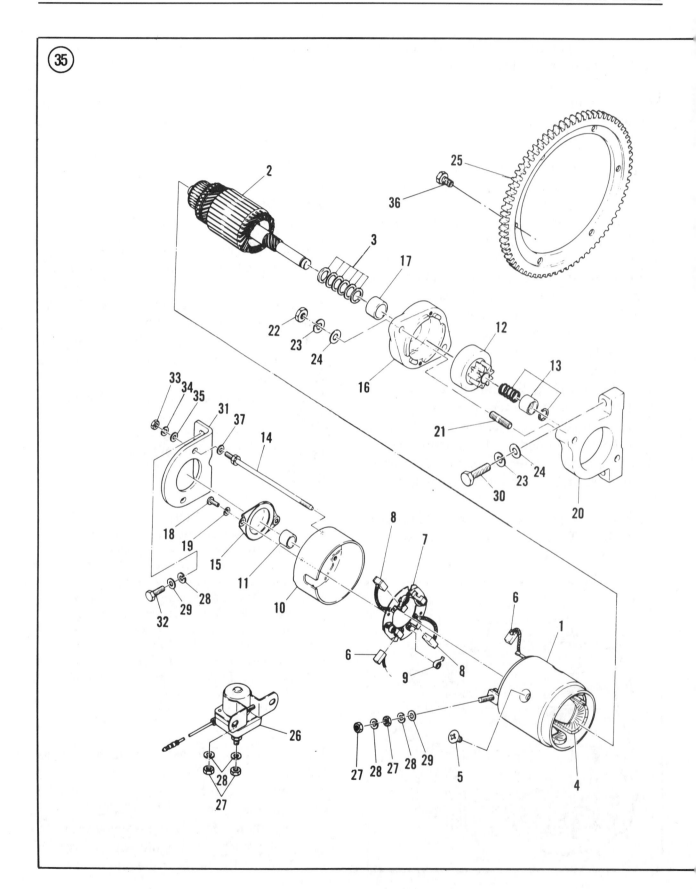

STARTER MOTOR AND SOLENOID

1. Armature housing
2. Armature
3. Washer assembly
4. Field coils
5. Screw
6. Positive brushes
7. Brush holder
8. Negative brushes
9. Brush springs
10. Rear cover
11. Rear bushing
12. Pinion gear assembly
13. Pinion stop assembly
14. Through bolt
15. Dust cover
16. Front cover
17. Front bushing
18. Screw
19. Lockwasher
20. Starter bracket
21. Stud
22. Nut
23. Lockwasher
24. Washer
25. Ring gear
26. Solenoid switch
27. Nuts
28. Lockwashers
29. Washer
30. Bolt
31. Starter bracket
32. Bolt
33. Nut
34. Lockwasher
35. Washer
36. Bolt
37. Washer

c. Hold the clip with Vise-grips as shown in **Figure 37**. Then pry the clip open with a screwdriver.
d. Remove the clip, pinion stopper and spring.
e. Before removing the pinion gear, check the clip groove for burrs that could damage the gear. Smooth the metal surface with a file.
f. Remove the pinion gear.
g. Remove the armature (**Figure 38**) and account for the washers on the armature shaft. See **Figure 35**.

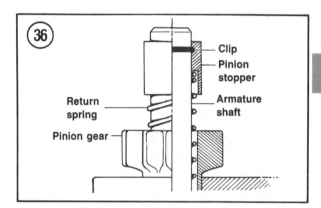

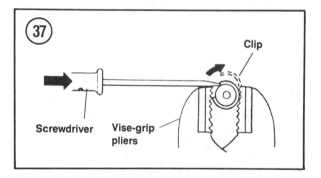

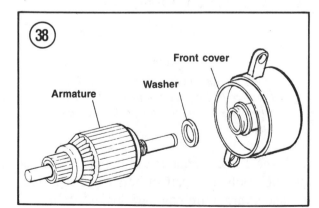

Starter Motor
Inspection

1. Clean all grease, dirt and carbon from the armature, case and end covers.

> *CAUTION*
> *Do not immerse brushes or the armature in solvent as the insulation may be damaged. Wipe the windings with a cloth lightly moistened with solvent and dry thoroughly.*

2. Check the metal thrust washers for damage. Replace if necessary.

3. Inspect the bushings for cracks, deep scoring or excessive wear. Replace the bushings if necessary.

4. Pull the brushes (**Figure 39**) out of the brush holder. Check the brush leads for fraying, cracks or other abnormal conditions. Replace all 4 brushes and brush springs if one lead is damaged.

5. Measure the length of each brush with a vernier caliper (**Figure 40**). If the length of any brush is less than 6.3 mm (1/4 in.), replace all 4 brushes and brush springs as a set. Brush replacement is described later in this section.

6. Inspect the commutator (**Figure 41**). The mica in a good commutator is below the surface of the copper bars. On a worn commutator, the mica and copper bars may be worn to the same level. See **Figure 42**. If necessary, remove buildup with a hacksaw blade, making sure that you do not cut deeper than original groove.

> *NOTE*
> *It will be necessary to grind a hacksaw blade to fit into the commutator grooves when removing buildup as described in Step 6.*

7. Inspect the commutator copper bars for discoloration. If a pair of bars are discolored, grounded armature coils are indicated.

> *NOTE*
> *An ohmmeter is required to perform the following checks.*

8. Use and ohmmeter and check the following:
   a. Switch an ohmmeter to the R × 100 scale. Touch the meter leads and zero the meter.

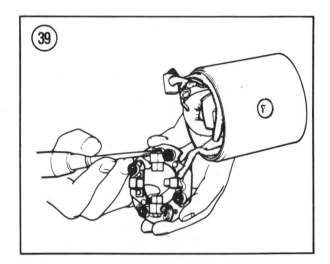

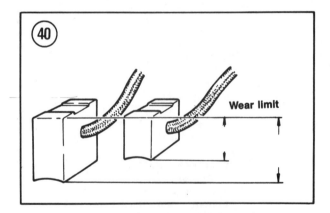

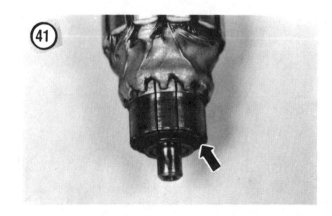

b. Check for continuity between the commutator bars and the shaft (**Figure 43**); there should be no continuity (infinite resistance).

c. Switch an ohmmeter to the R × 1 scale. Touch the meter leads and zero the meter.

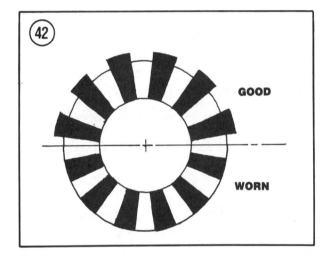

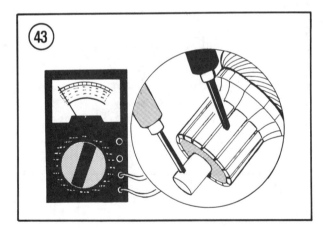

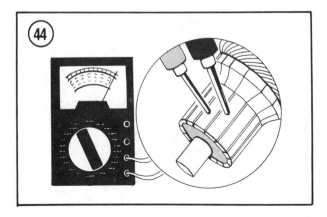

d. Check for continuity between the commutator bars (**Figure 44**); there should be continuity (indicated resistance) between pairs of bars.

e. If the armature fails either of these tests, the armature should be replaced.

Pinion Gear Inspection

CAUTION
If the pinion gear was not removed from the armature shaft, do not clean the gear in solvent as the armature insulation may be damaged.

1. Clean the pinion gear in solvent and dry thoroughly.

2. Check the pinion gear for cracks, chips, deep scoring, excessive wear or heat discoloration.

3. Check the armature shaft spiral gear for cracks, chips, excessive wear or heat discoloration.

4. Slide the pinion gear onto the armature shaft and work the gear back and forth by hand. The gear should move smoothly with no sign of roughness or binding.

5. Check the pinion spring for wear, damage or fatigue.

6. Replace worn or damaged parts as required.

Brush Replacement/Starter Assembly

1. Remove the brushes and springs from the brush holder.

2. Remove the cap screws securing the ground brushes to the end cap. Remove the brushes and brush holder from the brush plate.

3. Clean brush holder with compressed air.

4. Install new springs and brushes.

5. If the pinion gear was removed, install it as follows:

   a. Apply low-temperature grease to the front cover bushing.

   b. Install the washers (**Figure 35**) onto the armature shaft.

8

c. Install the armature into the front cover (**Figure 38**).

d. Apply low-temperature grease to the armature shaft worm gear.

e. Install the pinion gear onto the armature shaft. Then install the spring and pinion stopper.

f. Have an assistant hold the pinion stopper down while the clip is being installed in sub-step g.

g. Using pliers, open the new stopper clip to widen the clip ends (**Figure 45**).

h. Slip the clip into the groove in the armature shaft (**Figure 36**). Close the clip with pliers (**Figure 46**). Check that the clip seats in the groove completely.

6. Install the armature into the case housing and align housing marks made before disassembly.

7. Press the brushes into the holders and use a narrow strip of flexible metal or plastic as shown in **Figure 47** to keep them in place. Then align the commutator with the brushes and install the brush plate assembly. When the brushes are installed around the commutator, remove the metal or plastic strips.

8. Install the rear cover over the brush plate.

9. Install the 2 through bolts and tighten securely.

10. Install the starter bracket onto the through bolts. Secure the bracket with the flat washer, lockwasher and nut. Tighten nuts securely.

Starter Relay Replacement

The starter relay is mounted next to the battery (**Figure 2**).

1. Open the shroud.

2. Disconnect the negative battery lead, then the positive battery lead.

3. Label all wiring at the starter relay.

4. Slide off the rubber protective boots and disconnect the electrical wires from the terminals on the starter relay.

5. Disconnect the 2 small electrical wires from the starter relay.

6. Remove the screws holding the starter relay in place and remove the starter relay (**Figure 35**).

7. Replace by reversing these steps. Note the following.

8. Clean all electrical wire connectors with electrical contact cleaner.

9. Reconnect all electrical wires to the solenoid. Tighten the nuts on the large terminals securely and reposition the rubber boots.

10. Reconnect the positive battery lead, then the negative battery lead.

11. Close and secure the shroud.

Starter Relay Testing

Refer to starter testing in Chapter Two.

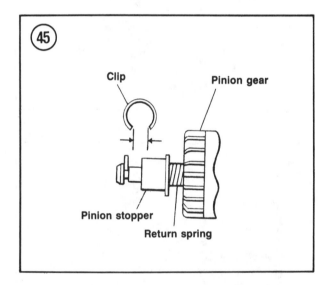

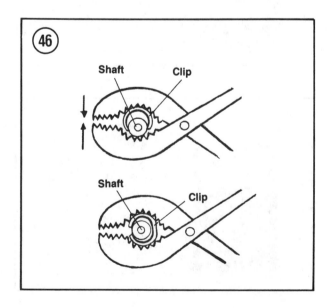

LIGHTING SYSTEM

The lighting system consists of the headlight, taillight/brakelight combination and meter illumination lights. If the event of trouble with any light, the first thing to check is the affected bulb itself. If the bulb is good, check all wiring and connections with a test light or ohmmeter.

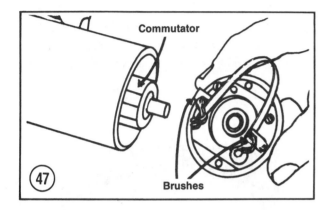

Commutator

Brushes

47

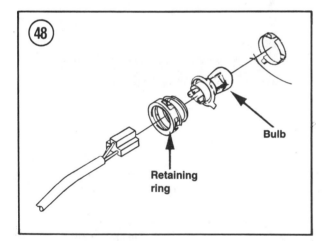

48

Bulb

Retaining ring

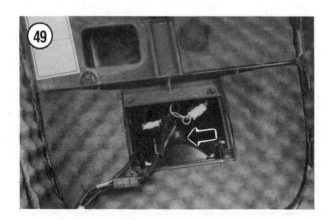

49

Headlight Bulb Replacement

The headlight on all models is equipped with a standard type bulb element. However, a quartz halogen bulb is optional from Polaris dealers when replacing the old bulb. Special handling of the quartz halogen bulb is required in order to prolong bulb life.

> *CAUTION*
> *When installing a quartz halogen bulb, do not touch the bulb glass with your fingers because of oil on your skin. Any traces of oil on the glass will drastically reduce the life of the bulb. Clean any traces of oil from the bulb with a cloth moistened in alcohol or lacquer thinner.*

> *WARNING*
> *If the headlight has just burned out or turned off, it will be **hot**. Don't touch the bulb until it cools off.*

Refer to **Figure 48** for this procedure.
1. Open the shroud.
2. Pull the rubber cover (**Figure 49**) away from the bulb connector.
3. The bulb is held in the headlight housing with a spring loaded retaining ring. Depress the retaining ring and turn it counterclockwise. When the retaining ring releases from the headlight housing, slide the ring down the wiring harness.
4. Remove the bulb (**Figure 48**) from the headlight housing with the wiring harness connector still attached. When the bulb is free of the housing, disconnect the bulb from the connector.
5. Check the electrical connector for contamination. If necessary, clean with electrical contact cleaner.
6. Make sure the retaining ring is installed on the wiring harness connector with the spring portion facing toward the headlight housing.

8

NOTE
If you are installing a quartz halogen bulb, read the information at the beginning of this procedure.

7. Reconnect the new bulb at the connector and insert the bulb into the headlight housing.

8. Align the retaining ring with the headlight housing so that the word "TOP" on the retaining ring faces up. Then depress the retaining ring and turn it clockwise to lock it into the headlight housing grooves.

9. Check headlight adjustment as described in this chapter.

Headlight Adjustment

1. Park the snowmobile on a level surface 25 ft. (8 m) from a vertical wall (**Figure 50**).

NOTE
A rider should be seated on the snowmobile when performing the following.

2. Measure the distance from the floor to the center of the headlight lens (**Figure 50**). Make a mark on the wall the same distance from the floor. For instance, if the center of the headlight lens is 2 feet above the floor, mark "A" in **Figure 50** should also be 2 feet above the floor.

3. Start the engine, turn on the headlight and set the beam selector to HIGH. Do not adjust the headlight beam with the selector set at LOW.

4. The most intense area of the beam on the wall should be 2 in. (50 mm) below the "A" mark (**Figure 51**) and in line with the imaginary vertical centerline from the headlight to the wall.

5. If the beam aim is incorrect, move the headlight up or down as required by turning the

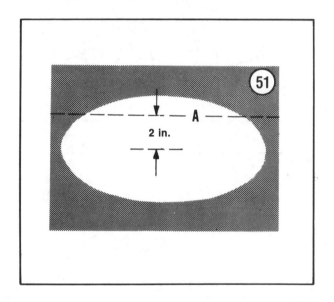

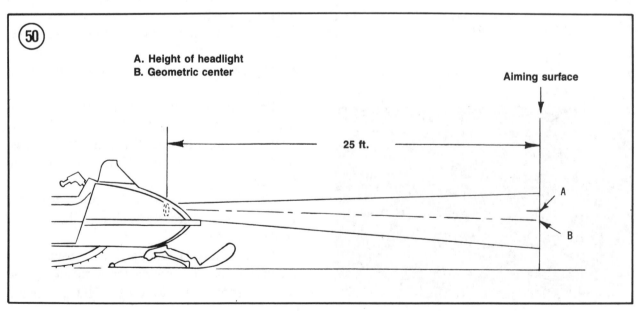

spring loaded adjuster screws in or out (A, **Figure 52**). Repeat until the adjustment is correct.

Taillight Bulb Replacement

1. Remove the taillight lens mounting screws and remove the lens and gasket.
2. Push the bulb in and turn it counterclockwise to remove it.
3. Clean the lens in a mild detergent and check for cracks.
4. Replace the lens gasket if it is torn or otherwise damaged.
5. Installation is the reverse of these steps.

METER ASSEMBLY

There are a number of different meter assemblies, depending on model.

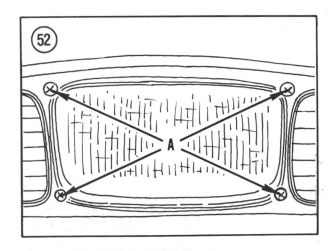

Meter Bulb Replacement

1. Open the shroud.
2. Locate the bulb socket at the affected meter assembly and pull the bulb socket out of the meter assembly. Replace the bulb.
3. Installation is the reverse of these steps.

Coolant Temperature Bulb Replacement (1984-1987 Liquid Cooled Models)

A coolant temperature bulb is mounted on the instrument panel. If the bulb should blow, replace it with the same wattage bulb.

Warning Indicator Light Replacement (1988-on)

Plug-in type indicator lights (A, **Figure 53**) are used on liquid cooled models. Depending on model, the indicator lights monitor the following circuits:

   a. Coolant temperature.
   b. Low oil level.
   c. Hi beam.
   d. Accessory.

To replace a blown indicator, pull the old indicator from the instrument panel. Reverse to install.

Meter Removal/Installation

1. Open the shroud.
2. Disconnect the electrical connectors at the affected meter.
3. Pull the bulb socket out of the meter assembly.
4. *Speedometer*: Disconnect the speedometer cable at the meter.
5. Remove the nuts securing the meter mounting bracket and remove the meter.
6. Installation is the reverse of these steps. Note the following.
7. Make sure to install a rubber cushion underneath each gauge assembly.

8

THROTTLE SAFETY SWITCH

All models are equipped with a throttle safety switch assembly that consists of 2 connected switches—off idle and idle. See **Figure 54**. For the system to operate properly, the throttle cable must be properly adjusted and both switches must have their plungers in the proper position (inward) in relation to throttle lever operation. **Figure 54** shows the position of both switches in the throttle block.

When the engine is at idle, the idle switch plunger is closed by the position of the throttle lever. The safety switch circuit is open and the engine can run.

When the engine moves from idle to off idle by the movement of the throttle lever, the idle switch plunger opens but at the same time the off idle switch plunger is closed. The safety switch circuit is still open and the engine can continue to run.

If the throttle lever is released and the carburetor slides (or cable) do not return to their closed position, the engine will turn off.

If the engine should stop when the throttle lever is released, perform the following:

1. Turn the ignition switch OFF.

2. Open the shroud and check the throttle cables for breakage or other abnormal conditions. If the throttle cables are okay, check the throttle cable entry into the carburetors. If these parts appear okay, check throttle cable operation by operating the throttle lever. The slides should open and close smoothly and the throttle lever should return to its closed position.

WARNING
Do not start the engine if the throttle lever fails to work properly.

3. When you are sure the throttle lever is operating correctly, attempt to restart the engine. If the engine will not start, perform the *Throttle Lever Free Play Adjustment* in Chapter Three. If it appears the free play is correct, test the safety switches as described in the following procedure.

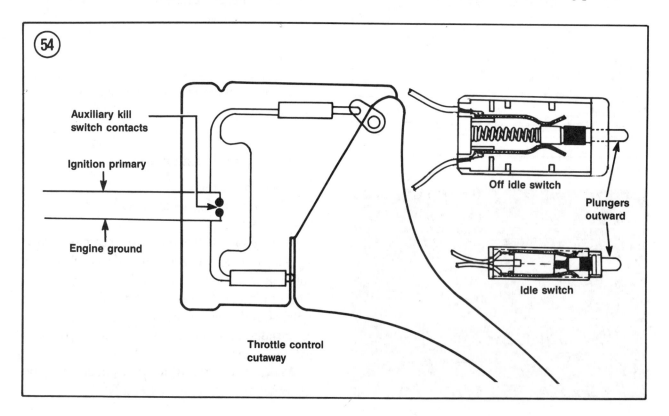

Auxiliary kill switch contacts

Ignition primary

Engine ground

Off idle switch

Plungers outward

Idle switch

Throttle control cutaway

Safety Switch Testing

1. Check and adjust the throttle cable assembly as described in Chapter Three. When the throttle cable is properly adjusted, proceed with Step 2.
2. Disconnect the safety switch 2-prong connector.
3. Set an ohmmeter to the R × 1 scale.
4. Connect the ohmmeter to the 2 terminals on the safety switch connector. Test the run circuit as follows:
   a. Set the kill switch (A, **Figure 55**) to its ON position (button pulled up).
   b. With the throttle lever (B, **Figure 55**) in the idle position, the ohmmeter should read infinity (high resistance).
   c. Slowly operate the throttle lever (B, **Figure 55**) and watch the ohmmeter; it should continue to show infinity as the throttle lever is moved from the idle to full open position.
   d. If the ohmmeter read infinity for sub-steps b and c, the safety switch open circuit is operating properly. Proceed to Step 5. If the ohmmeter needle shows continuity (low resistance) or if the needle fluctuated when performing sub-step c, replace the safety switch assembly.
5. Connect the ohmmeter to the 2 terminals on the safety switch connector. Test the kill circuit as follows:

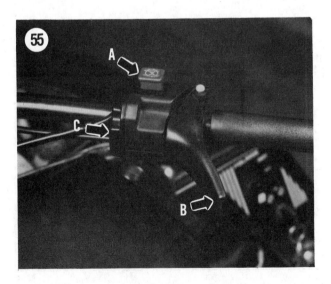

a. Set the kill switch (A, **Figure 55**) to its ON position (button pulled up).
   b. Pull the throttle lever away from the throttle block so that the plunger on both safety switches is out (**Figure 54**).
   c. The ohmmeter should read 0 ohms when the safety switch plungers are out.
   d. If the ohmmeter shows a high resistance reading when performing sub-step b or c, replace the safety switch assembly.
   e. Release the throttle lever assembly so that it rests in its idle position (B, **Figure 55**).
   f. If the throttle safety switch tested properly, perform Step 6.
6. Connect the ohmmeter to the 2 terminals on the safety switch connector. Test the kill button as follows:
   a. The ohmmeter should read 0 ohms when the kill switch (A, **Figure 55**) is in its ON position (button up).
   b. The ohmmeter should read infinity (high resistance) when the kill switch (A, **Figure 55**) is in its OFF position (button down).
   c. Replace the safety switch assembly if it failed to provide the readings in sub-step a or b.
7. Reconnect the safety switch 2-prong connector.

Safety Switch Replacement

The safety switch assembly is mounted at the back of the throttle block assembly (C, **Figure 55**).
1. Disconnect the safety switch 2-prong connector.
2. Remove the screws securing the cover to the back of the throttle block assembly. Slide the cover down the handlebar.
3. Remove the screws securing the 2 throttle switches to the throttle block. Then remove the 3 switches from the throttle block.
4. Secure the 2 safety switches with screws and install the kill switch into the throttle block assembly.

8

5. Install the cover to the back of the throttle block assembly.

6. Reconnect the safety switch 2-prong connector.

7. Check the safety switch operation as described in the following procedure.

Safety Switch Testing

The following procedure should be checked daily or after replacing the safety switch assembly.

1. Start the engine and allow to idle.

2. See **Figure 56**. Hold the throttle lever pin stationary by pushing it in the direction shown in **Figure 56**. Then open the throttle lever slightly to increase engine RPM; the engine should turn off.

3. If the engine should continue to run after performing Step 2, the safety switches are not operating properly. Test them as described in this chapter.

SWITCHES
(EXCEPT SAFETY SWITCH)

Switches can be tested for continuity with an ohmmeter (see Chapter One) or a test light at the switch connector plug by operating the switch in each of its operating positions and comparing results with the switch operation.

When testing switches, note the following:

a. When separating 2 connectors, pull on the connector housings and not the wires.

b. After locating a defective circuit, check the connectors to make sure they are clean and properly connected. Check all wires going into a connector housing to make sure each wire is properly positioned and that the wire end is not loose.

c. When joining connectors, push them together until they click into place.

d. When replacing handlebar switch assemblies, make sure the cables are routed correctly so that they are not crimped when the handlebar is turned from side-to-side.

**Handlebar Heater Switch
Replacement**

Refer to **Figure 57** for this procedure.

1. Open the shroud.

2. Disconnect the electrical connector at the toggle switch.

3. Unscrew the toggle switch nut from the toggle switch. Remove the nut and washer.

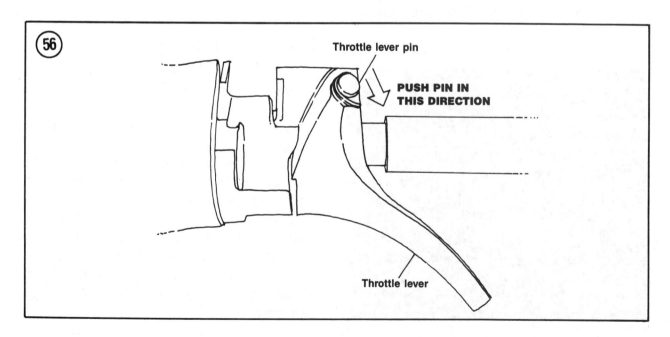

(56)

Throttle lever pin

**PUSH PIN IN
THIS DIRECTION**

Throttle lever

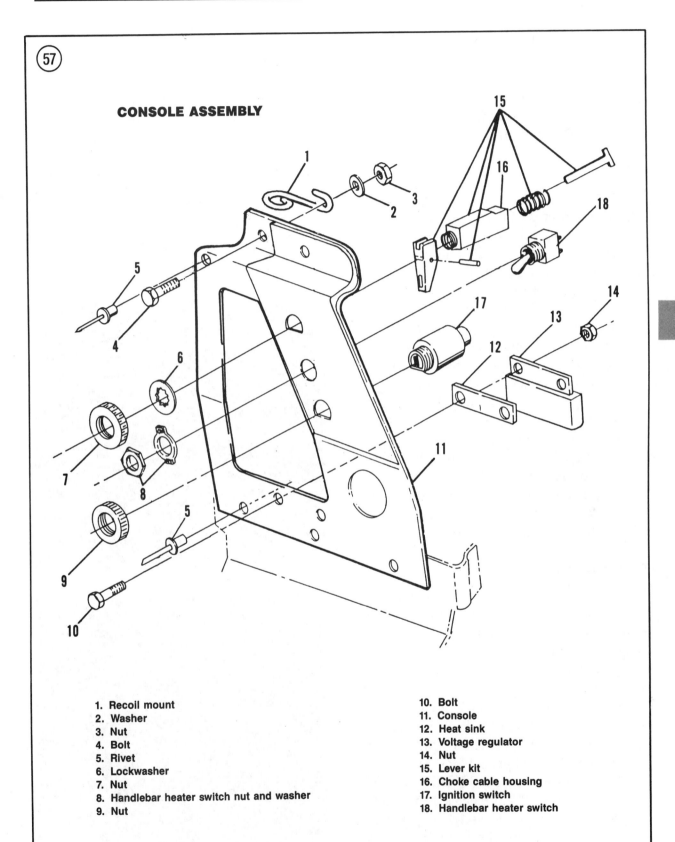

CONSOLE ASSEMBLY

1. Recoil mount
2. Washer
3. Nut
4. Bolt
5. Rivet
6. Lockwasher
7. Nut
8. Handlebar heater switch nut and washer
9. Nut
10. Bolt
11. Console
12. Heat sink
13. Voltage regulator
14. Nut
15. Lever kit
16. Choke cable housing
17. Ignition switch
18. Handlebar heater switch

4. Remove the toggle switch from the rear of the console.

5. Installation is the reverse of these steps. When installing the toggle switch washer (**Figure 57**), install it so that the word "ON" faces up when you are looking at the front of the console.

Ignition Switch
Testing/Replacement

Refer to **Figure 57** for this procedure.

1. Open the shroud.

2. Disconnect the electrical connector at the back of the ignition switch (**Figure 58**).

3. Test the ignition switch as follows:
  a. Set an ohmmeter to the R × 1 scale.
  b. Connect the ohmmeter leads to the 2 ignition switch terminals.
  c. Turn the ignition switch to the ON position. The ohmmeter should read infinity.
  d. If the ignition switch did not provide the reading as specified in sub-step c, the switch is shorted and must be replaced.

4. Replace the ignition switch as follows:
  a. Remove the key from the ignition switch. Then remove the nut from the front of the console and remove the switch.
  b. Installation is the reverse of these steps.

Dimmer Switch
Testing/Replacement

The dimmer switch (**Figure 59**) is mounted on the left-hand handlebar.

1. Open the shroud.

2. Disconnect the dimmer switch electrical connector.

3. Test the dimmer switch as follows:
  a. Set an ohmmeter to the R × 1 scale.
  b. Connect the ohmmeter leads to the dimmer switch yellow and green wires. Switch the dimmer switch to its LOW position. The ohmmeter should read 0 ohms. Now switch the dimmer switch to its HIGH position. The ohmmeter should show infinity (high resistance).

  c. Connect the ohmmeter leads between the yellow and red wires. Switch the dimmer switch to its HIGH position. The ohmmeter should read 0 ohms. Now switch the dimmer switch to its LOW position. The ohmmeter should show infinity (high resistance).

  d. If the dimmer switch does not provide the readings as specified in sub-step b and c, replace it as follows.

4. Replace the dimmer switch as follows:
  a. Remove any strap holding the dimmer switch wire harness to the handlebar.

b. Disconnect the clamp holding the dimmer switch (**Figure 59**) to the handlebar and remove the dimmer switch.

c. Reverse to install the dimmer switch. Make sure to route the dimmer switch wiring harness to prevent damage. Secure the wire harness to the handlebar with a cable tie. Spray the mating connector halves with electrical contact cleaner before reassembly.

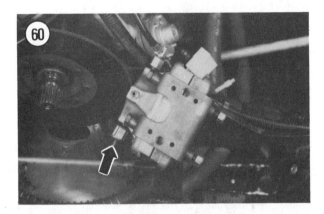

Brake Switch
Replacement

The brake switch (**Figure 60**) is mounted on the brake caliper.

1. Open the shroud.

> *NOTE*
> *On some liquid cooled models, it may be necessary to remove the brake caliper to gain access to the brake switch. Refer to Chapter Thirteen.*

2. Disconnect the electrical connector at the brake switch.

3. Loosen and remove the brake switch.

4. Installation is the reverse of these steps. Check brake switch operation with the engine running and applying the brake lever.

> *WARNING*
> *Do not ride the snowmobile until the rear brake light is operating correctly.*

Kill Switch

Refer to *Throttle Safety Switch* in this chapter.

Thermoswitch
Testing/Replacement

The thermoswitch is mounted in the cylinder head on 400 and 500 models (**Figure 61**) and in the water manifold on 600/650 models (**Figure 62**).

1. Open the shroud.

2. Disconnect the electrical connector at the thermoswitch.

3. Test the thermoswitch as follows:

   a. Switch an ohmmeter to the R × 1 scale.

   b. Connect one ohmmeter lead to the thermoswitch terminal pin and the other ohmmeter lead to a good engine ground. The ohmmeter should read infinity (high resistance).

8

c. If the thermoswitch does not provide the reading as specified in sub-step b, replace the switch.

3. Replace the thermoswitch as follows:
a. Drain the cooling system as described in Chapter Three.
b. Remove and replace the thermoswitch. See **Figure 61** or **Figure 62**.
c. Refill and bleed the cooling system as described in Chapter Three.

4. Reconnect the electrical connector at the thermoswitch.

OIL LEVEL SENSOR

Some models are equipped with an oil level sensor mounted in the oil tank (**Figure 63**). To replace the sensor, open the shroud and disconnect the oil level sensor electrical connector. Pull the oil sensor out of the oil tank and remove it. Reverse to install.

CIRCUIT BREAKER

Models with electric start are equipped with a circuit breaker.

WIRING DIAGRAMS

Wiring diagrams are located at the end of this book.

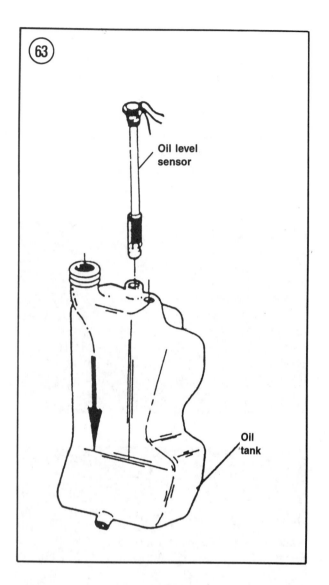

Table 1 BATTERY

Battery capacity	12 volt, 18 amp hour

Table 2 STATE OF CHARGE

Specific Gravity	State of Charge
1.110-1.130	Discharged
1.140-1.160	Almost discharged
1.170-1.190	One-quarter charged
1.200-1.220	One-half charged
1.230-1.250	Three-quarters charged
1.260-1.280	Fully charged

Chapter Nine

Oil Injection System

The fuel:oil ratio required by snowmobile engines depends upon engine demand. Without oil injection, oil must be hand-mixed with gasoline to assure that sufficient lubrication is provided at all operating speeds and engine load conditions. This ratio is adequate for high-speed operation, but contains more oil than required to lubricate the engine properly during idle.

With oil injection, the amount of oil provided with the fuel sent to the engine cylinders can be varied instantly and accurately to provide the optimum ratio for proper engine lubrication at any operating speed or engine load condition.

All models are equipped with an oil injection system. The system consists of a mechanical gear-driven pump, external oil tank, oil injection hoses and throttle/oil pump cable assembly. A junction box connects the throttle and pump cable so that they operate simultaneously.

This chapter covers complete oil injection system service.

SYSTEM COMPONENTS

The oil injection pump is mounted on the lower case half at the back of the engine. See **Figure 1**, typical. The oil pump is connected to the throttle by a cable. An oil reservoir tank is mounted in the engine compartment (**Figure 2**). Oil injection hoses connect the oil tank to the pump and connect the pump to the engine.

OIL PUMP SERVICE

Oil Pump Bleeding

The oil pump must be bled whenever one of the following conditions have been met:
  a. The oil tank ran empty.
  b. When any one of the oil injection hoses were disconnected.
  c. The machine was turned on its side.
  d. Pre-delivery service.

1. Check that the oil tank is full. See Chapter Three.

2. Check that all hoses are connected to the oil pump, oil reservoir tank and engine.

> *NOTE*
> *Place a shop rag underneath the oil pump when performing Step 3.*

3. Loosen the brass screw (**Figure 3**) on top of the oil pump and leave in this position until oil begins to flow from underneath the bolt head. When oil starts to flow, tighten the brass screw securely.

4. Remove the rag from underneath the oil pump and check for leaks.

Synchronizing Oil Pump

For proper operation, the oil pump must be synchronized with the carburetors. Synchronization is brought about by adjusting the oil pump cable at the oil pump. Perform the following:

1. Open and secure the shroud.

2. Check and adjust carburetor synchronization as described in Chapter Three.

3. Check and adjust engine idle speed as described in Chapter Three.

> *NOTE*
> *Step 4 must be performed with the engine OFF.*

4. With the throttle lever in its idle position, check the index mark on the pump lever (A, **Figure 4**) with the index mark on the pump housing (B, **Figure 4**); both marks should align. If the marks do not align, loosen the oil pump cable locknuts and adjust the oil pump cable until the index marks align. Tighten the cable locknuts and recheck the adjustment.

Oil Pump Operational Check

The following procedure should be checked whenever the engine is reassembled or when faulty oil pump operation is suspected.

1. To assure adequate protection to the engine when performing this procedure, the fuel should be pre-mixed at a 40:1 ratio.

> *NOTE*
> *Do not continue to use a 40:1 pre-mix mixture after checking the oil pump unless the engine is under break-in. The use of a 40:1 pre-mix mixture under normal operating conditions will lead to spark plug fouling and rapid carbon*

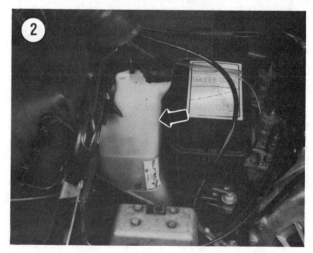

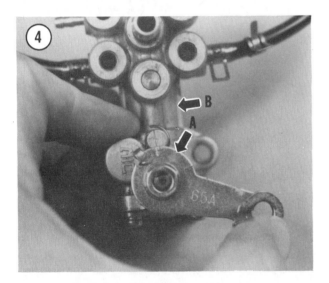

*build-up. Refer to **Correct Fuel Mixing** in Chapter Three for information on pre-mixing fuel.*

2. Bleed the oil pump as described in this chapter.

NOTE
The following procedure shows the engine removed from the frame for clarity. The engine must be installed and in running order when performing this procedure.

3. Disconnect the oil lines at the cylinder blocks (**Figure 5**) as follows. Refer to **Figures 6-12** for your model when performing the following.
4. Remove the banjo bolts and washers and disconnect the oil lines or disconnect the oil line at the fitting on the cylinder block.
5. Check the banjo bolts and oil lines for debris or other contamination.
6. On models with banjo bolts, screw the banjo bolts back into the cylinder blocks loosely. On models without banjo bolts, plug the oil hose fittings on the cylinder blocks. On all models, allow the oil hoses to hang loosely so that you can see the end of the check valves or hoses when performing the following.

WARNING
***Never** lean into the snowmobile's engine compartment while wearing a scarf or other loose clothing when the engine is running or when attempting to start the engine. If the scarf or clothing should catch in the drive belt or clutch, severe injury or death could occur. Make sure the belt guard is in place.*

7. Raise and support the rear of the snowmobile so that the track is clear of the ground and the skis are placed against a wall or other immovable barrier.

9

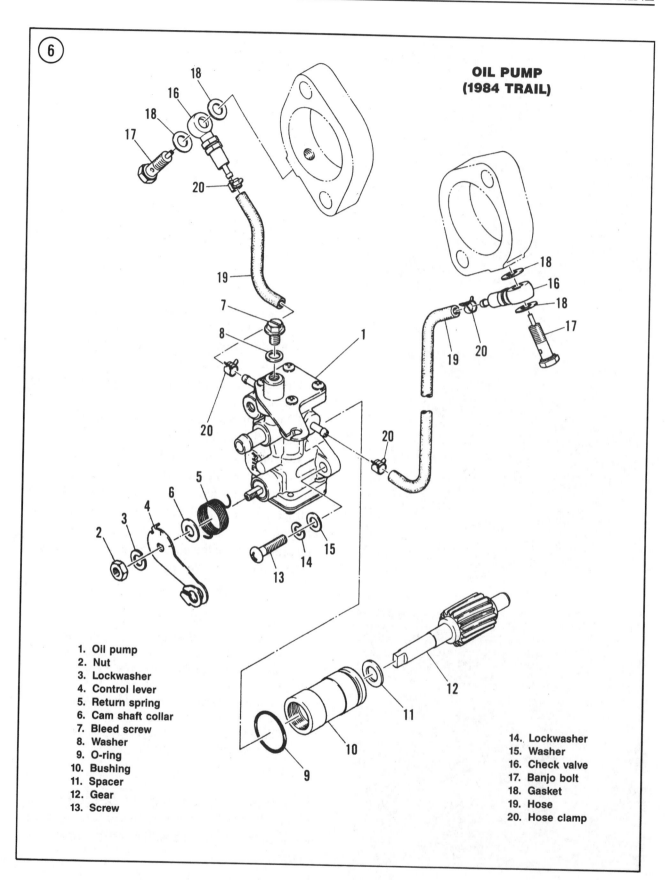

**OIL PUMP
(1984 TRAIL)**

1. Oil pump
2. Nut
3. Lockwasher
4. Control lever
5. Return spring
6. Cam shaft collar
7. Bleed screw
8. Washer
9. O-ring
10. Bushing
11. Spacer
12. Gear
13. Screw
14. Lockwasher
15. Washer
16. Check valve
17. Banjo bolt
18. Gasket
19. Hose
20. Hose clamp

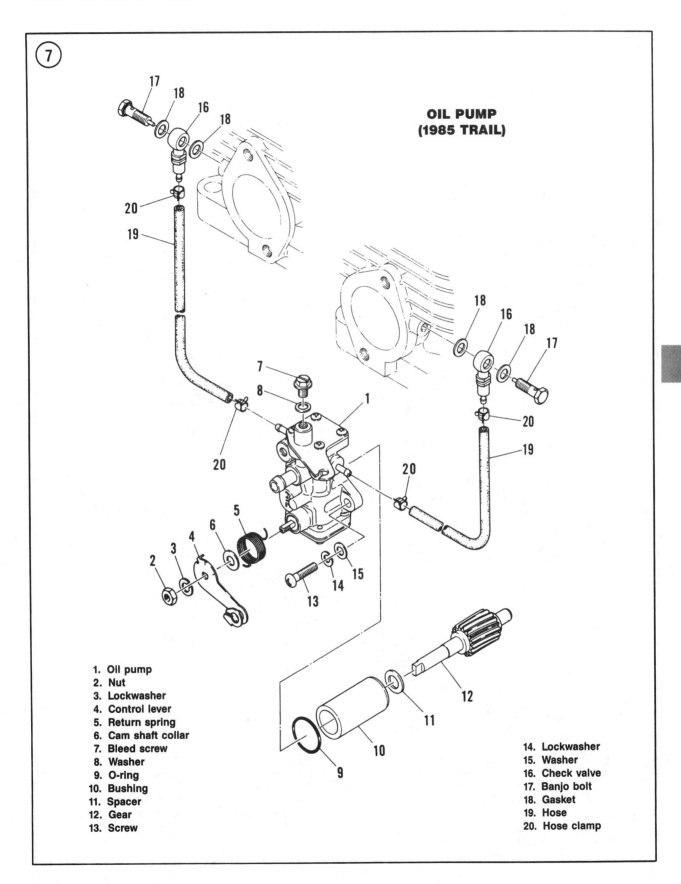

**OIL PUMP
(1985 TRAIL)**

1. Oil pump
2. Nut
3. Lockwasher
4. Control lever
5. Return spring
6. Cam shaft collar
7. Bleed screw
8. Washer
9. O-ring
10. Bushing
11. Spacer
12. Gear
13. Screw
14. Lockwasher
15. Washer
16. Check valve
17. Banjo bolt
18. Gasket
19. Hose
20. Hose clamp

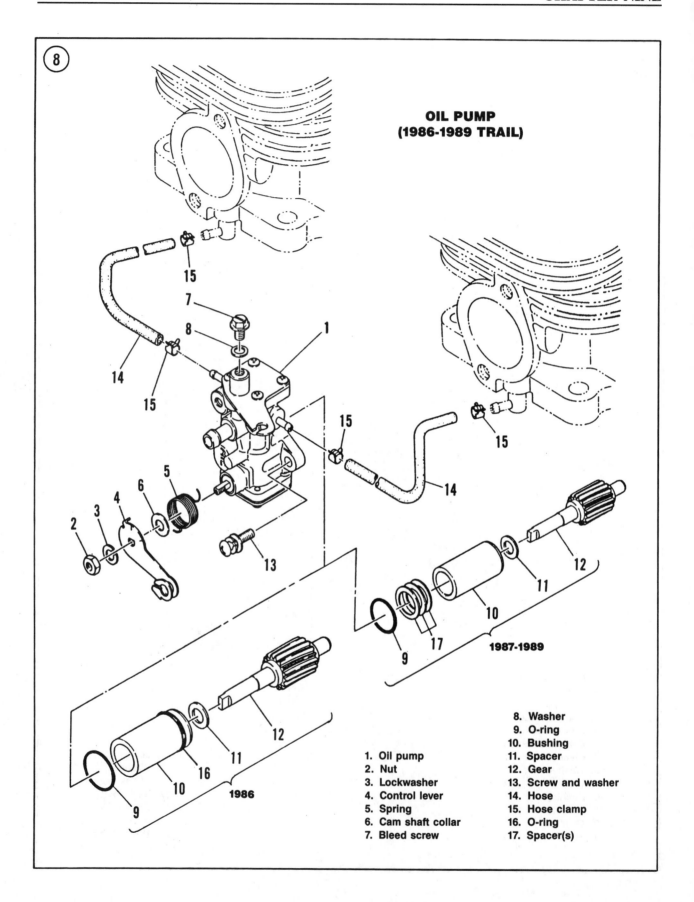

⑧

**OIL PUMP
(1986-1989 TRAIL)**

1987-1989

1986

1. Oil pump
2. Nut
3. Lockwasher
4. Control lever
5. Spring
6. Cam shaft collar
7. Bleed screw
8. Washer
9. O-ring
10. Bushing
11. Spacer
12. Gear
13. Screw and washer
14. Hose
15. Hose clamp
16. O-ring
17. Spacer(s)

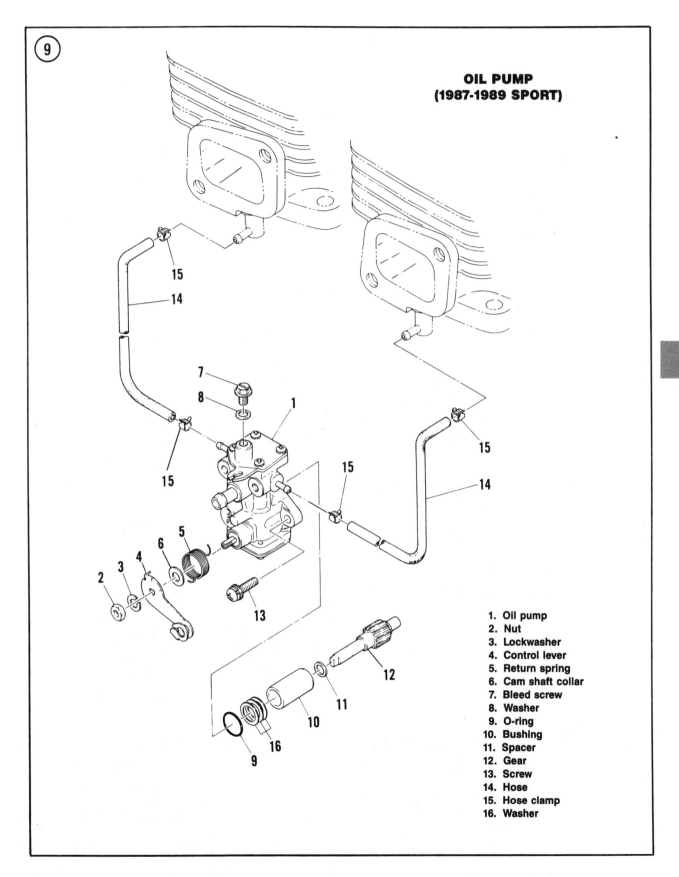

**OIL PUMP
(1987-1989 SPORT)**

1. Oil pump
2. Nut
3. Lockwasher
4. Control lever
5. Return spring
6. Cam shaft collar
7. Bleed screw
8. Washer
9. O-ring
10. Bushing
11. Spacer
12. Gear
13. Screw
14. Hose
15. Hose clamp
16. Washer

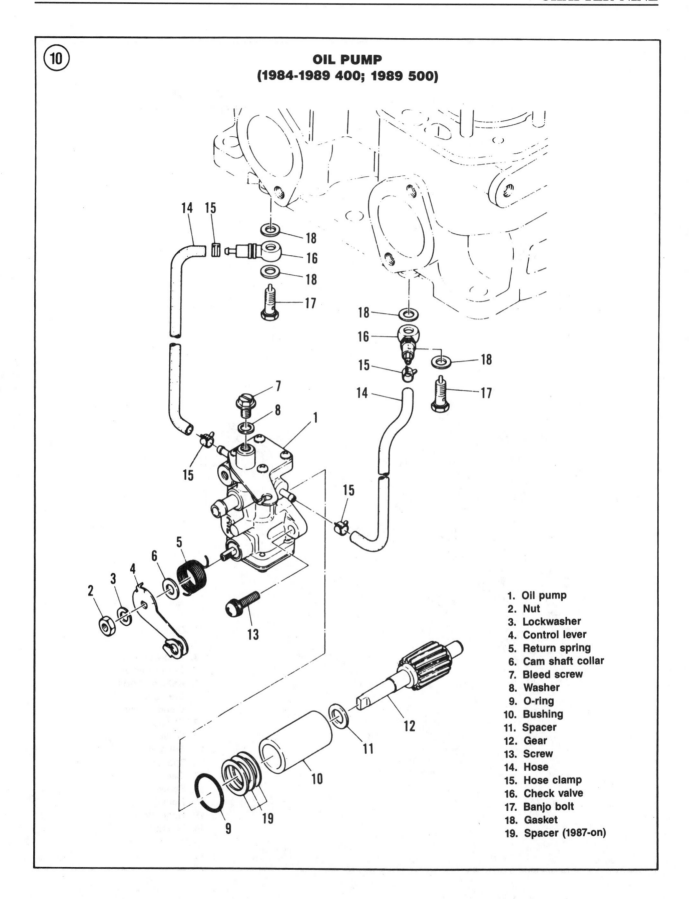

⑩

OIL PUMP
(1984-1989 400; 1989 500)

1. Oil pump
2. Nut
3. Lockwasher
4. Control lever
5. Return spring
6. Cam shaft collar
7. Bleed screw
8. Washer
9. O-ring
10. Bushing
11. Spacer
12. Gear
13. Screw
14. Hose
15. Hose clamp
16. Check valve
17. Banjo bolt
18. Gasket
19. Spacer (1987-on)

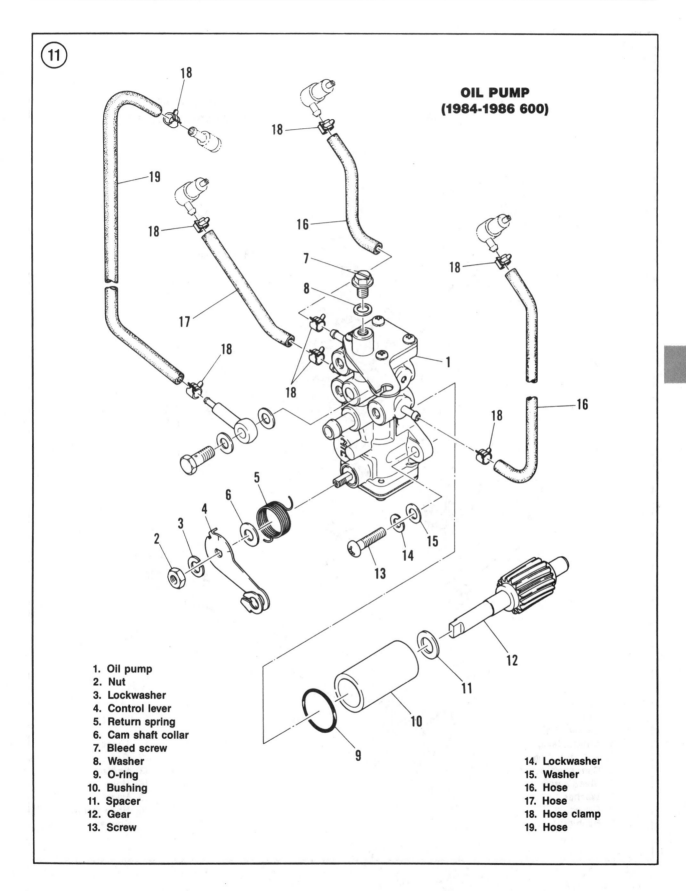

11

OIL PUMP
(1984-1986 600)

9

1. Oil pump
2. Nut
3. Lockwasher
4. Control lever
5. Return spring
6. Cam shaft collar
7. Bleed screw
8. Washer
9. O-ring
10. Bushing
11. Spacer
12. Gear
13. Screw
14. Lockwasher
15. Washer
16. Hose
17. Hose
18. Hose clamp
19. Hose

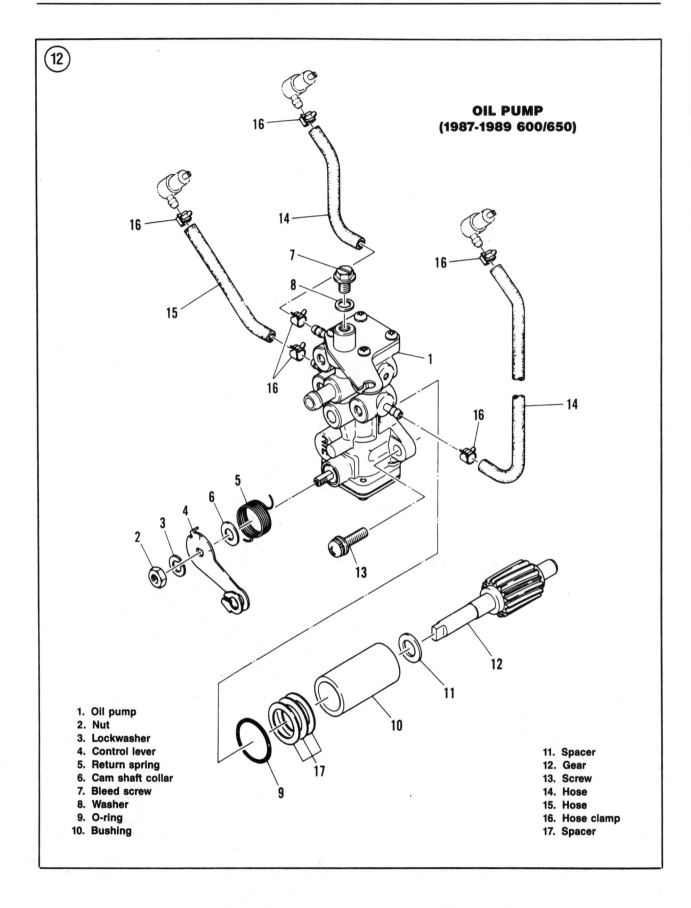

⑫

OIL PUMP
(1987-1989 600/650)

1. Oil pump
2. Nut
3. Lockwasher
4. Control lever
5. Return spring
6. Cam shaft collar
7. Bleed screw
8. Washer
9. O-ring
10. Bushing
11. Spacer
12. Gear
13. Screw
14. Hose
15. Hose
16. Hose clamp
17. Spacer

NOTE
When performing Step 8, it may be easier to use a piece of wire with a hook on one end to hold the control arm in its fully open position.

8. Start the engine and allow to idle. Then hold the pump control arm (A, **Figure 4**) in the fully open position. Drops of oil should be visible from the check valves or hoses after the engine has idled for approximately 1-2 minutes. If oil does not drip from the check valve, the check valve may be defective. Remove the check valve from the hose and recheck. If oil does not drip from the hose (with check valve removed), check for one of the following conditions:
   a. Plugged oil filter.
   b. Leaking or damaged oil feed lines.
   c. Defective oil pump.
9. Reinstall the check valves, banjo bolts and washers when you are sure the oil pump is operating correctly. On all other models, reconnect the oil hose to the cylinder block fitting and secure it with the hose clamp.
10. Drain the gas tank of the 40:1 mixture and refill with straight gas.

COMPONENT REPLACEMENT

Oil Reservoir Tank
Removal/Installation

Refer to **Figure 13** (typical) for this procedure.
1. Open the shroud.
2. Remove the air box, if necessary.
3. Label the hoses at the tank before removal.

NOTE
To prevent the oil tank from leaking, purchase a length of hose with the same ID as that used at the oil tank outlet port. Block one end of the hose with a bolt. The hose can then be used to plug the outlet port when the original hose is disconnected.

4. Disconnect the vent hoses at the oil tank.

5. Remove the oil level gauge from the oil tank, if used.
6. Remove the oil tank mounting bolts.
7. Lift the oil tank (**Figure 14**) up slightly and disconnect the hoses at the tank. Plug the outlet ports to prevent oil leakage.
8. Installation is the reverse of these steps. Note the following.
9. Bleed the oil pump as described in this chapter.

Oil Level Sensor

Refer to Chapter Eight.

Oil Hoses

Fresh oil hoses should be installed whenever the old hoses become hard and brittle. When replacing damaged or worn oil hoses, make sure to install transparent hoses with the correct ID. Non-transparent hoses will not allow you to visually inspect the hoses for air pockets (**Figure 15**) or other blockage that could cause engine seizure. When reconnecting hoses, secure each hose end with a clamp.

Oil Pump
Removal

Refer to **Figures 6-12** for your model when performing this procedure.
1. If the engine is installed in the frame, perform the following:
   a. Remove the carburetors as described in Chapter Seven.
   b. Disconnect the oil pump cable at the oil pump.
   c. Disconnect the hoses at the oil pump. Plug the oil supply hose to prevent oil leakage.

NOTE
The following steps are shown with the engine partially disassembled for clarity. Engine disassembly is not required to remove and install the oil pump and pump shaft.

9

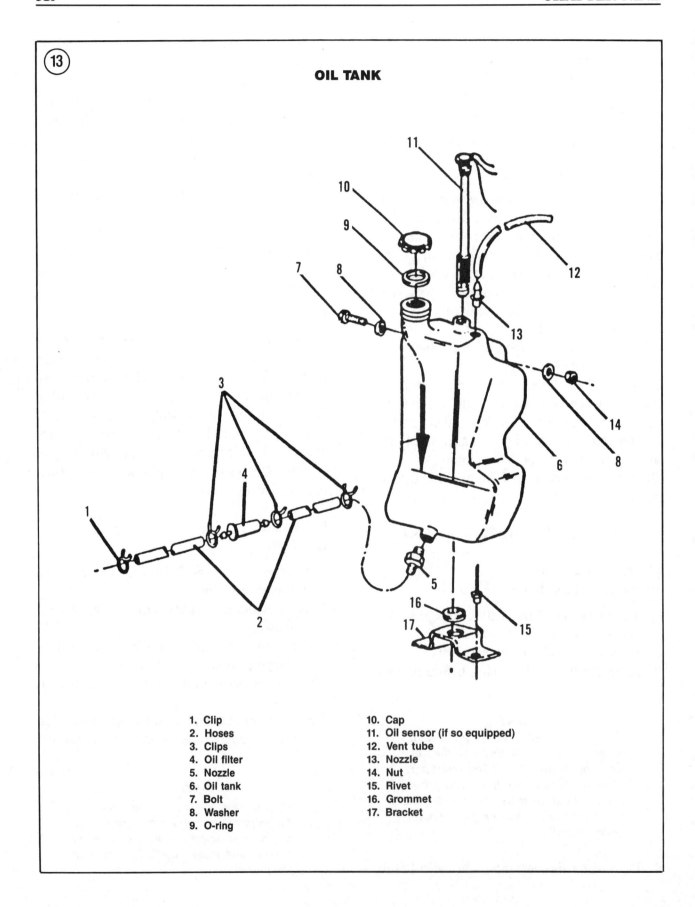

OIL TANK

1. Clip
2. Hoses
3. Clips
4. Oil filter
5. Nozzle
6. Oil tank
7. Bolt
8. Washer
9. O-ring
10. Cap
11. Oil sensor (if so equipped)
12. Vent tube
13. Nozzle
14. Nut
15. Rivet
16. Grommet
17. Bracket

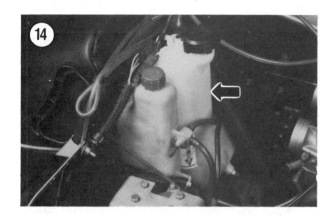

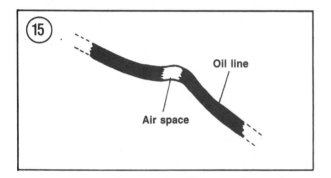

Oil line

Air space

2. Remove the screws and lockwashers holding the oil pump (**Figure 16**) to the crankcase. Remove the oil pump.

3. Remove the spacers (**Figure 17**), if used.

4. Remove the bushing (**Figure 18**) from the crankcase.

5. Remove the spacer (**Figure 19**) and the oil pump shaft (**Figure 20**).

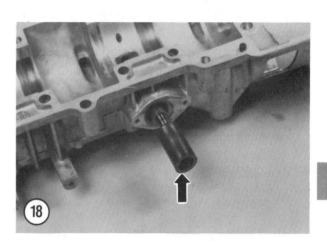

9

Inspection

1. The oil pump (A, **Figure 21**) is not rebuildable. If damaged, it must be replaced. The oil hoses, check valves, banjo bolts and washers are replaceable. In addition, the oil pump lever and spring assembly (B, **Figure 21**) are replaceable.

2. Check the O-ring on the back of the oil pump (**Figure 22**) for wear or damage; replace if necessary.

3. Check the oil pump shaft and gear (**Figure 23**) for cracks, excessive wear or gear damage. Replace if necessary.

NOTE
*If the gear is damaged, the mating gear on the crankshaft may also be damaged. Refer to **Crankshaft** in Chapter Five or Chapter Six.*

4. Check the bushing, washer(s) and spacer for cracks or damage. Replace if necessary.

5. If oil supply hoses are contaminated, remove check valves (if used) and flush valves and hoses.

Installation

1. Oil all parts with engine oil before installation.

2. Slide the washer (C, **Figure 24**) onto the pump shaft (D, **Figure 24**) and insert the shaft into the crankcase. See **Figure 20**. If the crankshaft is installed in the case half, turn the pump shaft as required to engage it with the crankshaft gear.

3. Slide the bushing over the pump shaft (**Figure 18**).

4. Install the washer(s) over the pump shaft (**Figure 17**).

5. See **Figure 25**. Align the notch in the end of the oil pump (A) with the end of the pump shaft (B) and install the oil pump assembly. If the oil pump does not fit into the crankcase easily, the

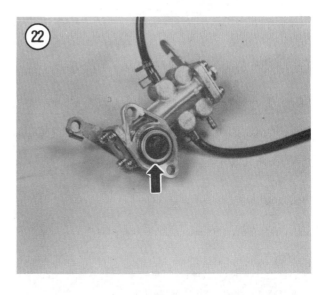

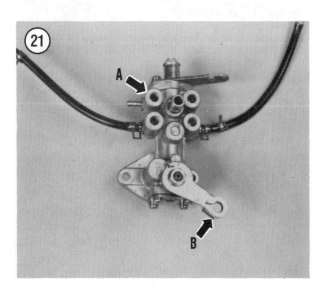

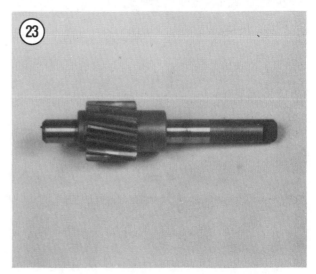

shaft and oil pump are not properly aligned. When the pump and shaft are aligned, install the pump screws and washers and tighten securely.

6. If the engine is installed in the frame, note the following:

a. Connect the oil supply hose at the oil pump.

b. Connect the 2 or 3 small oil hoses at the oil pump and secure with clamps.

c. Connect the oil pump cable at the oil pump.

d. Bleed the oil pump as described in this chapter.

e. Install the carburetors as described in Chapter Seven.

f. Adjust the carburetors as described in Chapter Three, then synchronize oil pump as described in this chapter.

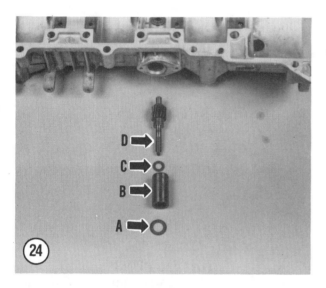

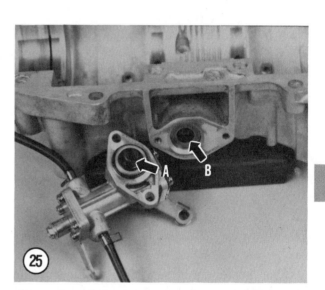

9

Chapter Ten

Liquid Cooling System

This chapter covers service procedures for the water pump, drive belt, connecting hoses and heat exchangers.

Cooling system flushing procedures are provided in Chapter Three.

The cooling system is a closed system. During operation, the coolant heats up and expands, thus pressurizing the system.

The liquid cooling system consists of a pressure cap, recovery tank, water pump, inlet and outlet manifolds, heat exchangers and cylinder heads. See **Figures 1-3**.

> *WARNING*
> *Do not remove the pressure cap (**Figure 4**) when the engine is hot. The coolant is very hot and is under pressure. Severe scalding could result if the coolant comes in contact with your skin.*

The cooling system must be allowed to cool prior to removing any component of the system.

WATER PUMP (400 AND 500)

Refer to **Figure 5** when performing procedures in this section.

Removal

1. Drain the cooling system as described in Chapter Three.
2. Disconnect the 2 hoses (**Figure 6**) at the water pump.
3. Remove the recoil starter housing as described in Chapter Eleven.
4. Remove the bolts holding the water pump onto the engine assembly. Remove the water pump (**Figure 7**) and drive belt.
5. If necessary, remove the lower pulley as follows:
    a. Hold the starter pulley with a holder and loosen the water pump pulley bolts.
    b. Remove the starter pulley (**Figure 8**) and the water pump pulley (**Figure 9**).

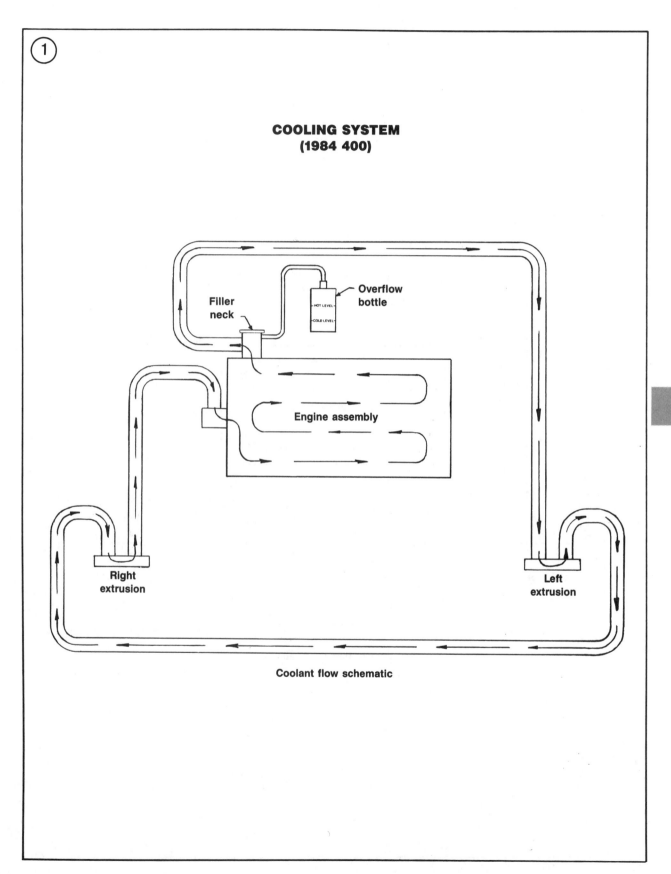

**COOLING SYSTEM
(1984 400)**

Coolant flow schematic

10

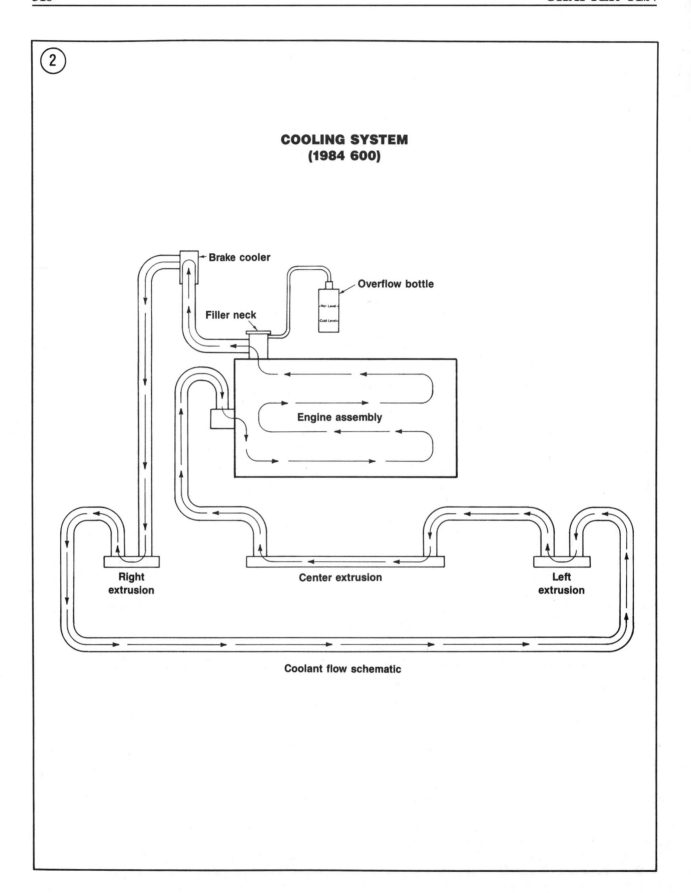

② COOLING SYSTEM
(1984 600)

Coolant flow schematic

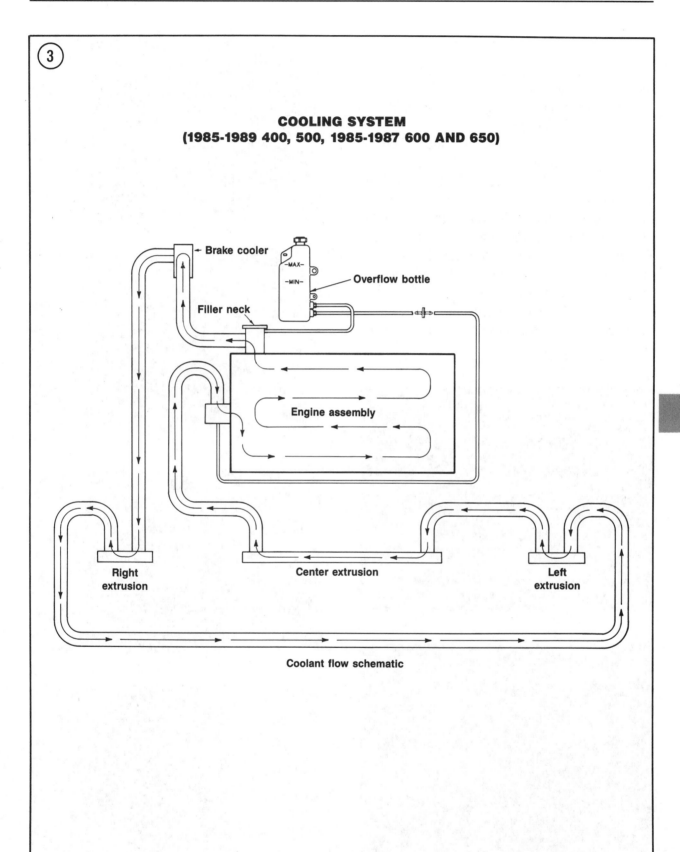

**COOLING SYSTEM
(1985-1989 400, 500, 1985-1987 600 AND 650)**

Brake cooler

Overflow bottle

Filler neck

Engine assembly

Right
extrusion

Center extrusion

Left
extrusion

Coolant flow schematic

10

Inspection

The water pump is non-rebuildable. If damaged, it must be replaced.

1. Check the water pump housing for cracks or other signs of damage.

2. Turn the water pump shaft by hand and check for excessive play or roughness. The shaft should turn smoothly without excessive play or any sign of tightness.

3. Inspect the check hole (A, **Figure 10**) in the bottom of the water pump for signs of coolant. If there is evidence of coolant leakage from the check hole, the pump's inner seal is damaged and the water pump must be replaced.

4. Check the O-ring (B, **Figure 10**) on the water pump spigot for cracks, wear or other damage; replace if necessary.

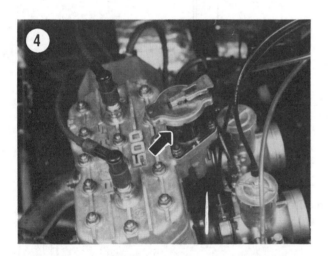

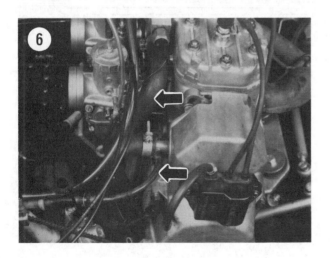

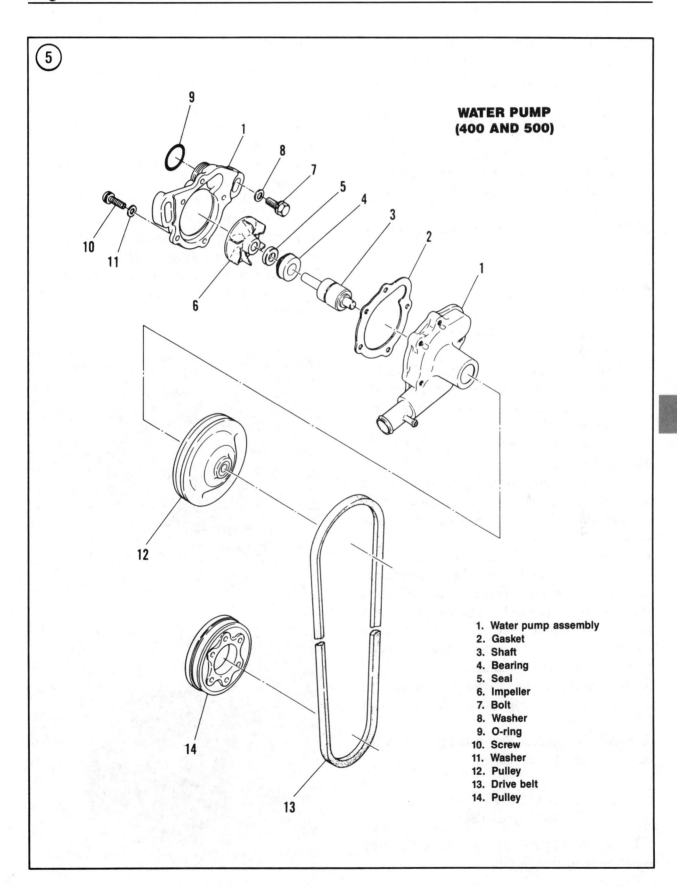

**WATER PUMP
(400 AND 500)**

1. Water pump assembly
2. Gasket
3. Shaft
4. Bearing
5. Seal
6. Impeller
7. Bolt
8. Washer
9. O-ring
10. Screw
11. Washer
12. Pulley
13. Drive belt
14. Pulley

10

5. Inspect the impeller as follows:
   a. Remove the screws (**Figure 11**) holding the water pump halves together and separate the halves.
   b. Check the impeller (**Figure 12**) for cracks or other damage. If the impeller is damaged, replace the water pump assembly.
   c. Using a new gasket (**Figure 13**), reassemble the water pump halves. Tighten the screws securely.

6. Check the pulleys for damage. Replace if necessary.

Assembly

1. Install the lower pulley as follows:
   a. Place the pulley (**Figure 9**) onto the flywheel.
   b. Place the starter pulley (**Figure 8**) against the water pump pulley and install the mounting bolts and washers. Tighten the bolts securely.

2. Apply grease to the water pump O-ring (B, **Figure 10**).

3. Fit the drive belt over the pulleys and place the water pump into position on the engine (**Figure 15**). Install the mounting bolts finger tight.

4. Adjust drive belt tension. Check belt deflection by pressing on the belt midway between the pulleys (**Figure 15**). Deflection should be 6 mm (1/4 in.). To adjust, pivot water pump as required. When belt deflection is correct, tighten the water pump mounting bolts securely. Recheck adjustment.

5. Reconnect the water hoses (**Figure 6**).

6. Reinstall the recoil starter housing. See Chapter Eleven.

7. Refill and bleed the cooling system as described in Chapter Three.

WATER PUMP
(600 AND 650)

Refer to **Figure 16** when performing procedures in this section.

Removal

1. Drain the cooling system as described in Chapter Three.

2. Remove the water pump cover (**Figure 17**).

3. Disconnect the 3 hoses (**Figure 18**) at the water pump.

4. Remove the recoil starter housing as described in Chapter Eleven.

5. Loosen the water pump mounting bolts and remove the drive belt (**Figure 19**).

6. Remove the starter pulley (**Figure 20**) and the water pump pulley (**Figure 21**).

7. Remove the bolts holding the water pump onto the engine assembly and remove the water pump (**Figure 22**).

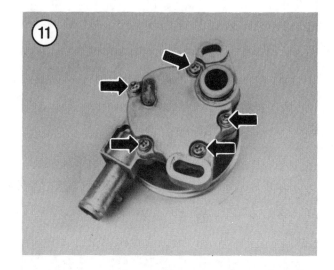

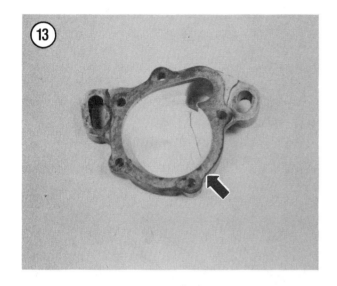

Inspection

The water pump is non-rebuildable. If damaged, it must be replaced.

1. Check the water pump housing for cracks or other signs of damage.

2. Turn the water pump shaft by hand and check for excessive play or roughness. The shaft should turn smoothly without excessive play or any sign of tightness.

3. Inspect the check hole (**Figure 23**) in the bottom of the water pump for signs of coolant. If there is evidence of coolant leaking from the check hole, the pump's inner seal is damaged and the water pump must be replaced.

4. Inspect the impeller as follows:

    a. Remove the nuts holding the water pump halves together and separate the halves (**Figure 24**).

    b. Check the impeller (**Figure 12**) for cracks or other damage. If the impeller is damaged, replace the water pump assembly.

    c. Using a new gasket, reassemble the water pump halves. Tighten the screws securely.

5. Check the pulleys for damage. Replace if necessary.

Assembly

1. Install the lower pulley as follows:

    a. Place the pulley (**Figure 21**) onto the flywheel.

    b. Place the starter pulley (**Figure 20**) against the water pump pulley and install the mounting bolts and washers. Tighten the bolts securely.

2. Fit the drive belt over the pulleys and place the water pump into position on the engine (**Figure 22**). Install the mounting bolts finger tight.

3. Adjust drive belt tension. Check belt deflection by pressing on the belt midway between the pulleys (**Figure 19**). Deflection should be 6 mm (1/4 in.). To adjust, pivot water pump as required. When belt deflection is

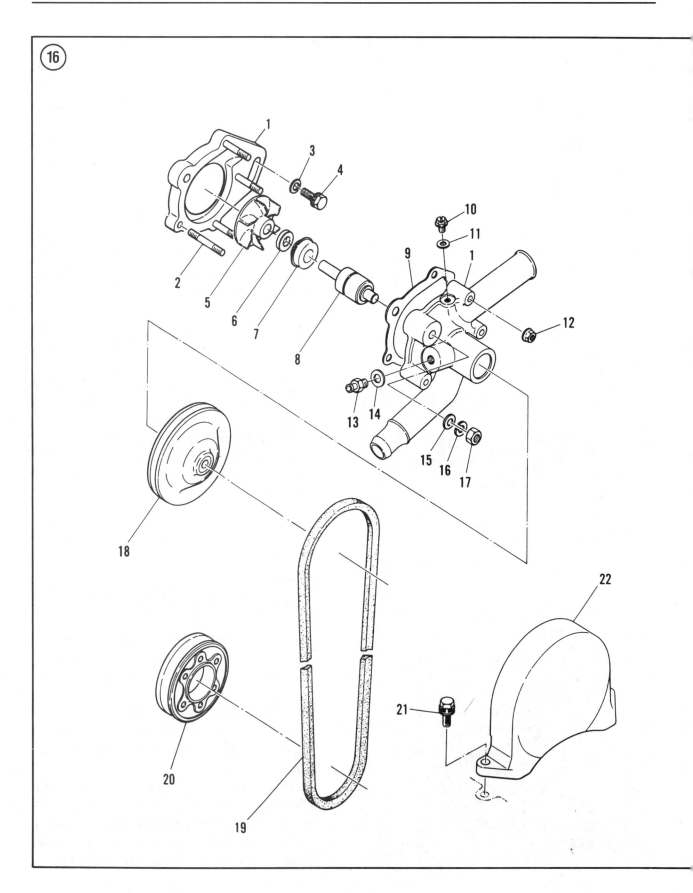

WATER PUMP
(600/650)

1. Water pump assembly
2. Stud
3. Lockwasher
4. Bolt
5. Impeller
6. Seal
7. Bearing
8. Shaft
9. Gasket
10. Bleed bolt
11. Washer
12. Nut
13. Union
14. Gasket
15. Washer
16. Lockwasher
17. Nut
18. Pulley
19. Drive belt
20. Pulley
21. Bolt
22. Cover

10

correct, tighten the water pump mounting bolts securely. Recheck adjustment.

4. Reconnect the water hoses (**Figure 18**).

5. Reinstall the recoil starter housing. See Chapter Eleven.

6. Install the water pump cover (**Figure 17**).

7. Refill and bleed the cooling system as described in Chapter Three.

HEAT EXCHANGER

Refer to **Figures 25-28** when servicing the heat exchangers.

Inspection/Replacement

The heat exchangers (**Figure 29**) should be inspected for damaged and replaced when necessary. Refer to **Figures 25-28** when replacing the heat exchanger(s).

HOSES

Hoses deteriorate with age and should be replaced periodically or whenever they show signs of cracking or leakage. To be safe, replace the hoses every 2 years. Loss of coolant will cause the engine to overheat and result in severe damage.

Whenever any component of the cooling system is removed, inspect the hoses(s) and determine if replacement is necessary.

Inspection

1. With the engine cool, check the cooling hoses for brittleness or hardness. A hose in this condition will usually show cracks and must be replaced.

2. With the engine hot, examine the hoses for swelling along the entire hose length. Eventually a hose will rupture at this point.

3. Check area around hose clamps. Signs of rust around clamps indicate possible hose leakage.

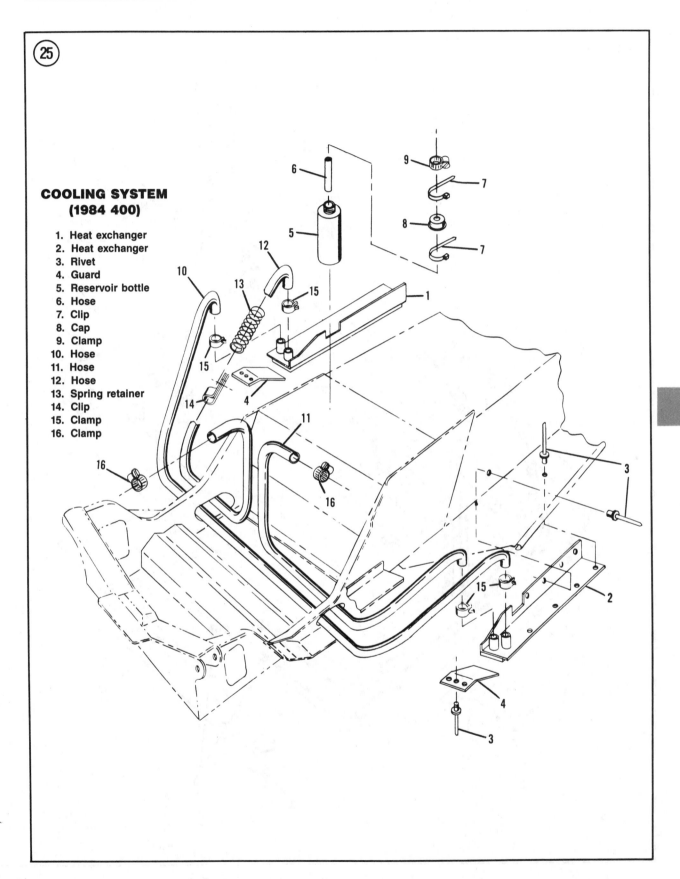

㉕

COOLING SYSTEM
(1984 400)

1. Heat exchanger
2. Heat exchanger
3. Rivet
4. Guard
5. Reservoir bottle
6. Hose
7. Clip
8. Cap
9. Clamp
10. Hose
11. Hose
12. Hose
13. Spring retainer
14. Clip
15. Clamp
16. Clamp

10

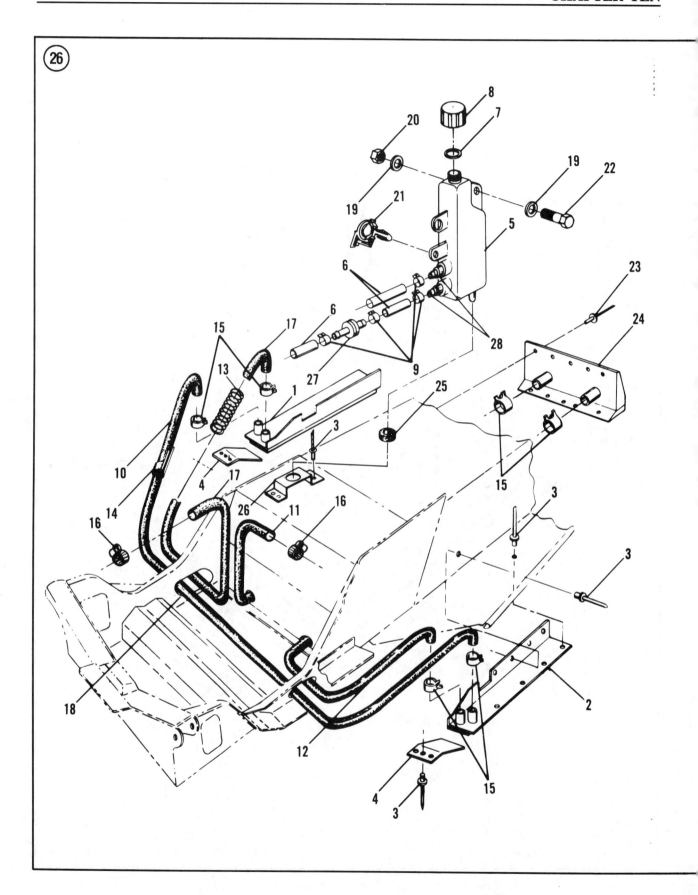

COOLING SYSTEM
(1985-1989 400; 1989 500)

1. Heat exchanger
2. Heat exchanger
3. Rivet
4. Guard
5. Reservoir bottle
6. Hoses
7. Gasket
8. Cap
9. Clamps
10. Hose
11. Hose
12. Hose
13. Spring
14. Clip
15. Clamps
16. Clamp
17. Hose
18. Edge trim
19. Washer
20. Nut
21. Clip
22. Bolt
23. Rivet
24. Heat exchanger
25. Grommet
26. Bracket
27. Check valve
28. Fittings

10

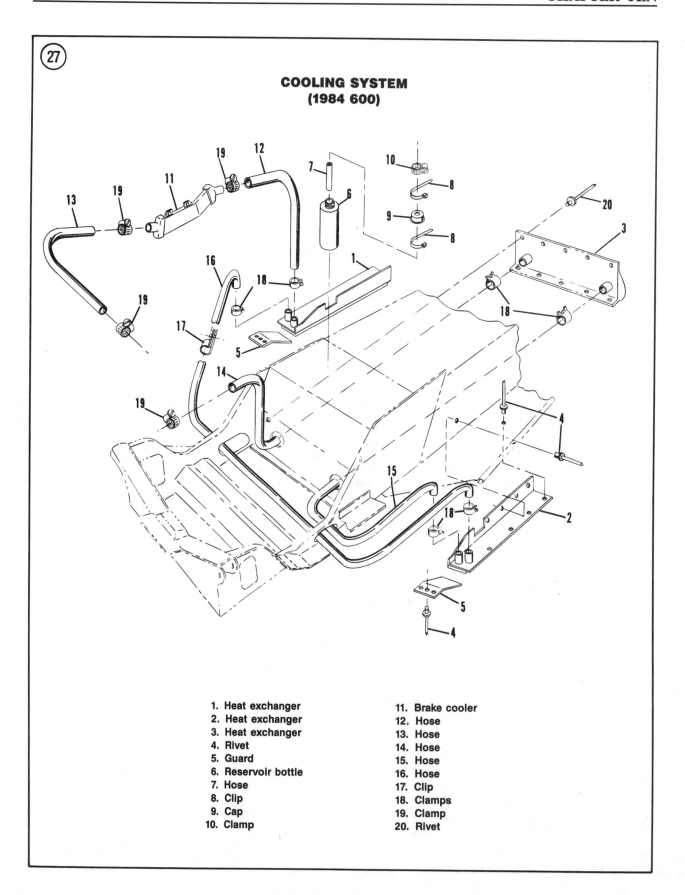

**COOLING SYSTEM
(1984 600)**

1. Heat exchanger
2. Heat exchanger
3. Heat exchanger
4. Rivet
5. Guard
6. Reservoir bottle
7. Hose
8. Clip
9. Cap
10. Clamp
11. Brake cooler
12. Hose
13. Hose
14. Hose
15. Hose
16. Hose
17. Clip
18. Clamps
19. Clamp
20. Rivet

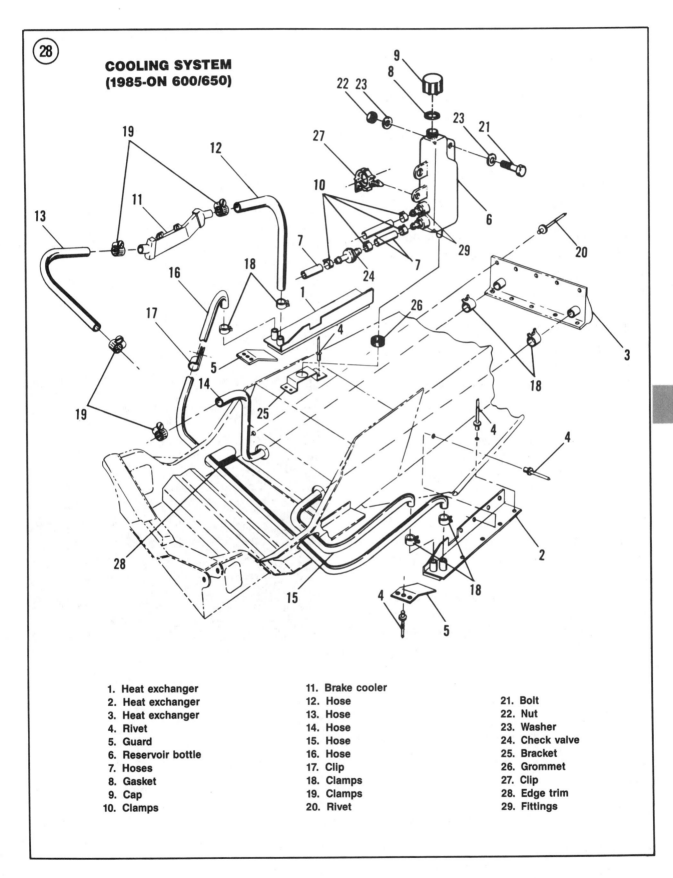

**COOLING SYSTEM
(1985-ON 600/650)**

1. Heat exchanger	11. Brake cooler	
2. Heat exchanger	12. Hose	21. Bolt
3. Heat exchanger	13. Hose	22. Nut
4. Rivet	14. Hose	23. Washer
5. Guard	15. Hose	24. Check valve
6. Reservoir bottle	16. Hose	25. Bracket
7. Hoses	17. Clip	26. Grommet
8. Gasket	18. Clamps	27. Clip
9. Cap	19. Clamps	28. Edge trim
10. Clamps	20. Rivet	29. Fittings

10

Replacement

Hose replacement should be performed when the engine is cool.

1. Drain the cooling system as described under *Coolant Change* in Chapter Three.

> *NOTE*
> *Make sure to note the routing and any clamps supported by the hoses.*

2. Loosen the hose clamps from the hose to be replaced. Slide the clamps along the hose and out of the way.

3. Twist the hose end to break the seal and remove from the connecting joint. If the hose has been on for some time, it may have become fused to the joint. If so, cut the hose parallel to the joint connections with a knife or razor. The hose then can be carefully pried loose with a screwdriver.

> *CAUTION*
> *Excessive force applied to the hose during removal could damage the connecting joint.*

4. Examine the connecting joint for cracks or other damage. Repair or replace parts as required. If the joint is okay, remove rust with sandpaper.

5. Inspect hose clamps and replace as necessary.

6. Slide hose clamps over outside of hose and install hose to inlet and outlet connecting joint.

Make sure hose clears all obstructions and is routed properly.

> *NOTE*
> *If it is difficult to install a hose on a joint, soak the end of the hose in hot water for approximately 2 minutes. This will soften the hose and ease installation.*

7. With the hose positioned correctly on joint, position clamps back away from end of hose slightly. Tighten clamps securely, but not so much that hose is damaged.

8. Refill cooling system as described under *Coolant Change* in Chapter Three. Start the engine and check for leaks. Retighten hose clamps as necessary.

Chapter Eleven

Recoil Starter

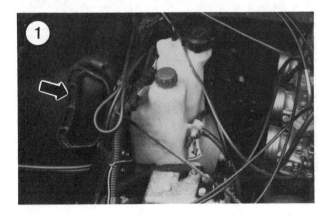

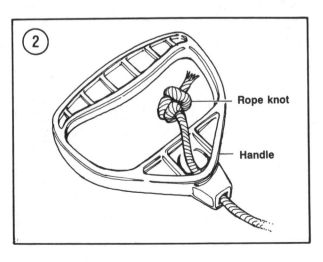

Rope knot

Handle

All models are equipped with a rope-operated recoil starter. The starter is mounted in a housing that is bolted onto the engine next to the flywheel. Pulling the rope handle causes the starter sheave shaft to rotate against spring tension, moving the drive pawl to engage the starter pulley on the flywheel and turn the engine over. When the rope handle is released, the spring inside the assembly reverses direction of the sheave shaft and winds the rope around the sheave.

Rewind starters are relatively trouble-free; a broken or frayed rope is the most common malfunction. This chapter covers removal and installation of the starter assembly, starter pulley, and rope and spring replacement.

RECOIL STARTER ASSEMBLY
(1984 600)

Removal/Installation

1. Open the shroud.
2. Remove the exhaust pipe. See Chapter Seven.
3. Pull the starter handle (**Figure 1**) out. Then untie the knot (**Figure 2**) and remove the

handle—don't release the rope yet. After removing the handle, tie a knot in the end of the rope to prevent the rope from retracting into the starter housing. Slowly release the rope.

4. Remove the water pump cover.

5. Remove the bolts holding the starter housing to the engine. Remove the starter housing (**Figure 3**).

6. Installation is the reverse of these steps. Note the following.

7. Align the starter housing with the starter pulley and install the housing (**Figure 3**). Install the starter housing mounting bolts and tighten securely.

8. Feed the starter rope back through its original path.

9. Untie the knot at the end of the rope and feed the rope through the handle. Tie a knot in the end of the rope as shown in **Figure 4**. Operate the starter assembly to make sure it is working properly. Check the path of the rope through the engine compartment to make sure it is not kinked or interfering with any other component.

10. Install the water pump cover and exhaust pipe.

11. Close the shroud.

Starter Pulley
Removal/Installation

The starter pulley is mounted onto the flywheel. The starter pulley can be removed with the engine installed in the frame. Refer to *Flywheel Removal* in Chapter Eight.

Starter Housing Disassembly

This procedure describes complete disassembly of the starter housing assembly. Refer to **Figure 5** when performing this procedure.

1. Remove the recoil starter housing as described in this chapter.

2. Invert the recoil assembly on a clean work bench. Untie the knot in the starter rope. Slowly

release the starter rope and allow the sheave to turn slowly until it stops.

3. Remove the friction plate nut, lockwasher and spring washer. Then remove the friction plate, return spring and slide plate. See **Figure 6**.

4. Remove the 3 return springs and the slide plate (**Figure 7**).

5. See **Figure 8**. Remove the 3 ratchets, friction spring and washer.

6. Lift the sheave drum (**Figure 9**) from the housing, making sure the drum is free from the recoil spring.

> *WARNING*
> *The recoil spring (**Figure 9**) is spring loaded. Failure to observe Step 7 may allow the spring to unwind violently and*

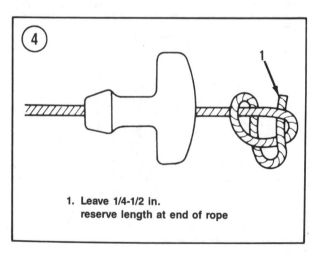

1. Leave 1/4-1/2 in. reserve length at end of rope

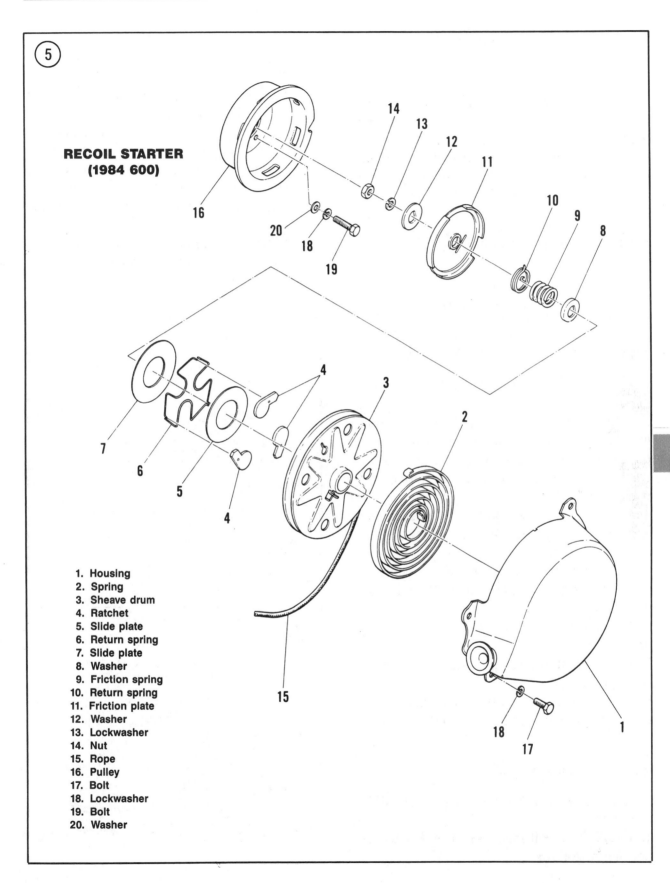

**RECOIL STARTER
(1984 600)**

1. Housing
2. Spring
3. Sheave drum
4. Ratchet
5. Slide plate
6. Return spring
7. Slide plate
8. Washer
9. Friction spring
10. Return spring
11. Friction plate
12. Washer
13. Lockwasher
14. Nut
15. Rope
16. Pulley
17. Bolt
18. Lockwasher
19. Bolt
20. Washer

11

may result in serious personal injury.
Wear safety glasses while removing and
installing the spring.

7. Place the starter assembly housing on the floor right side up. Tap lightly on the top of the housing while holding it tightly against the floor. The spring will fall out and unwind inside the housing. When the spring has unwound, pick the housing up off the floor along with the spring.

Starter Housing Inspection

NOTE
Before cleaning plastic components, make sure the cleaning agent is compatible with plastic. Some types of solvents can cause permanent damage.

1. Clean all parts thoroughly in solvent and allow to dry.
2. Visually check the starter post in the housing for cracks, deep scoring or excessive wear. Check the post threads for thread damage or cracks. Replace the housing if the starter post is damaged.
3. Check the ratchets, springs and washers for excessive wear or damage. Replace damaged parts as required.
4. Check the sheave drum for cracks or damage.
5. Check the recoil spring (**Figure 9**) for cracks or bending.
6. Check the starter rope for fraying, splitting or breakage.
7. Check the starter handle for cracks or other signs of damage.
8. If there is any doubt as to the condition of any part, replace it with a new one.

Starter Housing Assembly

Refer to **Figure 5** when performing this procedure.
1. Install the recoil spring (**Figure 9**) into the housing as follows:

WARNING
Safety glasses should be worn while installing the recoil spring.

a. Fit the outer spring loop around the spring lug in the starter housing.
b. Slowly wind the spring counterclockwise until it fits into the spring retainer in the starter housing. Check that the ends of the spring seat in the housing as shown in **Figure 9** and that the spring lays flat in the housing.
c. When the spring is installed correctly, lubricate the post and spring with a low-temperature grease.

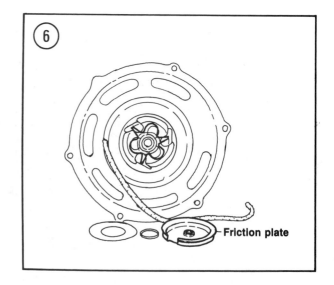

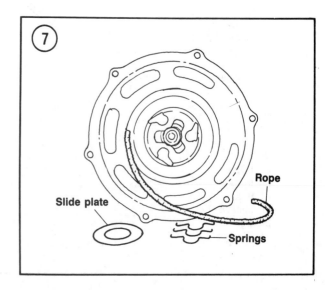

2. Install the starter rope as follows:

a. Check the ends of the starter rope for fraying. To tighten up the end of the rope, apply heat to slightly melt the end of the rope. Do not overheat.

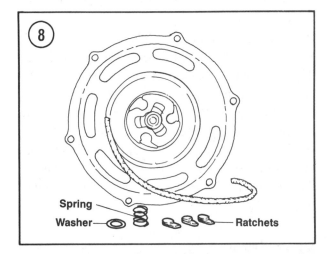

Spring

Washer———Ratchets

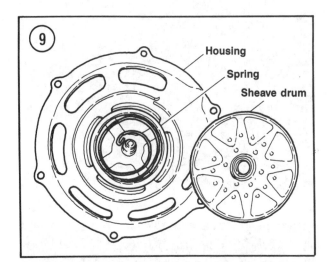

Housing

Spring

Sheave drum

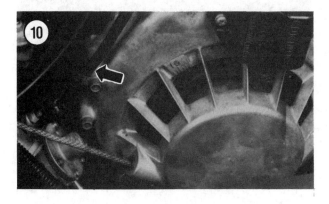

b. Then insert the rope through the sheave drum and pull the rope until the knot seats into the drum.

c. Place the sheave drum on the workbench in front of you so that it faces as shown in **Figure 9** and wind the rope counterclockwise around the sheave drum.

3. Insert the sheave drum into the housing.

4. Install the ratchets, spring and washer (**Figure 8**).

5. Install the slide plate and the 3 ratchet return springs (**Figure 7**). Install the slide plate, return spring and the friction plate.

6. Install the friction plate flat washer, lockwasher and nut.

7. With the starter rope engaged with the notch in the sheave drum, rotate the sheave drum 4-5 turns counterclockwise. Then release the rope from the notch and allow the rope to retract into the housing.

8. Pull on the rope to check tension. If tension is okay, tie a knot (**Figure 4**) in the end of the rope and release the rope.

9. Operate the recoil starter to make sure the sheave assembly works smoothly and returns properly.

11

RECOIL STARTER HOUSING (TRAIL, SPORT, 400, 500 AND 1985-ON 600/650)

Starter Housing
Removal/Installation

1. Open the shroud.

2. Remove the exhaust pipe. See Chapter Seven.

3. Pull the starter handle out (**Figure 1**). Then untie the knot (**Figure 2**) and remove the handle—don't release the rope yet. After removing the handle, tie a knot in the end of the rope to prevent the rope from retracting into the starter housing. Slowly release the rope.

4A. *Trail and Sport*: Remove the starter housing as follows:

a. Disconnect the ignition coil electrical connector (**Figure 10**, typical).

b. Disconnect the spark plug caps at the spark plugs.

c. Remove the bolts holding the starter housing to the engine and remove it. See **Figure 11**.

4B. *400 and 500*: Remove the starter housing as follows:

a. Disconnect the ignition coil electrical connector (A, **Figure 12**, typical).

b. Disconnect the spark plug caps at the spark plugs.

c. Remove the bolts holding the starter housing to the engine and remove it. See B, **Figure 12**.

4C. *1985-on 600/650*: Remove the starter housing as follows:

a. Remove the water pump cover.

b. Remove the bolts holding the starter housing to the engine. Remove the starter housing (**Figure 3**).

5. Installation is the reverse of these steps. Note the following.

6. Align the starter housing with the starter pulley and install the housing. Install the starter housing mounting bolts and tighten securely.

7. Reconnect all electrical connectors and spark plug caps, as required.

8. Feed the starter rope back through its original path and out the rear cowl hole.

9. Untie the knot at the end of the rope and feed the rope through the handle. Tie a knot in the end of the rope as shown in **Figure 4**. Operate the starter assembly to make sure it is working properly. Check the path of the rope through the engine compartment to make sure it is not kinked or interfering with any other component.

10. Close the shroud.

Starter Pulley
Removal/Installation

The starter pulley is mounted onto the flywheel. The starter pulley can be removed with the engine installed in the frame. Refer to *Flywheel Removal* in Chapter Eight.

Starter Housing Disassembly

This procedure describes complete disassembly of the starter housing assembly. Refer to **Figures 13-16** for your model when performing this procedure.

1. Remove the recoil starter housing as described in this chapter.

2. Invert the recoil assembly on a clean work bench. Untie the knot in the starter rope (**Figure 17**). Slowly release the starter rope and allow the sheave to turn slowly until it stops.

3. Turn the recoil assembly over so that the sheave assembly faces up (**Figure 18**).

4. Compress the sheave drum by hand and remove the nut and friction plate. See A & B, **Figure 18**. Note the spring on the fiction plate (**Figure 19**).

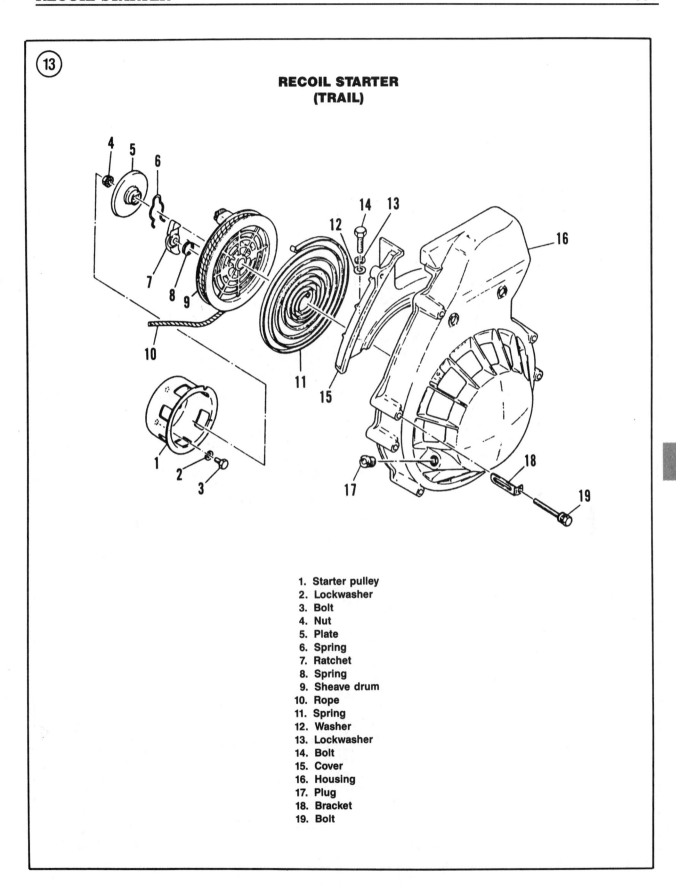

RECOIL STARTER
(TRAIL)

1. Starter pulley
2. Lockwasher
3. Bolt
4. Nut
5. Plate
6. Spring
7. Ratchet
8. Spring
9. Sheave drum
10. Rope
11. Spring
12. Washer
13. Lockwasher
14. Bolt
15. Cover
16. Housing
17. Plug
18. Bracket
19. Bolt

11

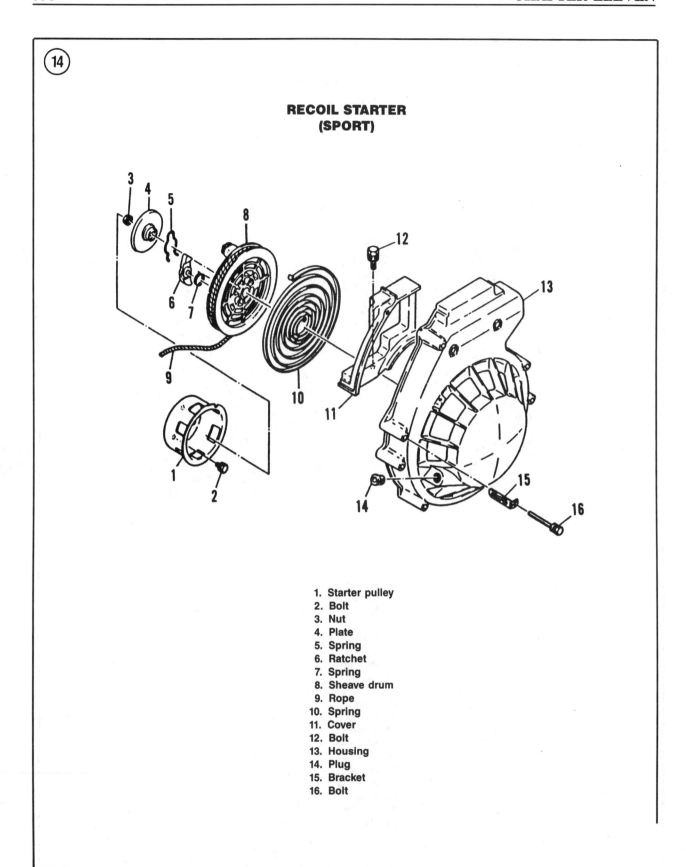

**RECOIL STARTER
(SPORT)**

1. Starter pulley
2. Bolt
3. Nut
4. Plate
5. Spring
6. Ratchet
7. Spring
8. Sheave drum
9. Rope
10. Spring
11. Cover
12. Bolt
13. Housing
14. Plug
15. Bracket
16. Bolt

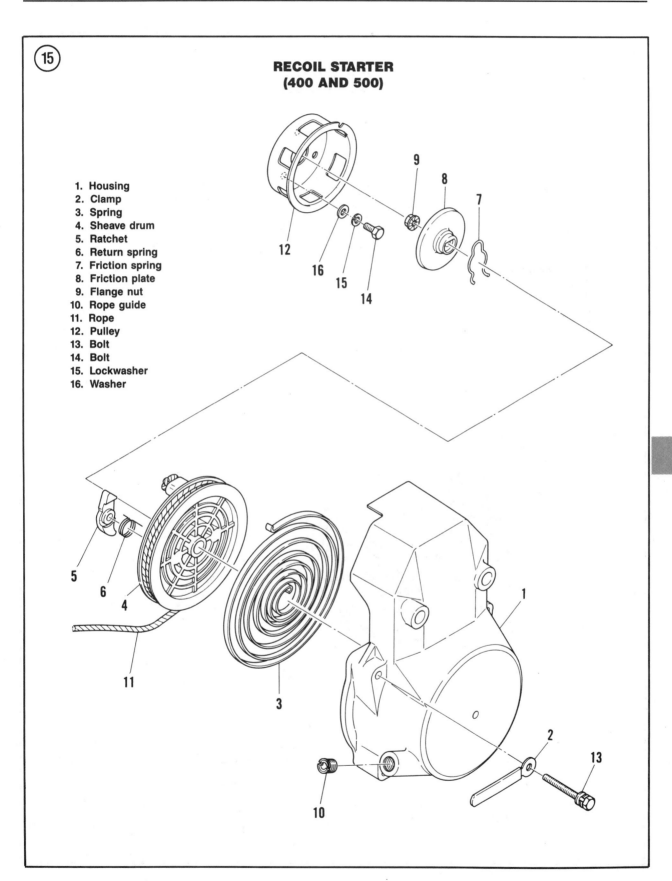

⑮

**RECOIL STARTER
(400 AND 500)**

1. Housing
2. Clamp
3. Spring
4. Sheave drum
5. Ratchet
6. Return spring
7. Friction spring
8. Friction plate
9. Flange nut
10. Rope guide
11. Rope
12. Pulley
13. Bolt
14. Bolt
15. Lockwasher
16. Washer

11

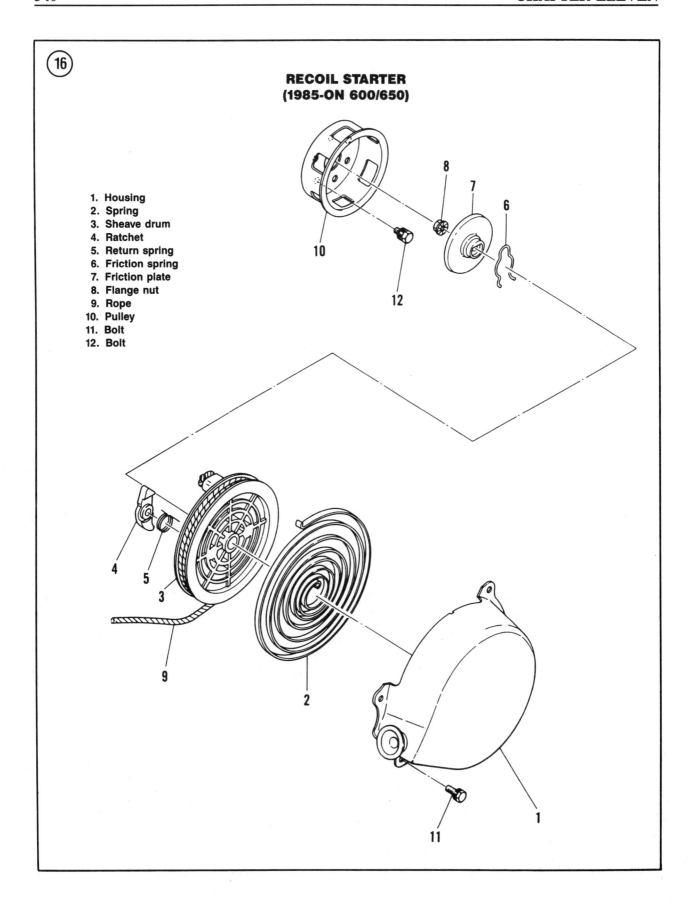

(16)

**RECOIL STARTER
(1985-ON 600/650)**

1. Housing
2. Spring
3. Sheave drum
4. Ratchet
5. Return spring
6. Friction spring
7. Friction plate
8. Flange nut
9. Rope
10. Pulley
11. Bolt
12. Bolt

5. Remove the ratchet (**Figure 20**) and spring (**Figure 21**).

6. Remove the sheave (**Figure 22**) and rope assembly from the starter housing.

11

*may result in serious personal injury.
Wear safety glasses while removing and
installing the spring.*

7. Place the starter assembly housing on the floor right side up. Tap lightly on the top of the housing while holding it tightly against the floor. The spring will fall out and unwind inside the housing. When the spring has unwound, pick the housing up off the floor along with the spring.

Starter Housing Inspection

NOTE
Before cleaning plastic components, make sure the cleaning agent is compatible with plastic. Some types of solvents can cause permanent damage.

1. Clean all parts thoroughly in solvent and allow to dry.
2. Visually check the starter post (**Figure 24**) in the housing for cracks, deep scoring or excessive wear. Check the post threads for thread damage or cracks. Replace the housing if the starter post is damaged.
3. Check the ratchet (**Figure 25**) and ratchet spring (A, **Figure 26**) for cracks or other damage. Replace damaged parts as required.
4. Install the spring onto the friction plate (B, **Figure 26**) and check that the spring fits securely.

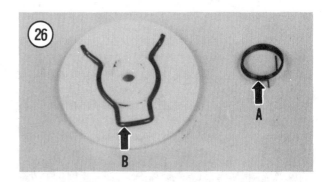

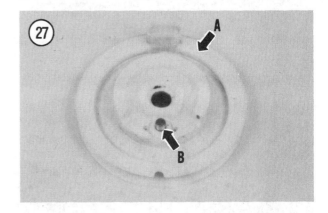

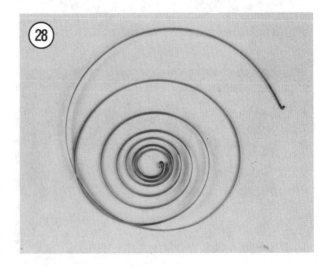

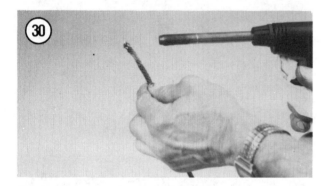

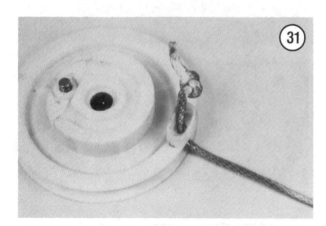

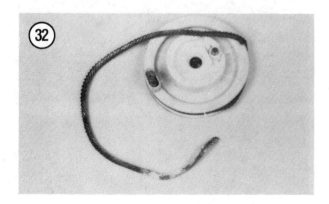

5. Check the sheave drum (A, **Figure 27**) for cracks or damage. Check the ratchet pin (B, **Figure 27**) in the drum for looseness or damage.

6. Check the recoil spring (**Figure 28**) for cracks or bending.

7. Check the starter rope for fraying, splitting or breakage.

8. Check the starter handle for cracks or other signs of damage.

9. If there is any doubt as to the condition of any part, replace it with a new one.

Starter Housing Assembly

Refer to **Figures 13-16** for your model when performing this procedure.

1. Install the recoil spring (**Figure 28**) into the housing as follows:

WARNING
Safety glasses should be worn while installing the recoil spring.

a. Fit the outer spring loop around the spring lug in the starter housing. See **Figure 29**.

b. Slowly wind the spring counterclockwise until it fits into the spring retainer in the starter housing. Check that the ends of the spring seat in the housing as shown in **Figure 23** and that the spring lays flat in the housing.

c. When the spring is installed correctly, lubricate the post and spring with a low-temperature grease.

2. Install the starter rope as follows:

a. Check the ends of the starter rope for fraying. To tighten up the end of the rope, apply heat to slightly melt the end of the rope (**Figure 30**). Do not overheat.

b. Then insert the rope through the sheave drum as shown in **Figure 31** and pull the rope until the knot seats into the drum.

c. Place the sheave drum on the workbench in front of you so that it faces as shown in **Figure 32** and wind the rope clockwise around the sheave drum.

11

3. Align the inner spring loop (**Figure 33**) with the guide notch in the sheave drum (**Figure 34**) and install the sheave drum assembly (**Figure 22**). Twist the sheave drum slightly to make sure the drum and spring are engaged.

> *NOTE*
> *Make sure the rope engages the notch in the sheave drum as shown in A, **Figure 35**.*

4. Install the ratchet spring (B, **Figure 35**) into the sheave drum.

5. Install the ratchet so that it engages with the ratchet spring as shown in **Figure 36**.

6. Install the spring onto the friction plate (**Figure 19**) so that the top of the spring is opposite the flat on the drive plate. Lubricate the drive plate pivot hole with a low-temperature grease. Then slide the drive plate over the post in the housing (B, **Figure 18**) so that the flat on the friction plate aligns with the flat on the threaded post in the starter housing.

7. Apply Loctite 242 (blue) onto the drive plate nut threads and install the washer and nut (A, **Figure 18**). Tighten securely.

8. With the starter rope engaged with the notch in the sheave drum, rotate the sheave drum 4-5 turns counterclockwise. Then release the rope from the notch and allow the rope to retract into the housing.

9. Pull on the rope to check tension. If tension is okay, tie a knot in the end of the rope and release the rope (**Figure 17**).

10. Operate the recoil starter to make sure the sheave assembly works smoothly and returns properly.

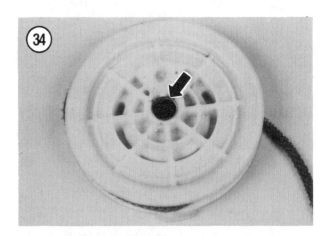

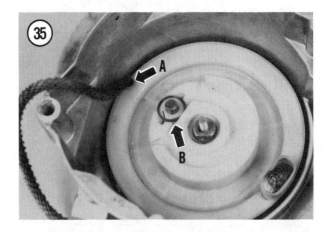

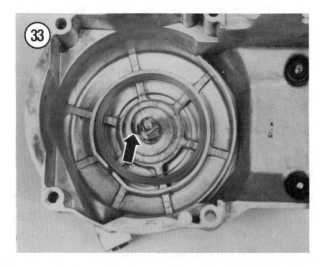

Chapter Twelve

Drive System

The drive train consists of a drive sheave on the end of the crankshaft, connected to a driven sheave on the chaincase by a belt, a drive chain and sprockets inside the chaincase, a drive shaft with 2 drive sprockets and a brake. This chapter describes complete procedures for the primary and secondary sheave components. Service to the chaincase, chain and sprockets and jackshaft are described in Chapter Fourteen.

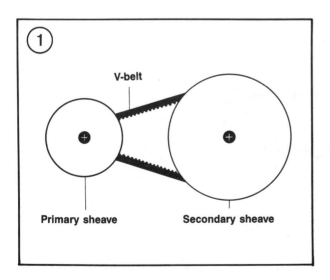

V-belt

Primary sheave   Secondary sheave

Service specifications are listed in **Tables 1-6**. **Table 7** lists tightening torques. **Tables 1-7** are at the end of the chapter.

> *WARNING*
> *Never lean into a snowmobile's engine compartment while wearing a scarf or other loose clothing when the engine is running or when attempting to start the engine. If the scarf or clothing should catch in the drive belt or drive system, severe injury or death could occur. Make sure the drive belt guard is in place.*

DRIVE UNIT

Torque is transferred from the engine crankshaft to the rear track unit by a centrifugally actuated sheave-type transmission. The transmission or drive unit automatically selects the proper drive ratio to permit the machine to move from idle to maximum speed. Major components are the primary sheave assembly, secondary sheave assembly and drive belt. See **Figure 1**.

The primary and secondary sheaves are basically 2 sets of variable diameter pulleys that automatically vary the gear ratio. Gear changes are brought about by the primary and secondary sheave drums changing their diameter to correspond to prevailing track conditions. See **Figure 2**.

The shift sequence is determined by engine torque instead of engine rpm. When track resistance increases, such as when going up hill, the sheaves change the gear ratios; engine rpm will remain the same but the vehicle's speed drops. When track resistance decreases, the sheaves automatically shift toward a higher ratio; engine rpm remains the same but the vehicle speeds up.

PRIMARY SHEAVE ASSEMBLY

Major components of the primary sheave assembly (A, **Figure 3**) are the sliding sheave, fixed sheave, shift weights, primary spring and V-belt. The V-belt connects the primary and secondary sheaves (**Figure 1**).

Fixed and Sliding Sheaves

The sheaves are precision made of mild steel and are balanced to prevent vibration induced problems with the crankshaft and its bearings. The sheave belt surfaces are machined to a smooth tapered surface. The tapered surface on both sheaves matches the V-belt gripping surface (**Figure 4**).

The primary sheave assembly is mounted onto the engine crankshaft. When the engine is stopped or at idle, the fixed and sliding sheaves are held apart by the primary spring. This allows the V-belt to drop down between both sheaves. There is no engagement because the belt diameter is *less* than the space between the sheaves.

When the engine is started and allowed to idle, the primary spring is still pushing the sliding sheave away from the fixed sheave; the V-belt is not contacting either sheave. The engine will stay at idle until it reaches a specific engine rpm. As the engine approaches its engagement rpm, the shift weights mounted on the sliding sheave push or swing out against the rollers mounted on the spider. At this point, the weights begin to overcome the strength of the primary spring. As engine rpm increases, the weights force the sliding sheave to move closer toward the fixed sheave, thus increasing sheave diameter. The free space between the sheaves is reduced and both sheaves begin to grip the V-belt. As engine rpm

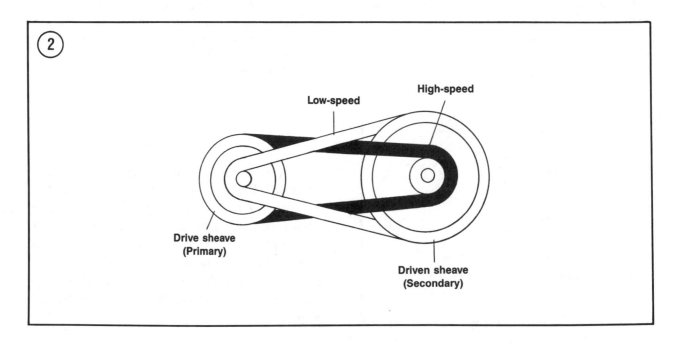

increases from engagement rpm to shift rpm, the weights swing out further and force the sliding sheave closer to the fixed sheave (the sheave diameter is getting increasingly larger). As the primary sheaves come closer together, the V-belt is forced toward the sheaves outer edge.

Primary Sheave Spring

The primary sheave spring controls engagement speed. If a lighter spring is installed, a lower engine rpm will be required for engagement. If a heavier spring is installed,

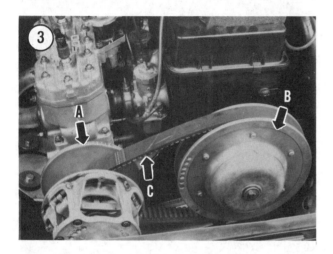

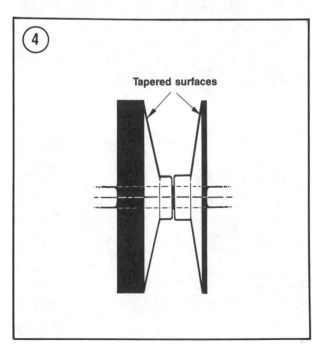

Tapered surfaces

engine rpm must be increased to overcome spring pressure and allow engagement.

Shift Weights

As noted in the previous section, weights react to engine rpm to move the sliding sheave. Until engine rpm is increased, the weights cannot move or pivot out and thus the sliding sheave cannot move. Weight movement is controlled by engine rpm. The faster the crankshaft rotates, the farther the weights pivot out.

SECONDARY SHEAVE

Major components of the secondary sheave assembly (B, **Figure 3**) are the sliding sheave, fixed sheave, secondary spring and helix. The sheaves are precision made of mild steel. The sheave belt surfaces are machined to a smooth tapered surface. The tapered surface on both sheaves matches the V-belt gripping surface (**Figure 4**).

The secondary sheave assembly is mounted onto a jackshaft that couples the sheave assembly to a gear and chain assembly that connects to the track. When the engine is stopped or at idle, the secondary sheave assembly is held in its low speed position by tension from the secondary spring.

The secondary sheave is a torque sensitive unit. If the snowmobile encounters an increased load condition while at wide open throttle, the helix forces the secondary sheaves to downshift (secondary sheave halves move closer together). While the snowmobile will be running slower, the engine is still running at a high rpm. By sensing load conditions and shifting accordingly, engine rpm remains at its peak power range.

Secondary Sheave Spring

The secondary sheave spring determines engine rpm during the shift pattern. On the secondary sheave, spring tension can be adjusted

12

by repositioning the spring position in holes drilled in the helix. Note the following:

a. Increasing secondary spring tension increases engine rpm. If the secondary sheave is shifting into a higher ratio than the engine can pull, engine rpm will drop below the peak power range. Increasing spring tension will prevent the secondary sheave from upshifting under the same riding conditions. By not shifting up, the engine will stay in a lower ratio at the peak power range.

b. Decreasing secondary spring tension decreases engine rpm. If the engine rpm is higher than the peak power range, decreasing spring tension will allow the clutch to shift into a higher ratio. Under the same load conditions, engine rpm will drop.

Helix Cam Angle

The secondary sheave spring and the helix cam angle work together to control how easily the driven clutch will shift up. For example, if the spring tension remains the same but the cam angle is decreased, engine rpm will increase. When increasing the helix angle, engine rpm will decrease.

Polaris has 3 helix cam angles available for the Indy clutches described in this manual. Note the following:

a. 30° helix: No identification.

b. 34° helix: The number "8" is stamped on the top of one ramp.

c. 36° helix: The letters "S/A" is stamped on the internal face of the helix.

DRIVE BELT

The drive belt (C, **Figure 3**) transmits power from the primary sheave to the secondary sheave and should be considered a vital link in the operation and performance of the snowmobile. To ensure top performance, the correct size drive belt must be matched to the primary and secondary sheaves. Belt width dimensions are

critical. Belt wear affects clutch operation and its shifting characteristics. Because belt width will change with use, it must be monitored and adjusted as described in this chapter.

With general use, there is no specific mileage or time limit on belt life. However, belt life is

directly related to maintenance and snowmobile operation. The belt should be inspected at the intervals listed in Chapter 3. Premature belt failure (200 miles or less) is abnormal and the cause must be determined to prevent secondary damage.

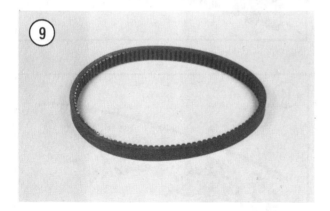

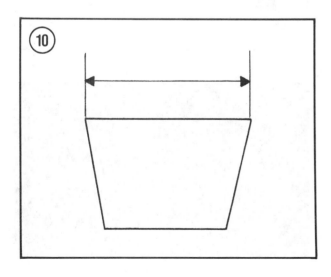

Removal/Installation

1. Open the shroud.
2. Remove the drive belt guard (**Figure 5**).
3. Check the drive belt (C, **Figure 3**) for markings so that during installation it will run in the same direction. If the belt is not marked, draw an arrow on the belt facing forward.
4. Push against the secondary sheaves in opposite directions (**Figure 6**) to open the clutch unit. Then roll the belt over the outer sheave and remove it. See **Figure 7**.
5. Inspect the drive belt as described in this chapter.
6. Perform the *Drive Belt Alignment* as described in this chapter.
7. Reverse Steps 1-4 and install the drive belt (**Figure 8**). If installing the original belt, make sure to install it so that the arrow mark on the belt (made before removal) faces in the same direction (forward). When installing a new belt, install it so that you can read the belt identification marks while standing on the left-hand side of the vehicle and looking into the engine compartment.

Inspection

1. Remove the drive belt (**Figure 9**) as described in this chapter.
2. Measure the width of the drive belt at its widest point (**Figure 10**). Replace the belt if the width meets or exceeds the wear limit in **Table 2**.
3. Visually inspect the belt for the following conditions:

12

a. *Frayed edge*: Check the sides of the belt for a frayed edge cord (**Figure 11**). This indicates drive belt misalignment. Drive belt misalignment can be caused by incorrect sheave alignment and loose engine mounting bolts.

b. *Worn narrow in one section*: Examine the belt for a section that is worn narrower in one section (**Figure 12**). This condition is caused by excessive belt slippage due to a stuck track or a too high engine idle speed.

c. *Belt disintegration*: Drive belt disintegration (**Figure 13**) is caused by severe belt wear or misalignment. Disintegration can also be caused by the use of an incorrect belt.

d. *Sheared cogs*: Sheared cogs shown in **Figure 14** are usually caused by violent drive sheave engagement. This is an indication of a defective or improperly installed drive sheave.

5. Replace a worn or damaged belt immediately. Always carry a spare belt on your snowmobile (**Figure 5**) for emergency purposes.

Drive Belt Tension Adjustment

For a 1:1 sheave ratio to occur, the drive belt must move to the outer secondary sheave edges (**Figure 15**) when the sheaves have closed together (neutral position). However, because friction gradually reduces the width of the drive belt, a narrow belt will not move to the outer edges, thus preventing the sheaves from shifting

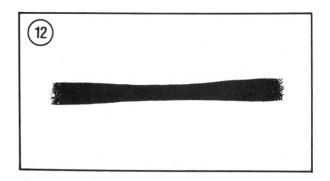

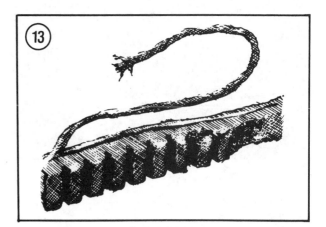

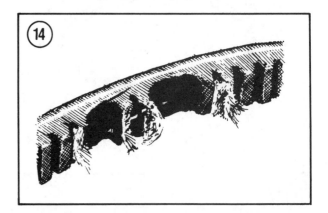

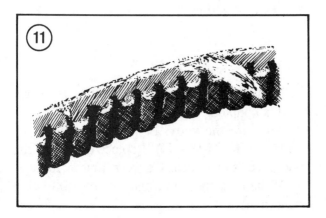

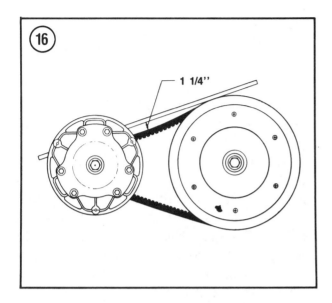

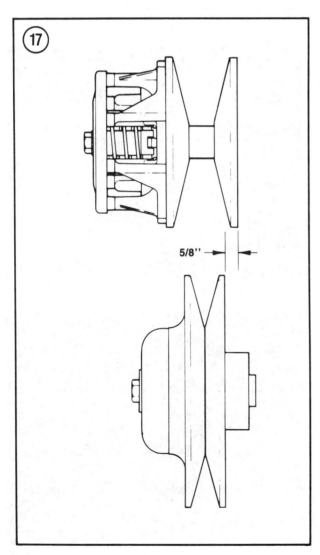

to a 1:1 ratio. To compensate for belt wear, washers of variable thicknesses can be installed between the secondary sheaves. A thin washer will increase belt tension whereas a thicker washer will decrease tension.

1. Open the shroud.

2. Lay a straightedge across the drive belt as shown in **Figure 16** and compress the belt midway across the sheaves with your finger to remove belt slack; measure belt deflection from top of belt surface to bottom of straightedge (**Figure 16**). The correct belt tension is 31.75 mm (1 1/4 in).

3. If belt tension is incorrect, remove and disassemble the secondary sheave assembly as described in this chapter to change washer thickness. Washers can be purchased through a Polaris dealer.

Primary and Secondary Sheave Offset Check and Adjustment

Primary and secondary sheave offset (**Figure 17**) must be correctly maintained for good clutch performance and long belt life.

The Polaris Offset Clutch Alignment Tool (part No. 2870426) will be required to perform this procedure. See **Figure 18**. Sheave alignment specifications are listed in **Table 3**.

NOTE
If you are using an accessory manufactured alignment bar, make sure it is machined for your clutch.

12

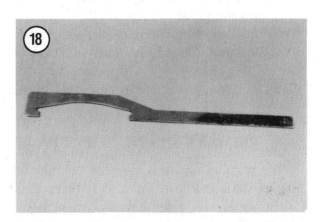

1. Remove the drive belt as described in this chapter.

2. Place the alignment tool between the primary sheave and against the secondary sheave as shown in **Figure 19**. For the offset to be correct, the alignment tool will contact *all* sheave surfaces equally as shown in **Figure 18**. If sheave offset is incorrect, proceed to Step 3.

3. Sheave offset is adjusted by repositioning the engine (those with slotted motor mounts) or by repositioning the secondary sheave laterally on the jackshaft (all models). The following steps describe adjustment by repositioning the secondary sheave. If you wish to reposition the engine instead, refer to Chapter Five or Chapter Six.

NOTE
Do not intermix the washers installed on the outside and inside of the secondary sheave. These washers are used for offset purposes.

  a. Remove the secondary sheave as described in this chapter.

  b. Adjust sheave offset by adding or removing shims (**Figure 20**) located behind the secondary sheave.

NOTE
Adding shims increases offset; removing shims decreases it.

  b. Reinstall the secondary shim and recheck the offset. If the offset is incorrect, repeat until the offset is correct.

  c. After adjusting sheave offset, perform the *Secondary Sheave Free Play* as described in this chapter.

4. Reinstall the drive belt as described in this chapter.

PRIMARY SHEAVE SERVICE

The primary sheave is mounted onto the left end of the crankshaft. Refer to **Figures 21-23** when performing procedures in this section.

Removal

The Polaris clutch puller (**Figure 24**) and an air gun will be required to remove the primary sheave. If you do not have access to air tools, a strap wrench (**Figure 25**) will be required to hold the sheave during removal and installation procedures.

1. Remove the drive belt as described in this chapter.

2. Secure the primary sheave with a strap wrench and loosen the sheave bolt. Remove the bolt, washer and bushing assembly (**Figure 26**).

3. Apply grease to the end of the puller (**Figure 24**) and oil the puller threads before installation. Then insert the puller (**Figure 27**) through the plug hole in the belly pan and into the primary

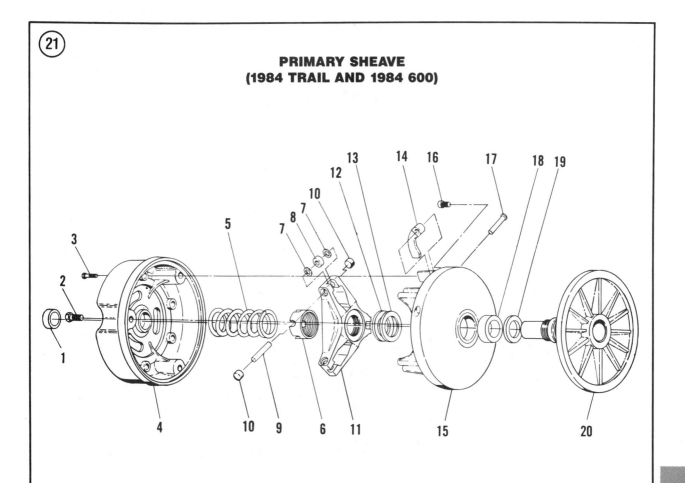

**PRIMARY SHEAVE
(1984 TRAIL AND 1984 600)**

1. Spacer
2. Bolt
3. Bolt
4. Cover
5. Spring
6. Jam nut
7. Fiber washer
8. Roller
9. Pin
10. Button
11. Spider
12. Washer
13. Washer
14. Weight
15. Movable sheave
16. Screw
17. Pin
18. Bearing
19. Insert
20. Fixed sheave

12

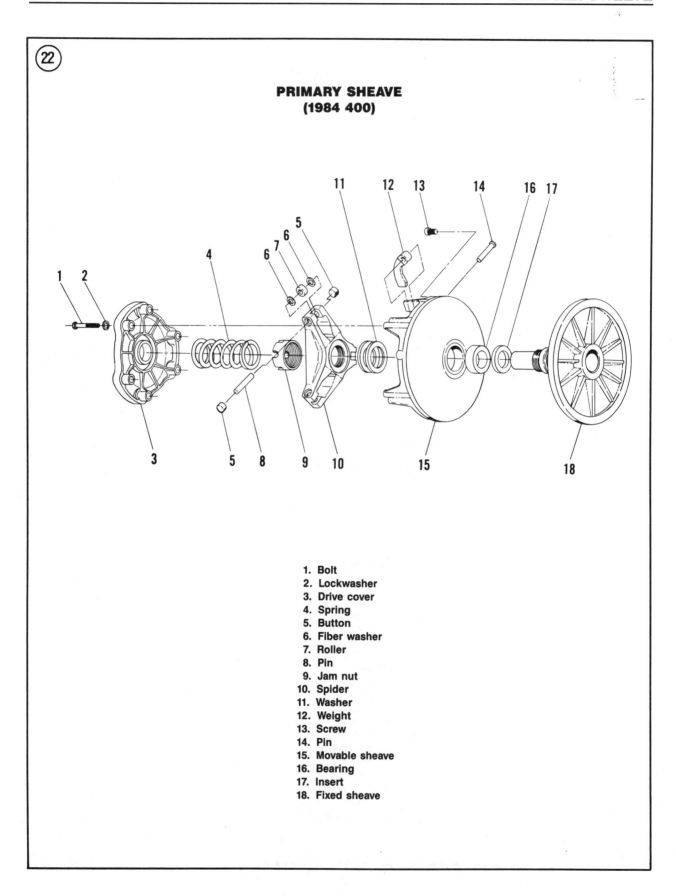

**PRIMARY SHEAVE
(1984 400)**

1. Bolt
2. Lockwasher
3. Drive cover
4. Spring
5. Button
6. Fiber washer
7. Roller
8. Pin
9. Jam nut
10. Spider
11. Washer
12. Weight
13. Screw
14. Pin
15. Movable sheave
16. Bearing
17. Insert
18. Fixed sheave

PRIMARY SHEAVE
(ALL 1985-1989 MODELS)

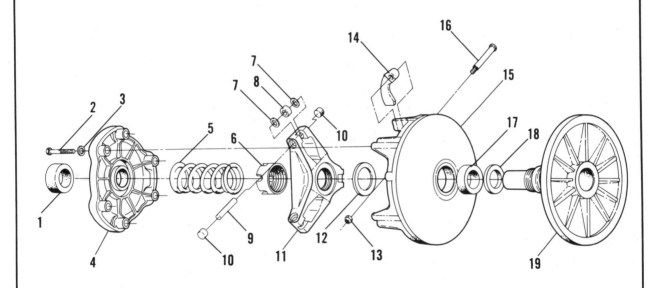

1. Bearing (1989 only)
2. Bolt
3. Washer
4. Drive plate
5. Spring
6. Jam nut
7. Fiber washer
8. Roller
9. Pin
10. Button
11. Spider
12. Spacer
13. Nut
14. Weight
15. Sliding sheave
16. Pin
17. Bearing
18. Insert
19. Fixed sheave

12

sheave. Make sure the puller threads engage with the primary sheave threads correctly.

4. Hold the sheave with the strap wrench and tighten the clutch puller with an impact wrench or breaker bar.

NOTE
It may be necessary to rap sharply on the head of the puller with a brass hammer to shock the primary sheave loose from the crankshaft.

5. When the primary sheave is loose, remove the puller.

6. Remove the primary sheave (**Figure 28**) from the crankshaft.

Disassembly

When servicing the primary sheave, a number of special tools are required. Do not attempt disassembly or reassembly without these tools:

WARNING
The primary sheave is under spring pressure. Attempting to disassemble or reassemble the primary sheave without the use of the specified special tools may cause severe personal damage. If you do not have access to the necessary tools, have the service performed by a dealer or other snowmobile mechanic.

a. Polaris drive clutch spider nut driver or equivalent. This tool is used to remove and install the spider jam nut. See A, **Figure 29**.

b. Polaris drive clutch holding fixture and holding tabs or equivalent. This tool is used to hold the drive clutch securely when loosening and tightening the spider jam nut. See B, **Figure 29**.

c. Polaris spider removal tool or equivalent. This tool is used to loosen and tighten the spider assembly. See C, **Figure 29**.

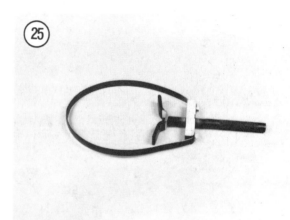

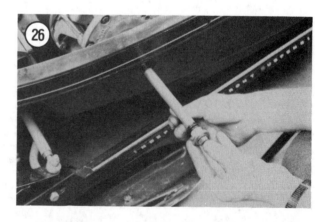

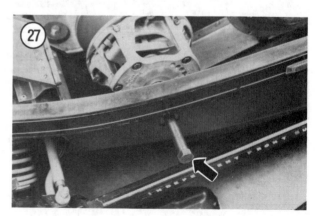

1. Scribe an alignment mark across each clutch component (**Figure 30**) with a felt-tip marker. These marks will be used during reassembly to ensure proper clutch alignment.

WARNING
*The cover plate (**A, Figure 31**) is spring loaded. To prevent it from flying off during removal, the plate must be removed as described in Step 2.*

2. Loosen the cover plate bolts (B, **Figure 31**) in small increments in a crisscross pattern until the plate is free from all spring tension. Remove the cover plate, bolts and washers (if used).
3. Remove the primary spring (**Figure 32**).

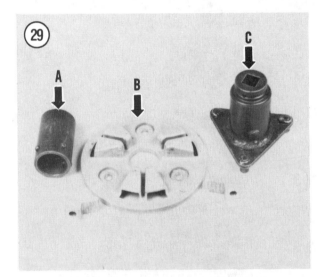

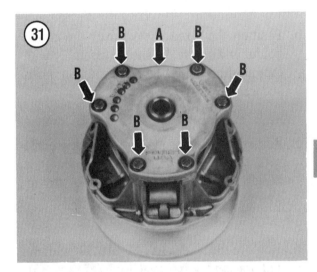

12

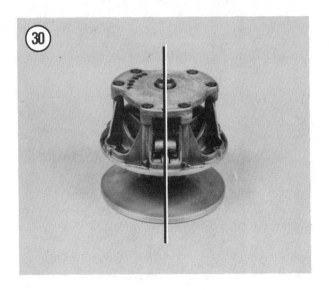

4. Remove the spider (A, **Figure 33**) as follows:

> *CAUTION*
> *The spider (**A, Figure 33**) is secured tightly onto the fixed sheave with Loctite and a high torque reading. To prevent primary sheave damage, make sure the fixed sheave is held securely when loosening the spider assembly as described in this procedure.*

a. Bolt the Polaris drive clutch holding fixture (B, **Figure 29**) onto a workbench or stand so that it is secure.

b. Position the fixed sheave over the clutch holder and secure the sheave assembly with the holding tabs. Tighten the tab bolts securely.

c. Remove the spider jam nut (B, **Figure 33**) with the Polaris drive clutch spider nut driver (**Figure 34**). Turn the driver *counterclockwise* to remove the nut.

d. Remove the spider (A, **Figure 33**) with the Polaris spider removal tool (**Figure 35**). Turn the removal tool *counterclockwise* to loosen the spider.

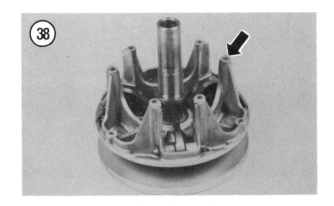

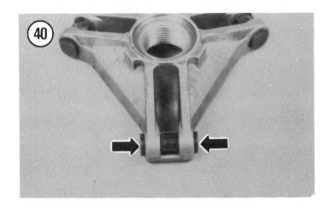

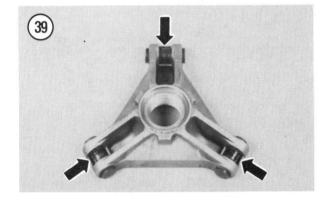

e. Remove the spider (**Figure 36**).

f. Remove the spacer washer(s) (**Figure 37**).

4. Remove the movable sheave assembly (**Figure 38**).

5. Disassemble the spider roller assembly (**Figure 39**) as follows:

   a. Remove the 2 buttons (**Figure 40**) from one spider arm with pliers. See **Figure 41**.

   b. Support the spider arm with a socket or similar tool and drive the pin (**Figure 42**) out of the spider.

   c. Remove the 2 washers (**Figure 43**) and roller (**Figure 44**).

12

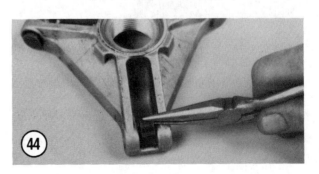

d. Store the roller and bushing assembly as a set until inspection and reassembly. Do not intermix parts.

e. Repeat for the 2 remaining roller assemblies.

6. Remove the shift weights (**Figure 45**) from the sliding sheave as follows:

a. *1984 models*: Remove the pins securing the weights and remove the weights.

b. *1985-on*: Remove the weight nut (A, **Figure 46**) and Allen bolt (B, **Figure 46**). Remove the weight (**Figure 47**).

c. Repeat for the 2 remaining weight assemblies.

Inspection

Refer to **Table 4** for shift weight specifications. High altitude clutch specifications are listed in Chapter Four.

CAUTION
Step 1 describes cleaning of the primary sheave assembly. Do not clean the rollers and weights with solvent as solvent will damage the bushings.

NOTE
When cleaning parts in Step 1, do not remove the alignment marks made before disassembly.

1. All parts, except the rollers and weights, should be cleaned thoroughly in solvent. Wipe the rollers and weights with a rag.

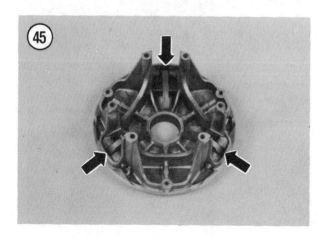

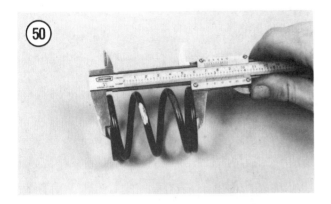

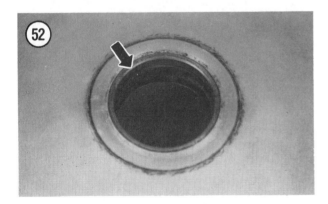

2. Remove all Loctite residue from all threads.

3. Check the sliding and fixed sheaves, spider and cover plate for cracks or damage.

4. Check the sheave drive belt surfaces (A, **Figure 48**) for rubber or rust buildup. For proper operation, the sheave surfaces must be *clean*. Remove rubber or rust debris with a course grade sandpaper and finish with #400 wet-or-dry sandpaper.

5. Check the fixed sheave threads (B, **Figure 48**) for strippage or other damage. Have damaged threads repaired by a machine shop.

6. Check the primary sheave spring (**Figure 49**) for cracks or distortion. If spring appears okay, measure its free length with a vernier caliper (**Figure 50**). Replace the spring if its free length is shorter than the length specified in **Table 5**. Replace the spring with the same color code and wire diameter.

7. Check the cover plate bushing (**Figure 51**) for wear, cracks or other signs of damage. If bushing is worn or damaged, replace it with a suitable size socket.

8. Check the sliding sheave bushing (**Figure 52**) for wear, cracks or other signs of damage. If bushing is worn or damaged, replace it. When installing a new bushing, drive it into the sheave with a suitable size socket.

9. Check the face of each shift weight (**Figure 53**) for wear, cracks or scoring. Then check the bushing (**Figure 54**) for excessive wear or

12

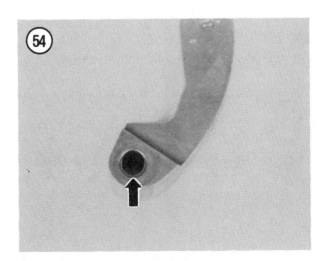

damage. Each weight must rotate freely at its pivot point. Insert the bolt (**Figure 53**) into the weight and check the bushing for wear. Replace the weight if the bushing is severely worn or damaged.

> *NOTE*
> *Make sure to replace shift weights with same weight. Refer to identification mark on weight (**Figure 53**). See **Table 4** for shift weight specifications.*

10. See **Figure 55**. Check the spider roller assembly for cracks, deep scoring or excessive wear. Then fit the roller onto its bushing and check for excessive wear or looseness.

> *NOTE*
> *If the outer roller surface is worn unevenly, the roller may have seized onto the bushing. If this condition is noted, check both parts for damage.*

11. Check the spider buttons (**Figure 55**) for excessive wear or damage. Then check the spider button contact surface on the sliding sheave for galling or excessive wear. Replace spider buttons as a set.

12. If there is any doubt as to the condition of any part, replace it with a new one.

Reassembly

1A. *1984*: Install the weights as follows:
a. Place the new weight into position.

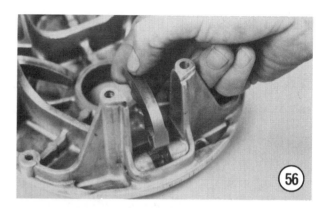

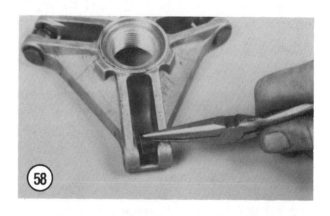

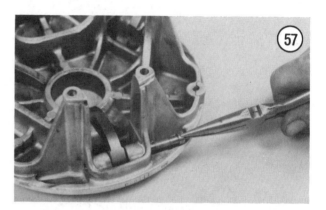

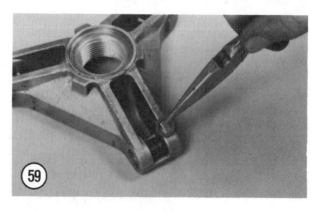

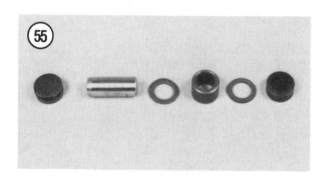

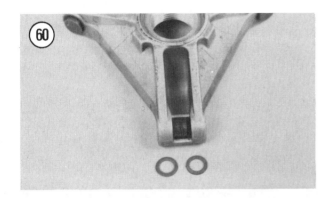

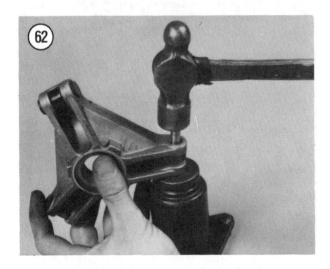

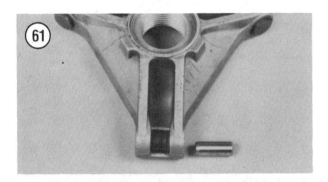

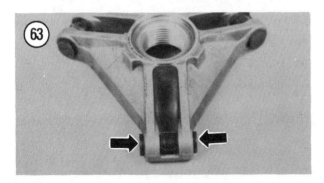

b. Install the pin into the moveable sheave and drive it through the weight with a punch.

c. Lubricate the weight pins with Polaris Snowmobile Drive Clutch Lubricant.

1B. Install the weights (**Figure 53**) as follows:

> *NOTE*
> *The 3 Allen bolts used to secure the shift weights must be installed so that each bolt head faces to the right of its weight as shown in **B, Figure 46**.*

a. Fit the weight (**Figure 56**) into the sliding sheave and slide the Allen bolt through the weight as shown in **Figure 57**.

b. Apply Loctite 242 (blue) onto the threads of a *new* locknut and install the locknut (A, **Figure 46**). Tighten the locknut until it bottoms against the aluminum sliding sheave tower; do not overtighten the locknut.

c. Pivot the weight back and forth to make sure it pivots smoothly.

> *NOTE*
> *If a weight does not pivot smoothly, the locknut is too tight.*

d. Repeat for each weight assembly.

2. Assemble the spider roller assembly (**Figure 55**) as follows:

a. Place a bushing into the middle of the tower (**Figure 58**).

b. Insert a washer (**Figure 59**) on both sides of the bushing. See **Figure 60**.

c. Install the pin (**Figure 61**) as follows. Place the spider over a socket or suitable tool and drive the pin through the bushing (**Figure 62**). Maintain the same amount of clearance on both sides of the pin.

d. Install guide buttons (**Figure 63**) on spider by tapping into place with a soft-faced hammer.

e. Repeat for each roller assembly.

12

3. Install the sliding sheave over the fixed sheave (**Figure 64**).

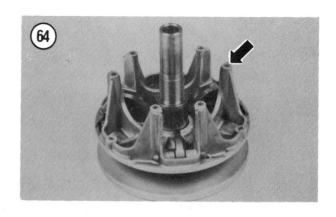

4. Install the spacer washer(s) over the fixed sheave shaft. See **Figure 65**A.

5. Apply Loctite 242 (blue) to the threaded portion on the fixed sheave.

6. Install the spider (**Figure 65B**) over the fixed sheave shaft, making sure to align the marks made on the spider and sliding sheave. Turn the spider *clockwise* when installing it.

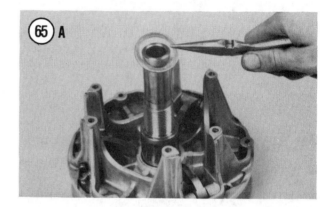

7. Secure the sheave assembly with the same tools used during disassembly. Then align the tool with the spider (**Figure 66**) and turn the wrench *clockwise* to tighten the spider. Tighten the spider to the tightening torque listed in **Table 7**.

> *NOTE*
> *After tightening the spider, work the sliding sheave up and down by hand. The sheave should slide freely without any sign of binding. If binding is noted, the shift weight Allen bolts may be too tight. Retighten bolts as described in Step 1B.*

8. Install the jam nut (**Figure 67**) by turning it *clockwise* on the fixed sheave shaft. Then tighten the jam nut using the same tool used during removal (**Figure 68**) to the torque specification listed in **Table 7**.

> *NOTE*
> *Place the cover plate (**Figure 69**) over the fixed sheave shaft (without primary spring) and check plate movement along shaft. If binding is noted during plate movement, the fixed sheave shaft is out-of-round. Repair shaft with a fine cut file until plate moves smoothly.*

9. Install the primary sheave spring (**Figure 70**).

10. Align the mark on the cover plate with the mark on the spider and sliding sheave and install the cover plate (**Figure 69**). Then apply Loctite 242 (blue) to the cover plate bolts and install the 6 bolts (**Figure 71**).

11. Tighten the bolts evenly in a crisscross pattern to the torque specification in **Table 7**.

Primary Sheave Bolt
Inspection

The primary sheave bolt assembly (**Figure 72**) should be checked whenever it is removed. Note the following:

1. Remove the bushing, washers, spacers and O-rings (**Figure 73**, typical) from the bolt.

2. Check the spacers and bushings for severe wear or damage. Check the O-rings (if used) for cracks or other damage. See **Figure 74**. Check the washers for cupping, cracks or other abnormal conditions.

3. Check the primary sheave bolt for cracks, bending or other damage. Check for worn or deformed threads. Check the hex head for rounding at the corners.

4. Replace worn or damaged parts as required.

5. Reassemble the bolt in the order shown in **Figure 72**. On models which use a long spacer, install the spacer so that the hole in spacer faces toward the bolt head. See **Figure 75**.

12

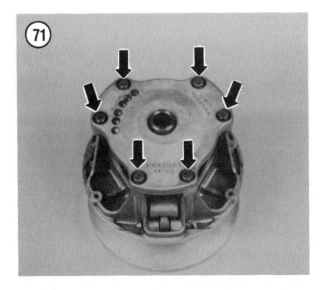

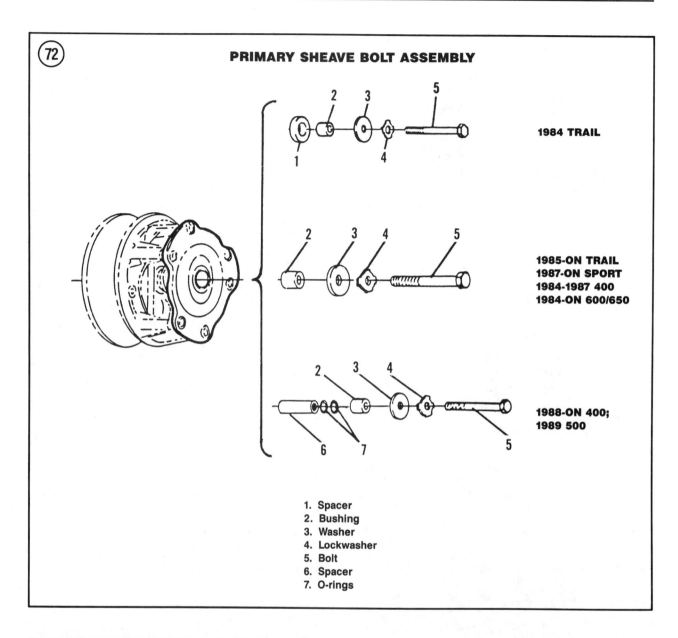

PRIMARY SHEAVE BOLT ASSEMBLY

1984 TRAIL

1985-ON TRAIL
1987-ON SPORT
1984-1987 400
1984-ON 600/650

1988-ON 400;
1989 500

1. Spacer
2. Bushing
3. Washer
4. Lockwasher
5. Bolt
6. Spacer
7. O-rings

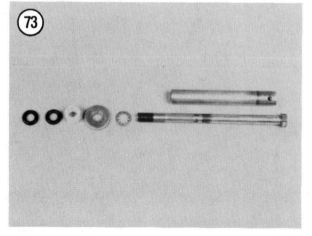

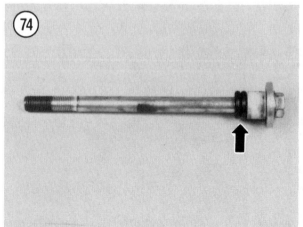

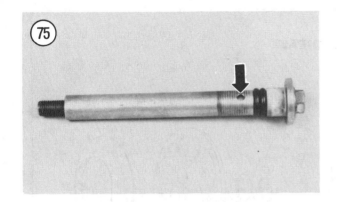

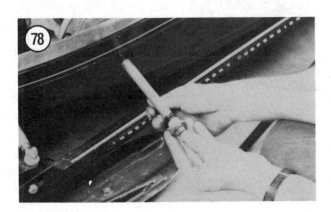

Installation

1. Clean the crankshaft taper with lacquer thinner or electrical contact cleaner.

2. Before installing the primary sheave onto the crankshaft, true the sheave taper bore with the Polaris tapered reamer (part No. 2870576). See **Figure 76**. Mount the primary sheave in a vise and apply a cutting oil to the sheave's tapered bore. Turn reamer clockwise by hand. Remove reamer and clean bore with lacquer thinner or electrical contact cleaner.

3. Slide the primary sheave (**Figure 77**) onto the crankshaft.

4. Apply a light weight oil onto the primary bolt threads. Then install the bolt assembly (**Figure 78**). Assemble bolt for your model in order shown in **Figure 72**.

> *WARNING*
> *Because of the tapered fit used between the crankshaft and primary sheave bore, **do not** use an air impact wrench to tighten the primary sheave bolt. In addition, an air impact wrench may loosen the jam nut and spider assembly.*

5. Secure the primary sheave with a strap wrench and tighten the bolt to the torque specification listed in **Table 7**. Remove the holder.

6. Check sheave alignment as described in this chapter.

7. Reinstall the drive belt and drive belt cover as described in this chapter.

8. Run snowmobile engine for 5 minutes and turn engine off. Retorque primary sheave belt to the torque specification listed in **Table 7**.

SECONDARY SHEAVE SERVICE

The secondary sheave is mounted onto the left-hand side of the jackshaft. Refer to **Figure 79** when performing procedures in this section.

12

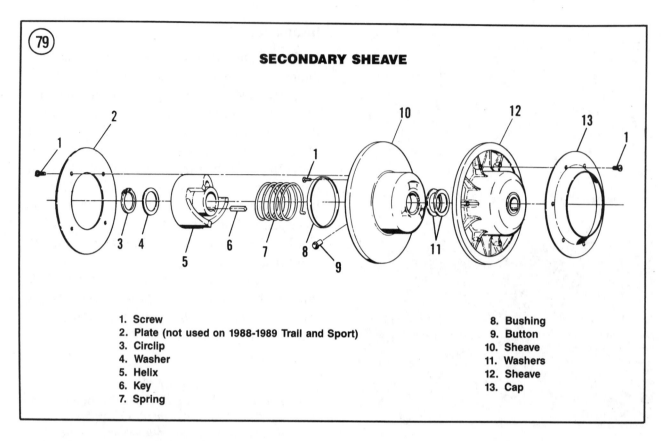

SECONDARY SHEAVE

1. Screw
2. Plate (not used on 1988-1989 Trail and Sport)
3. Circlip
4. Washer
5. Helix
6. Key
7. Spring

8. Bushing
9. Button
10. Sheave
11. Washers
12. Sheave
13. Cap

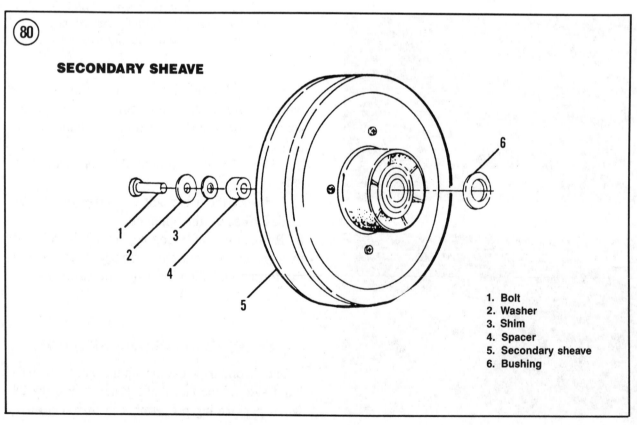

SECONDARY SHEAVE

1. Bolt
2. Washer
3. Shim
4. Spacer
5. Secondary sheave
6. Bushing

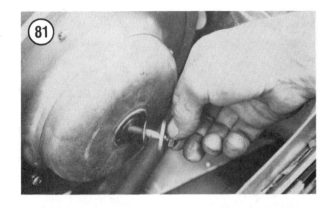

Removal

Refer to **Figure 80**.

1. Remove the drive belt as described in this chapter.

2. Loosen and remove the secondary sheave bolt and washers (**Figure 81**).

3. Pull the sheave out slightly and remove the spacer (**Figure 82**).

4. Remove the secondary sheave (**Figure 83**).

5. Remove the key (A, **Figure 84**) from jackshaft.

NOTE
*The washers (**B, Figure 84**) installed on the jackshaft behind the secondary sheave are used for sheave offset. Do not remove the washers unless jackshaft service is required. Make sure to install the same washers before installing the secondary sheave.*

Installation

1. Make sure the offset washers are installed on the jackshaft (B, **Figure 84**).

2. Place the key (A, **Figure 84**) in the jackshaft keyway.

3. Apply Loctite Anti-Seize Compound to the jackshaft.

4. Align the keyway in the secondary sheave with the key installed in the jackshaft and install the secondary sheave (**Figure 83**).

5. Install the spacer (**Figure 82**) into the end of the secondary sheave.

6. Install the secondary sheave bolt and washers (**Figure 80**). Tighten the bolt securely.

7. Perform the *Secondary Sheave Free Play Adjustment* in this chapter.

8. Install the drive belt as described in this chapter.

Secondary Sheave Free Play Adjustment

1. Remove the drive belt as described in this chapter.

12

2. The secondary sheave must have side clearance when installed on the jackshaft and with its bolt tightened securely. Check free play by moving the sheave back and forth by hand. The sheave should have approximately 0.5-1.0 mm (0.002-0.004 in.) axial clearance.

3. If the axial clearance is incorrect, proceed to Step 4.

4. Loosen and remove the secondary sheave bolt (**Figure 81**).

5. Adjust the secondary sheave free play by adding or subtracting the number of shims on the shaft. See 3, **Figure 80**. Adding shims increases free play while removing shims decreases it.

6. Install the secondary sheave bolt and the new number of shims. Tighten bolt securely.

7. Recheck secondary sheave free play. If the free play is not within specifications, repeat this procedure until free play is correct.

Disassembly

> *WARNING*
> *The secondary sheave is under spring pressure. Do not allow the helix to fly off when removing the circlip in Step 1.*

1. Compress the helix (A, **Figure 85**) by hand and remove the circlip (B, **Figure 85**).

2. Remove the washer (**Figure 86**).

> *NOTE*
> *The helix is equipped with 4 holes for spring adjustment (**Figure 87**). Record the hole number before removing the spring from the helix in Step 3. **Table 6** lists stock helix angle specifications.*

3. Lift the helix and spring off of the secondary sheave (**Figure 88**). Then record the spring hole number and remove the spring from the helix.

4. Separate the sliding and fixed sheaves (**Figure 89**).

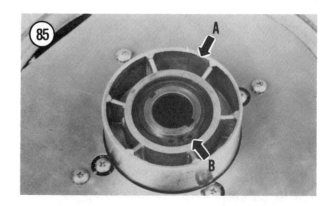

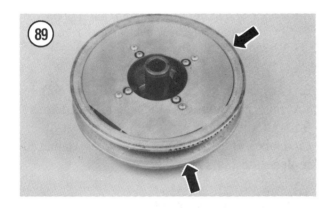

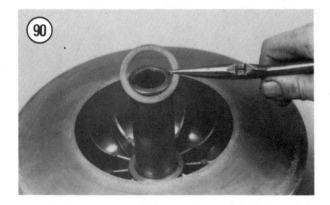

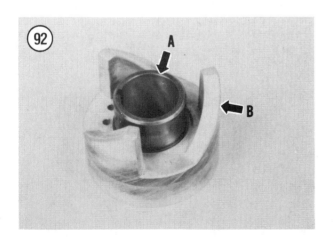

*The washers installed between the sliding sheaves are used to compensate for drive belt wear. If you are going to install a new drive belt during reassembly, refer to **Drive Belt Tension Adjustment** in this chapter.*

5. Remove the washer(s) (**Figure 90**).

Inspection

1. Clean the secondary sheave assembly in solvent.

2. Check the sheave surfaces (**Figure 91**) for cracks, deep scoring or excessive wear.

3. Check the sheave drive belt surfaces for rubber or rust buildup. For proper operation, the sheave surfaces must be *clean*. Remove rubber or rust debris with a course grade sandpaper and finish with #400 wet-or-dry sandpaper.

4. Check the movable sheave and helix mating surfaces (A, **Figure 92**) for cracks or deep scoring.

5. Check the keyway (**Figure 93**) in the helix for cracks or damage.

6. The nylon ramp shoes (**Figure 94**) provide a sliding surface between the fixed and sliding secondary sheaves. Because the nylon pads are rubbing against aluminum, wear is normally minimal. However, if a ramp shoe is gouged or damaged, smooth the surface with emery cloth;

12

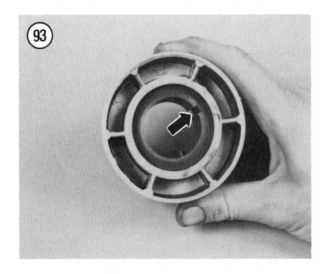

do not use a file. If a ramp shoe cannot be repaired, replace all the ramp shoes as a set.

7. Check the secondary sheave spring (**Figure 95**) for cracks or distortion. Replace the spring if necessary.

> *NOTE*
> *Secondary sheave spring fatigue is due mainly to metal fatigue from the constant twisting action. As the spring weakens, the secondary sheave will open quicker. This condition can be noticed when riding in mountain areas or deep snow; the machine will drive slower and have quite a bit less pulling power. Because it is difficult to gauge spring wear, you should count on replacing the spring every 2 years or 2,000 miles, whichever comes first. In addition, if you have noticed that the sheave shift point is 500 rpm below the maximum torque rpm, replace the spring.*

8. Check the helix ramps (B, **Figure 92**) for scoring, gouging or other signs of damage. Smooth the ramp area with #400 wet-or-dry sandpaper. If the ramp area is severely damaged, replace the helix.

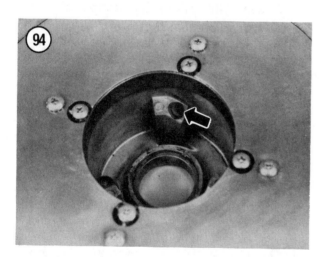

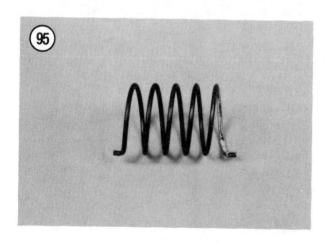

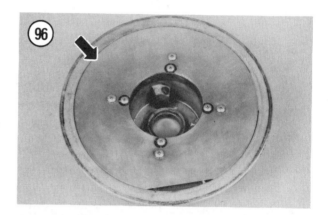

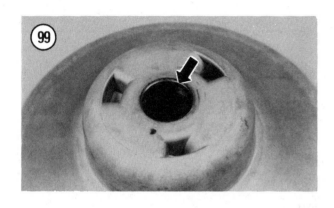

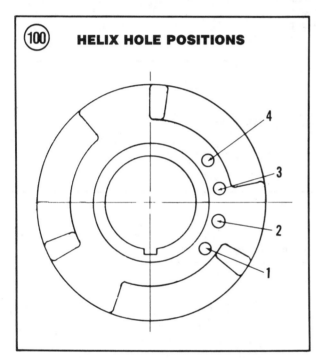

HELIX HOLE POSITIONS

4
3
2
1

9. Remove the plate (**Figure 96**) from the movable sheave, if used. Inspect the bushing (**Figure 97**) for looseness, severe wear or damage. Check bushing wear by inserting helix into bushing (**Figure 98**) and check play. Replace bushing if necessary. When installing a new bushing, apply Loctite Depend Activator and Adhesive to outside of bushing. Install bushing and allow 24 hours of cure time before using clutch. Reinstall plate (if used); tighten screws securely.

10. Check the smaller bushing (**Figure 99**) for wear or damage; replace bushing if necessary.

Assembly

1. Place the fixed sheave on the workbench so that the shaft faces up.

2. Install the shim(s) (**Figure 90**) over the shaft.

3. Install a low-temperature grease to the inside of the sliding sheave spline bore.

4. Install the sliding sheave (**Figure 89**) over the fixed sheave shaft so that the belt surfaces face together.

5. Install the spring into the secondary sheave as follows:

    a. Hook the end of the secondary spring into the hole recorded before removal. See **Figure 87**.

    b. If you did not identify the hole position during removal, set the spring to the No. 2 hole position (standard setting). See **Figure 100**.

6. Insert end of spring so that it engages the hole in the sliding sheave (**Figure 88**). Then turn helix 1/3 turn and align keyway in helix with keyway in fixed sheave; install key (**Figure 101**).

7. Compress helix by hand and install washer (**Figure 86**) and circlip (B, **Figure 85**).

8. Install the secondary sheave as described in this chapter.

12

Tables are on the following pages.

Table 1 OPERATING RPM

Model	RPM
Trail	
1984-1985	6,800
1986-1988	7,000
1989	6,600
Sport	6,500
400	
1984	*
1985	7,800
1986-1987	7,700
1988-1989	8,000
500	7,800
600	7,800
650	7,800
* Not specified.	

Table 2 DRIVE BELT SPECIFICATIONS

	mm	in.
Belt width limit	30.16	1 3/16
Belt tension	31.75	1 1/4

Table 3 CLUTCH ALIGNMENT

	mm	in.
Offset	15.88	5/8
Center-to-center distance	304.8	12.0

Table 4 PRIMARY SHEAVE SHIFT WEIGHT SPECIFICATIONS

Model	Shift weight— gram
Trail	
1984	M1 Mod—46.0
1985	N1 Mod—51.5
1986	10—51.5
1987-1989	10—51.7
Sport	10—51.7
400	
1984	*
1985	N-1 Mod—51.5
1986	10 Mod—50.0
1987	10M—49.5
1988-1989	10M Blue—47.5
(continued)	

Table 4 PRIMARY SHEAVE SHIFT WEIGHT SPECIFICATIONS (continued)

Model	Shift weight— gram
500	10M—49.5
600	
1984	05—53.5
1985	05—53.5
1986	15—55.5
1987	10A—55.0
650	10A—55.0
* Not specified.	

Table 5 PRIMARY SHEAVE SPRING

Model	Color code	Wire diameter in.	Free length* in.
Trail			
1984	Green	0.177	3.05
1985	Brown	0.200	3.06
1986-1989	Red/white	0.192	3.59
Sport	Red/white	0.192	3.59
400			
1984	**	**	**
1985	Gold	0.207	3.25
1986-1987	Blue/gold	0.207	3.50
1988-1989	Blue	0.207	**
500	Silver	0.208	3.12
600			
1984-1985	Gold	0.207	3.25
1986-1987	Blue/gold	0.207	3.50
650	Gold	0.207	3.25
* ± 0.125 in.			
** Not specified.			

12

Table 6 SECONDARY SHEAVE HELIX ANGLE SPECIFICATIONS

Model	Helix angle
Trail	
1984	36°
1985	34°
1986	36°
1987-1989	36°
Sport	34°
400	
1984	*
1985	34°
1986	34°
1987	34°
1988-1989	34°
	(continued)

Table 6 SECONDARY SHEAVE HELIX ANGLE SPECIFICATIONS (continued)

Model	Helix angle
500	36°
600	
1984	34°
1985	34°
1986	34°
1987	34°
650	34°
* Not specified.	

Table 7 CLUTCH TIGHTENING TORQUES

	N·m	ft.-lb.,
Spider	272	200
Spider jam nut	320	235
Primary sheave bolt	54-61	40-45
Primary sheave cover plate	11	100 in.-lb.

Chapter Thirteen

Disc Brake

All Indy models are equipped with a front disc brake. The Polaris brake system consists of the following components:

a. Brake lever.
b. Master cylinder.
c. Brake caliper assembly.
d. Brake pads.
e. Brake disc (mounted on the jackshaft).
f. Hydraulic hoses.
g. Brake fluid.

The brake is actuated by hydraulic fluid and controlled by a hand lever on the master cylinder. When the brake lever is pulled in, hydraulic pressure is created within the master cylinder. This hydraulic pressure is then transferred through a hydraulic hose to move the brake caliper piston which activates the brake pads to press against the brake disc. As the brake pads wear, the caliper piston moves out automatically to keep the brake lever free play constant. Over a period of time, the gradual repositioning of the piston will drop the level of the hydraulic fluid in the reservoir. The brake fluid level in the master cylinder reservoir should be checked frequently and topped off as required with DOT 3 brake fluid.

Table 1 is at the end of the chapter.

Periodic Adjustment

Refer to Chapter Three for procedures dealing with brake fluid level and brake lever free play.

Disc Brake System Service

Consider the following when servicing the front disc brake.

1. When working on hydraulic brake systems, it is necessary that the work area and all tools be absolutely clean. Any tiny particles of foreign matter and grit in the caliper assembly or master cylinder can damage the components.

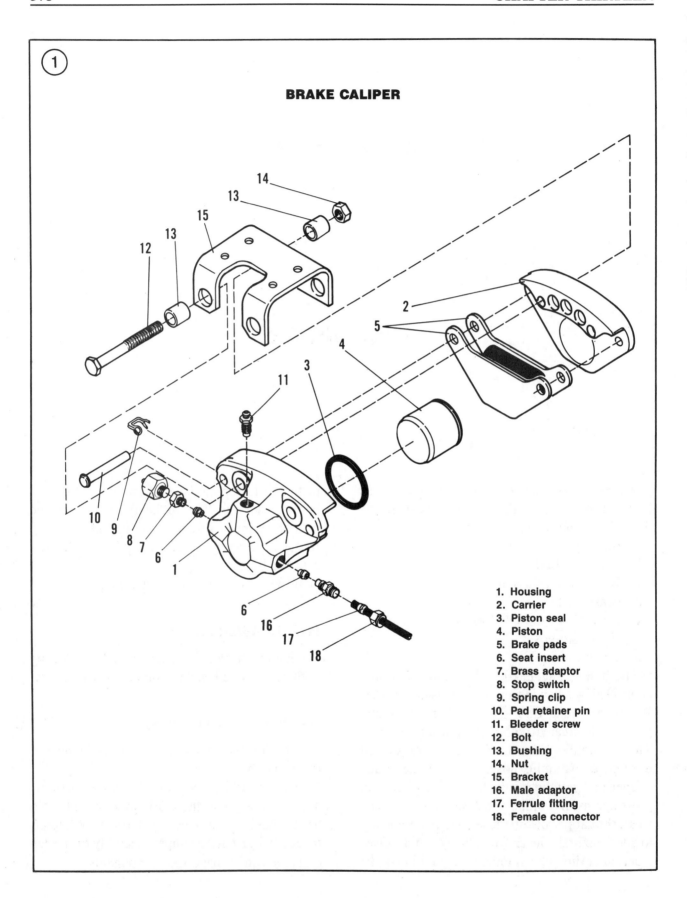

BRAKE CALIPER

1. Housing
2. Carrier
3. Piston seal
4. Piston
5. Brake pads
6. Seat insert
7. Brass adaptor
8. Stop switch
9. Spring clip
10. Pad retainer pin
11. Bleeder screw
12. Bolt
13. Bushing
14. Nut
15. Bracket
16. Male adaptor
17. Ferrule fitting
18. Female connector

2. Use only DOT 3 brake fluid from a sealed container.

WARNING
Do not intermix silicone based (DOT 5) brake fluid as it can cause brake component damage leading to brake system failure.

3. Do not allow disc brake fluid to contact any plastic parts or painted surfaces as damage will result.

4. Always keep the master cylinder reservoir and spare cans of brake fluid closed to prevent dust or moisture from entering. This would result in brake fluid contamination and brake problems.

5. Use only disc brake fluid (DOT 3) to wash parts. Never clean any internal brake components with solvent or any other petroleum-based cleaners. Solvents will cause the seals to sell and distort and require replacement.

6. Whenever *any* component has been removed from the brake system, the system is considered "opened" and must be bled to remove air bubbles. Also, if the brake feels "spongy," this usually means there are air bubbles in the system and it must be bled. For safe brake operation, refer to *Bleeding the System* in this chapter for complete details.

BRAKE PAD REPLACEMENT

There is no recommended time interval for changing the friction pads in the front disc brake. Pad wear depends greatly on riding habits and conditions.

To maintain an even brake pressure on the disc, always replace both pads in the caliper at the same time.

Refer to **Figure 1** for this procedure. The brake caliper is mounted inside the chaincase (A, **Figure 2**).

NOTE
*It is not necessary to disconnect the brake hose (**B, Figure 2**) from the inner caliper half when replacing the front brake pads.*

13

1. Open and secure the shroud.
2. Remove the chaincase as described in Chapter Fourteen.

NOTE
Do not operate the brake lever after removing the brake caliper in Step 3. Operating the brake lever will force the piston out of the caliper. If the brake caliper is going to be removed from the brake disc, place a wood shim or piece of hose between the brake pads to prevent the piston from unseating due to the accidental use of the brake lever.

3. Lift the brake caliper assembly (**Figure 3**) off of the brake disc.

4. Turn the brake caliper over and remove the 2 spring clips (**Figure 4**) from the pad retaining pins. See **Figure 5**.

5. Remove the 2 pad retaining pins (**Figure 6**) from the brake caliper.

6. Remove the 2 brake pads from between the brake caliper halves (**Figure 7**).

7. Check the brake pad friction surface for scoring, cracks or excessive wear. Check the brake pad holders for cracks or damage. Replace both brake pads as a set as required.

8. When new brake pads are installed in the caliper, the master cylinder brake fluid will rise as the caliper piston is repositioned. Clean the top of the master cylinder of all dirt. Remove the cap (**Figure 8**) and diaphragm from the master cylinder and slowly push the caliper piston into the caliper. Constantly check the reservoir to make sure brake fluid does not overflow. Remove fluid, if necessary, prior to it overflowing. The piston should move freely in the caliper bore. If the piston doesn't move smoothly and there is evidence of it sticking in the caliper, the caliper should be removed and serviced as described under *Brake Caliper Rebuilding* in this chapter.

9. Install brake pads as follows:

   a. Install the 2 brake pads into the caliper assembly so that the friction surface on both pads face each other (**Figure 7**).

   b. Install the pad retaining pins through the caliper and brake pad assembly. Note that one end of each pin has a shoulder. Install the retaining pins so that the shoulder faces against the inner caliper half (**Figure 6**).

   c. Secure both retaining pins with a spring clip (**Figure 5**). See **Figure 4**.

   d. Place the brake caliper assembly over the brake disc so that a brake pad is placed on each side of the disc (**Figure 3**). Do not force the brake caliper over the disc or you may damage the brake pads. If the caliper is difficult to install, the piston needs to be pushed further into the caliper bore.

10. Install the chaincase as described in Chapter Fourteen.

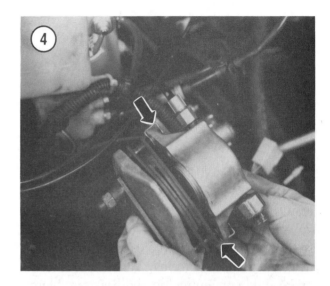

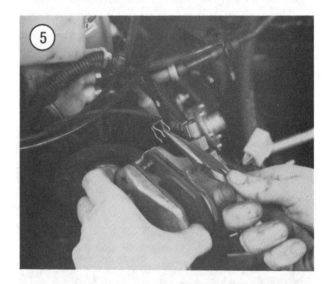

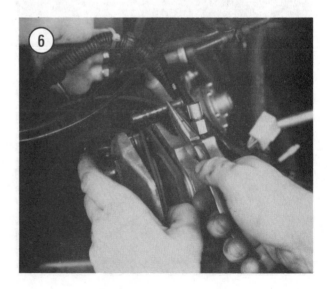

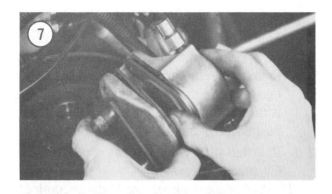

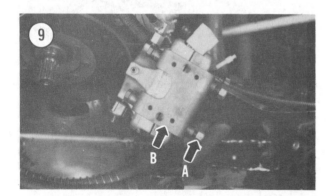

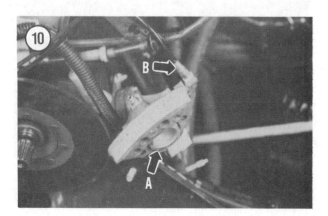

11. Activate the brake lever as many times as necessary as it takes to refill the cylinder in the caliper and correctly locate the pads.

12. Refill the master cylinder reservoir, if necessary, to maintain the correct brake fluid level. Install the diaphragm and top cover (**Figure 8**).

> *WARNING*
> *Use brake fluid clearly marked DOT 3 from a sealed container. Other types may vaporize and cause brake failure. Always use the same brand name; do not intermix as many brands are not compatible.*

> *WARNING*
> *Do not operate the snowmobile until you are sure the brake is operating correctly with full hydraulic advantage. If necessary, bleed the brake system as described in this chapter.*

BRAKE CALIPER

Removal

1. Open and secure the shroud.

> *NOTE*
> *If the brake caliper is going to be serviced, perform Step 2. If the brake caliper is not going to be serviced, perform Step 3.*

2. Disassemble and remove the brake caliper (**Figure 1**) as follows:
   a. Remove the brake pads as described in this chapter.
   b. Remove the 2 nuts and bolts (A, **Figure 9**) securing the bracket to the caliper halves. Then remove the bracket (B, **Figure 9**) and the outer caliper half.
   c. Place a rag over the end of the caliper piston (A, **Figure 10**) and operate the brake lever to force the piston out of the caliper bore. Remove the piston and wrap a rag around the caliper to prevent brake fluid from leaking into the engine compartment.

13

d. Loosen the brake hose connector at the caliper and disconnect the hose (B, **Figure 10**). Remove the inner brake caliper half. Tie a small plastic bag around the end of the brake hose to prevent brake fluid from dripping into the engine compartment.

e. Overhaul the brake caliper as described in this chapter.

3. Remove the brake caliper as follows:

a. Remove the chaincase as described in Chapter Fourteen.

b. Lift the brake caliper off of the brake disc (**Figure 3**).

c. If it is unnecessary to disconnect the brake hose at the caliper, insert a spacer (wood or vinyl tubing) between the brake pads. That way if the brake lever is accidently squeezed, the piston will not be forced out of the caliper bore. If this does happen, the caliper may have to be disassembled to reseat the piston and the system will have to be bled. After securing the spacer between the brake pads, place the caliper in the engine compartment safely out of the way. Make sure the caliper is not supported by the brake hose; use a Bunjee cord to hold the caliper if necessary.

d. If caliper removal is required, loosen the brake hose connector at the caliper and disconnect the hose (B, **Figure 10**). Remove the brake caliper assembly. Tie a small plastic bag around the end of the brake hose to prevent brake fluid from dripping into the engine compartment. Then plug the brake hose opening on the caliper to prevent the entry of dirt or moisture.

Installation

1. If the brake caliper was not disassembled, install it as follows:

a. Reconnect the brake hose at the brake caliper (if previously disconnected).

b. Remove the spacer from between the brake pads, if used.

c. Place the brake caliper assembly over the brake disc so that a brake pad is placed on each side of the disc (**Figure 3**). Do not force the brake caliper over the disc or you may damage the brake pads. If the caliper is difficult to install, the piston needs to be pushed further into the caliper bore.

d. Reinstall the chaincase as described in Chapter Fourteen.

2. If the brake caliper was disassembled, assemble and install it as follows:

a. Reconnect the brake hose at the outer caliper half (B, **Figure 10**).

b. Assemble the caliper halves, spacers, through bolts and bracket (B, **Figure 9**) as shown in **Figure 1**.

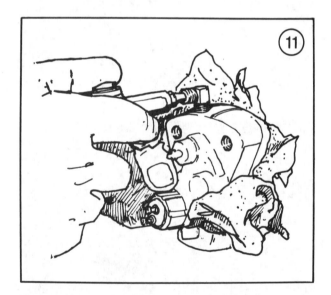

c. Apply Loctite 242 (blue) over the end of the through bolts (A, **Figure 9**) and install the nuts. Tighten the nuts to the torque specification listed in **Table 1**.

d. Reinstall the brake pads as described in this chapter.

3. After the brake caliper and chaincase assemblies are installed, bleed the brake as described in this chapter.

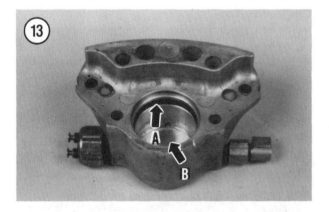

Caliper Overhaul

This procedure describes complete overhaul of the brake caliper assembly. Polaris does not provide service specifications for any caliper component except brake pads; replace any part that appears worn or damaged.

Refer to **Figure 1** when performing this procedure.

1. If the caliper piston was not removed as described under caliper removal, remove it as follows:

a. Either wrap the caliper half and piston with a heavy cloth or place a shop cloth or piece of soft wood over the end of the piston.

b. Perform this step over and close down to a workbench top. Hold the caliper half with the piston facing away from you.

WARNING
In the next step, the piston may shoot out of the caliper half like a bullet. Keep your fingers out of the way. Wear shop gloves and apply air pressure gradually.

c. Apply air pressure in short spurts to the hydraulic fluid port (**Figure 11**) and force the piston out of the caliper half. Remove the piston (**Figure 12**).

CAUTION
In the following step, do not use a sharp tool to remove the piston seal from the caliper bore. The bore can be easily scratched.

2. Carefully remove the piston seal (A, **Figure 13**) from the seal groove with a piece of wood or plastic. Discard the piston seal.

3. Check the piston (**Figure 14**) and the piston bore (B, **Figure 13**) for deep scratches or other obvious wear marks. If either part is less than perfect, replace it.

4. Remove the bleed screw (A, **Figure 15**) and check it for wear or damage. Clean the screw with compressed air.

13

5. Inspect the caliper body halves for damage; replace damaged caliper halves as required.

6. Inspect the hydraulic fluid passageway in the caliper bore. Apply compressed air to the opening to clean it.

7. Remove the brake switch (B, **Figure 15**) and hose connection joint (C, **Figure 15**) from the caliper half. Clean the hose connection assembly with compressed air.

8. Clean all parts (except brake pads) with DOT 3 brake fluid.

9. Reinstall the brake switch and hose connection.

10. Soak the new piston seal in fresh brake fluid. Coat the inside of the cylinder with fresh brake fluid prior to the assembly of parts.

> *WARNING*
> *Never reuse a piston seal that has been removed. Very minor damage or age deterioration can make the seals useless.*

11. Carefully install the new piston seal into the caliper groove (A, **Figure 13**). Make sure the seal seats evenly all the way around the groove.

12. Coat the piston with DOT 3 brake fluid.

13. Align the piston (**Figure 16**) with the caliper half so that the *open end faces out*. Then push the piston evenly into the caliper bore. Make sure that you do not cock the piston to one side during installation; push the piston in until it bottoms out (**Figure 17**).

14. Install the brake caliper assembly as described in this chapter.

MASTER CYLINDER

Removal/Disassembly

1. Remove the left-hand grip (A, **Figure 18**) from the handlebar as described in Chapter Fifteen.

> *CAUTION*
> *Cover the fuel tank and instrument panel with a heavy cloth or plastic tarp to protect them from accidental brake fluid*

spills. Wash spilt brake fluid off of all surfaces immediately. Use soapy water and rinse completely.

2. Disconnect the brake hose at the master cylinder. Then wrap the end of the hose with a small plastic bag, taped in place; this will prevent fluid spillage and at the same time prevents the entry of moisture and contaminants into the brake line.

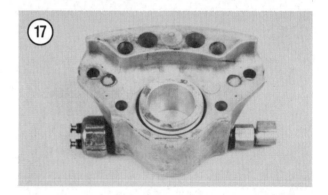

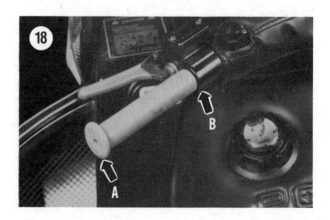

3. Loosen the Allen screw at the bottom of the master cylinder housing and slide the master cylinder (B, **Figure 18**) off of the handlebar.

4. Install by reversing these steps. Note the following.

5. Slide the master cylinder onto the handlebar and secure it by tightening the Allen screw at the bottom of the master cylinder housing.

6. Install the brake hose onto the master cylinder. Tighten the hose securely.

7. Bleed the brake as described in this chapter.

8. Reinstall the left-hand grip as described in Chapter Fifteen.

> *WARNING*
> *Do not operate the snowmobile until the*
> *front brake is working correctly.*

Disassembly

Refer to **Figure 19** for this procedure.

1. Remove the master cylinder as described in this chapter.

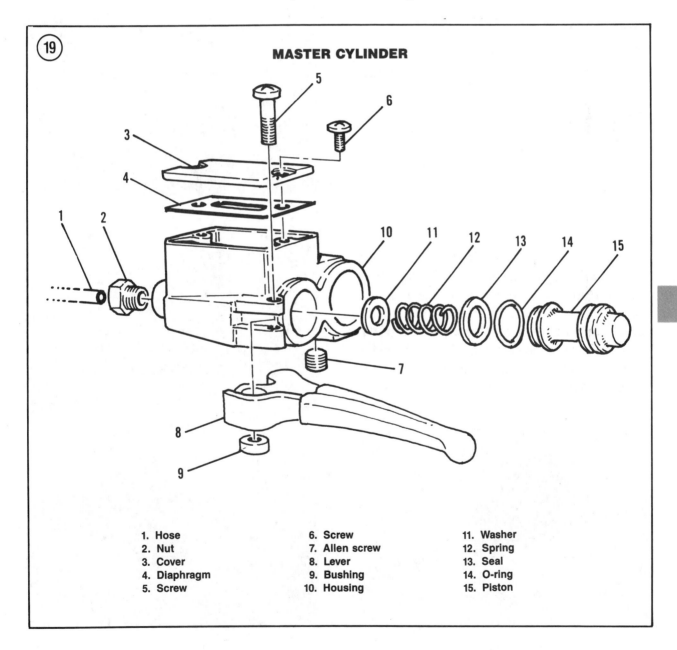

MASTER CYLINDER

1. Hose	6. Screw	11. Washer
2. Nut	7. Allen screw	12. Spring
3. Cover	8. Lever	13. Seal
4. Diaphragm	9. Bushing	14. O-ring
5. Screw	10. Housing	15. Piston

13

2. Remove the cover and diaphragm from the master cylinder and empty the old fluid into a disposable container.

3. Remove the nut and screw holding the brake lever in place. Then remove the brake lever (**Figure 20**) and collar.

4. Pull the piston (**Figure 21**) out of the master cylinder. Then remove the spring (**Figure 22**) and washer (**Figure 23**) from the master cylinder.

5. If necessary, disconnect the brake hose at the master cylinder (**Figure 24**).

Inspection

Polaris does not provide wear limit specifications on any of the master cylinder components. Replace worn or damaged parts as required.

Refer to **Figure 19** for this procedure.

1. Clean all parts in fresh DOT 3 brake fluid.

2. The piston assembly is identified in **Figure 25**:

   a. Seal.

   b. O-ring.

   c. Piston.

Check the seal (A) and O-ring (B) on the piston (**Figure 25**). Replace the seal and O-ring if they are worn, softened, swollen or damaged. Check the piston surface (C, **Figure 25**) for cracks, scoring or other damage. Check the end of the piston where it contacts the hand lever for excessive wear or damage. Replace worn or damaged parts as required.

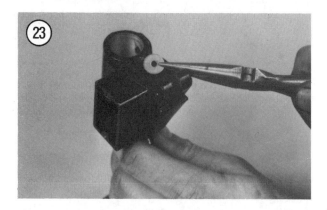

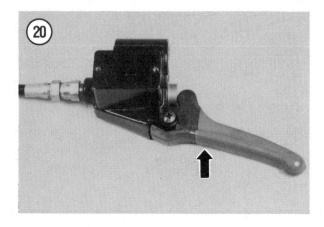

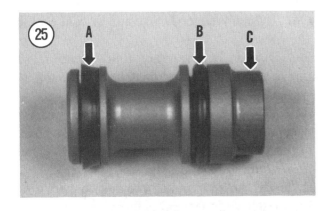

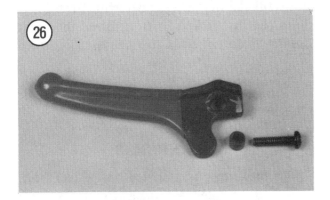

3. Check the piston spring and washer for damage. Replace damaged parts as required.

4. Check the pivot hole in the hand lever (**Figure 26**) and master cylinder for cracks, elongation or damage. Check the hand lever collar (**Figure 26**), screw and nut for damage.

5. Make sure the passages in the bottom of the brake fluid reservoir are clear (**Figure 27**). A plugged relief port will cause the pads to drag on the disc.

6. Check the reservoir cap and diaphragm (**Figure 28**) for damage and deterioration and replace as necessary.

Reassembly

Refer to **Figure 19** for this procedure.

1. Install a new seal (A, **Figure 25**) and O-ring (B, **Figure 25**) onto the piston as required.

2. Soak the piston assembly in fresh DOT 3 brake fluid. Coat the inside of the cylinder with fresh brake fluid prior to the assembly of parts.

NOTE
*When installing the piston assembly in Step 3, do not allow the seal (**A, Figure 25**) to turn inside out. This will cause seal damage and allow brake fluid to leak within the cylinder bore.*

3. Install the piston assembly (**Figure 29**) as follows:

   a. Install the washer (**Figure 23**) into the cylinder bore.

13

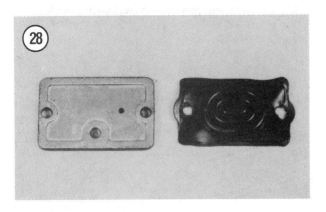

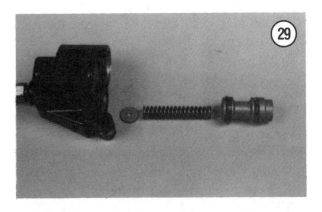

b. Install the spring (**Figure 22**).

c. Install the piston into the cylinder bore so that the seal end (A, **Figure 25**) faces the spring. See **Figure 30**.

NOTE
Do not allow the piston to fly out after its installation.

4. Assemble the brake lever assembly onto the master cylinder. Make sure to install the collar into the brake lever before installing the lever onto the master cylinder. Install the screw and nut. Tighten the nut and check the operation of the brake lever. The lever must operate smoothly with no sign of binding or tightening.

5. Install the diaphragm and top cap until the master cylinder is reinstalled onto the handlebar and the brakes bled.

6. Reinstall the master cylinder as described in this chapter.

FRONT BRAKE HOSE REPLACEMENT

There is no factory-recommended replacement interval, but it is a good idea to replace the flexible brake hose every 2-4 years or when it shows signs of cracking or damage.

CAUTION
Cover the fuel tank and instrument panel with a heavy cloth or plastic tarp to protect them from accidental brake fluid spills. Wash spilt brake fluid off of all surfaces immediately. Use soapy water and rinse completely.

1. Drain the hydraulic brake fluid from the front brake system as follows:

a. Attach a hose to the bleed valve on the caliper assembly.

b. Place the loose end of the hose in a clean container to catch the brake fluid (**Figure 31**).

c. Open the bleed valve on the caliper and apply the front brake lever until the brake fluid is pumped out of the system.

d. Tighten the bleed valve and disconnect the hose.

e. Dispose of this brake fluid—*never* reuse brake fluid. Contaminated brake fluid may cause brake failure.

2. Disconnect the brake hose at the master cylinder and caliper. Remove the old brake hose, noting the routing so that you can install the new hose along the same path.

3. Install the new brake hose and connect it to the master cylinder and caliper.

4. Refill the master cylinder with fresh brake fluid clearly marked DOT 3. Bleed the brake system as described in this chapter.

WARNING
Do not ride the snowmobile until you are sure that the brakes are operating properly.

BRAKE DISC

The brake disc is mounted on the jackshaft. Refer to Chapter Fourteen for all service procedures related to the brake disc.

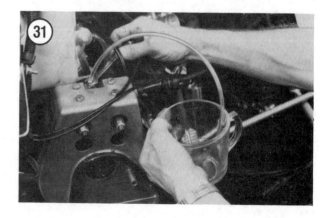

BLEEDING THE SYSTEM

This procedure is necessary only when the brake feels spongy, there is a leak in the hydraulic system, a component has been replaced or the brake fluid has been replaced.

1. Connect a length of clear hose to the bleeder valve on the caliper (**Figure 31**). Place the other end of the hose into a clean container. Fill the container with enough fresh brake fluid to keep the end submerged. The hose should be long enough so that a loop can be made higher than the bleeder valve to prevent air from being drawn into the caliper during bleeding.

CAUTION
Cover the fuel tank and instrument panel with a heavy cloth or plastic tarp to protect them from accidental brake fluid spills. Wash spilt brake fluid off of all surfaces immediately. Use soapy water and rinse completely.

2. Clean the top of the master cylinder of all dirt and foreign matter. Remove the cap and diaphragm (**Figure 8**). Fill the reservoir to about 6.3-7.9 mm (1/4-5/16 in.) from the top. Reinstall the diaphragm to prevent the entry of dirt and moisture.

WARNING
Use brake fluid clearly marked DOT 3 only. Others may vaporize and cause brake failure. Always use the same brand name; do not intermix the brake fluids, as many brands are not compatible.

3. Slowly apply the brake lever several times. Hold the lever in the applied position and open the bleeder valve about 1/2 turn (**Figure 31**). Allow the lever to travel to its limit. When this limit is reached, tighten the bleeder screw. As the brake fluid enters the system, the level will drop in the master cylinder reservoir. Maintain the level at about 6.3-7.9 mm (1/4-5/16 in.) from the top of the reservoir to prevent air from being drawn into the system.

4. Continue to pump the lever and fill the reservoir until the fluid emerging from the hose is completely free of air bubbles. If you are replacing the fluid, continue until the fluid emerging from the hose is clean.

5. Hold the lever in the applied position and tighten the bleeder valve. Remove the bleeder tube.

6. If necessary, add fluid to correct the level in the master cylinder reservoir. It must be above the level line.

7. Install the cap and tighten the screws. See **Figure 8**.

8. Test the feel of the brake lever. It should feel firm and should offer the same resistance each time it's operated. If it feels spongy, it is likely that air is still in the system and it must be bled again. When all air has been bled from the system and the brake fluid level is correct in the reservoir, double-check for leaks and tighten all fittings and connections.

WARNING
Before operating the snowmobile, make certain that the front brake is working correctly by operating the lever several times. Then make the test ride a slow one at first to make sure the brake is working correctly.

13

Table 1 BRAKE TIGHTENING TORQUES

	N·m	ft.-lb.
Brake caliper through bolt	34-41	25-30
Caliper bracket cap screws	11	8

Chapter Fourteen

Chaincase, Jackshaft and Driveshaft

This chapter describes complete service procedures for the chaincase, jackshaft and driveshaft. **Table 1** and **Table 2** are at the end of the chapter.

DRIVE CHAIN AND SPROCKETS

Removal

Refer to **Figure 1** when performing this procedure.

1. Open the shroud.
2. Remove the exhaust pipe as described in Chapter Seven.
3. Place a number of shop rags underneath the chaincase cover.

> *NOTE*
> *The chaincase is filled with oil and the cover is not equipped with a drain plug. When the chaincase cover is removed, try to absorb as much of the oil on the rags as possible.*

4. *Liquid cooled models:* Disconnect the coolant hose clamp at the chaincase (**Figure 2**).

5. Remove the bolts and washers holding the chaincase cover (**Figure 3**) to the chaincase. Remove the cover and gasket. Discard the gasket.

6. Wipe up as much oil as possible with the rags.

7. Loosen the chain adjuster locknut and loosen the chain adjuster bolt (A, **Figure 4**). Loosen the adjuster until the chain is completely loose.

8. Mark the outer face of the sprockets for reference during installation. Have an assistant apply the front brake and loosen the upper and lower sprocket bolts (B, **Figure 4**). Remove the bolts and lockwashers.

9. Remove the chain and sprockets (**Figure 5**) as an assembly.

10. Remove the spacer from the jackshaft (**Figure 6**).

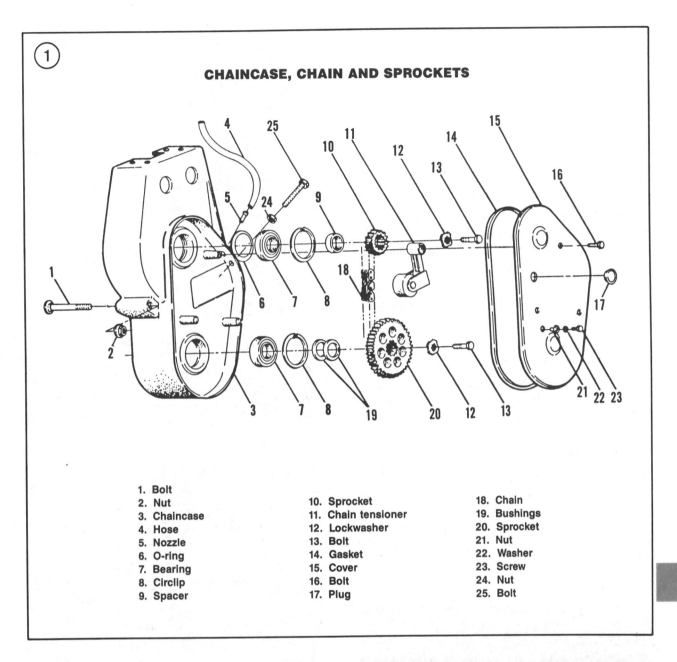

CHAINCASE, CHAIN AND SPROCKETS

1. Bolt
2. Nut
3. Chaincase
4. Hose
5. Nozzle
6. O-ring
7. Bearing
8. Circlip
9. Spacer
10. Sprocket
11. Chain tensioner
12. Lockwasher
13. Bolt
14. Gasket
15. Cover
16. Bolt
17. Plug
18. Chain
19. Bushings
20. Sprocket
21. Nut
22. Washer
23. Screw
24. Nut
25. Bolt

14

11. Remove the washers from the driveshaft (**Figure 7**).

12. Remove the chain tensioner assembly from the chaincase housing.

Inspection

Refer to **Figure 1** for this procedure. Drive chain specifications are listed in **Table 1**.

1. Clean all components thoroughly in solvent. Remove any gasket residue from the cover and housing machined surfaces.

2. Visually check the drive and driven sprockets (**Figure 8**) for cracks, deep scoring, excessive wear or tooth damage. Check the splines for the same abnormal conditions.

3. Check the chain (**Figure 9**) for cracks, excessive wear or pin breakage.

4. If the sprockets and/or chain are severely worn or damaged, replace all 3 components as a set.

5. Visually check the chain tensioner roller (**Figure 10**) for cracks, deep scoring or excessive wear. Replace the chain tensioner assembly if necessary.

6. Check the chain tensioner pivot stud (A, **Figure 11**) in the chaincase housing for cracks or excessive wear.

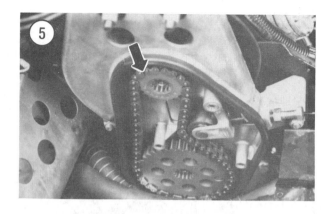

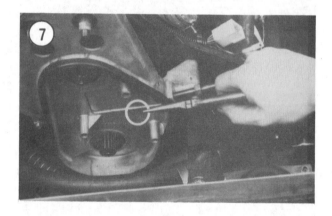

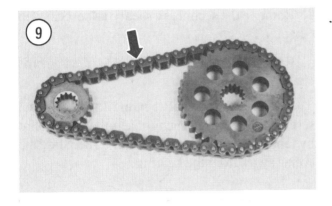

Installation

Refer to **Figure 1** for this procedure

1. Lubricate the chain tensioner pivot stud (A, **Figure 11**) and the ends of both shafts (B, **Figure 11**) with Polaris Chaincase Oil or equivalent.

2. Install the chain tensioner.

3. Install the washers onto the driveshaft (**Figure 7**). See **Figure 12**.

4. Install the spacer onto the jackshaft so that the shoulder side faes out. See **Figure 6**.

NOTE
Before installing the chain and sprocket assembly in Step 5, install the upper and lower sprockets without the chain. Then place a straightedge across the face of each sprocket and check that the sprockets are perfectly aligned. If the face of the sprockets are not aligned properly, add to or subtract from the driveshaft shims installed during Step 3. See **Figure 7** *and* **Figure 12**. *Recheck until sprocket alignment is correct.*

5. Fit the sprockets onto the drive chain (**Figure 13**). Make sure the marks placed on the sprockets during removal are facing outward. Then install the chain and sprockets. See **Figure 5**.

NOTE
If the sprockets were not marked for direction during removal, install the sprockets so the numbered side of the upper sprocket faces out and the shoul-

14

der (bevel) on the inner diameter of the
bottom sprocket faces in.

6. Install the upper and lower sprocket lock-washer and bolt (B, **Figure 4**). Apply the front brake and tighten the bolts securely.

7. Referring to **Figure 14**, adjust chain tension as follows:

    a. Turn brake disc a few degrees counter-clockwise as indicated in A, **Figure 14**.

    b. Turn the chain adjuster bolt (D) until the chain has approximately 6.3-9.5 mm (1/4-3.8 in.) free play at point B.

    c. Tighten the adjuster locknut (C) and check free play.

8. Install a new chaincase cover gasket.

9. Install the chaincase and secure it with its mounting bolts.

10. Remove the plug (B, **Figure 15**) and fill the chaincase with the amount and type of gear oil specified in Chapter Three. Reinstall the plug.

11. *Liquid cooled models:* Reconnect the coolant hose clamp at the chaincase cover (**Figure 2**).

12. Reinstall the exhaust pipe assembly. See Chapter Seven.

13. Close the shroud.

CHAINCASE AND JACKSHAFT

A hydraulic disc brake is installed on all models. The brake disc rides on the jackshaft. The jackshaft is installed horizontally at the rear of the engine compartment directly over the driveshaft. The secondary sheave is mounted onto the left-hand side of the jackshaft. Refer to Chapter Thirteen for disc brake service procedures. Refer to **Figure 1** and **Figure 16** when performing procedures in this section.

Removal

1. Remove the chain and sprockets as described in this chapter.

2. Remove the air box as described in Chapter Seven.

3. Remove the secondary sheave as described in Chapter Twelve.

4. Remove the driveshaft as described in this chapter.

5. Loosen the jackshaft bearing lock collar set screw (**Figure 17**). Then loosen the lock collar by tapping it with a punch (**Figure 18**). Slide the collar away from the bearing.

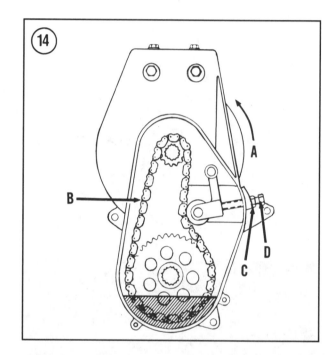

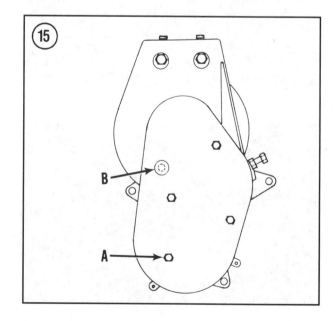

JACKSHAFT

1. Bolt
2. Bearing flange
3. Set screw
4. Bearing
5. Washer
6. Nut
7. Jackshaft
8. Woodruff key
9. Circlip
10. Brake disc
11. Seal

14

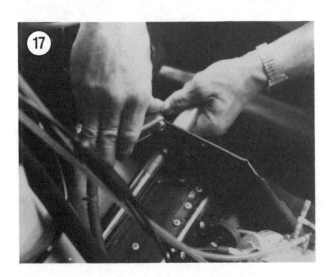

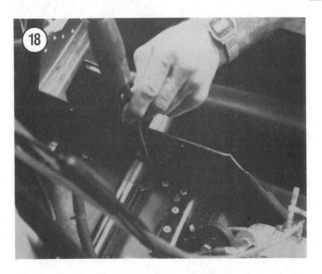

6. Remove the nuts, washers and bolts securing the bearing flanges to the bulkhead (**Figure 19**). Remove the bearing assembly in the following order:

   a. Remove the outer bearing flange (**Figure 20**).

   b. Remove the bearing (**Figure 21**).

   c. Remove the inner bearing flange (**Figure 22**).

   d. Remove the lock collar (**Figure 23**).

7. Remove the bolts securing the chaincase (A, **Figure 24**) to the bulkhead.

8. Remove the 4 brake caliper-to-chaincase mounting bolts (B, **Figure 24**).

9. Thread a 5/16 × 3/4 in. bolt into the jackshaft and tap the jackshaft out of the chaincase bearing. Remove the chaincase assembly (A, **Figure 24**).

10. Lift the brake caliper (**Figure 25**) off of the brake disc. Set the caliper assembly aside so that there is no strain on the brake hose. Do not allow

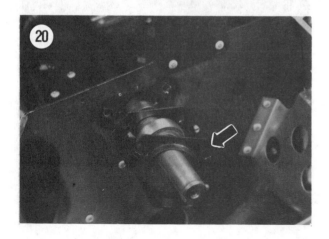

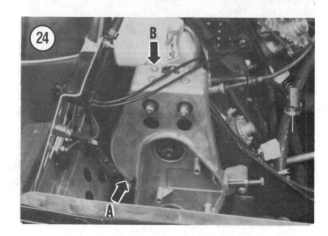

the caliper to hang so that all of its weight is put onto the brake hose.

NOTE
*While the caliper is removed from the brake disc, place a wood or plastic pad between the brake pads (**Figure 26**). That way, if the brake lever is accidently squeezed, the piston cannot be forced out of its bore. Disassembly of the brake caliper may be necessary to reseat the piston if the brake lever is squeezed.*

11. Remove the circlip from the end of the jackshaft (**Figure 27**). Then remove the spacer (A, **Figure 28**) and brake disc (B, **Figure 28**).

12. Remove the Woodruff key (A, **Figure 29**) from the jackshaft.

13. Remove the jackshaft (**Figure 30**) from the left-hand side.

14

Chaincase Inspection

Refer to **Figure 1** when performing this procedure.

1. Clean all components in solvent and thoroughly dry.

2. Remove all gasket residue from the chaincase and cover machined surfaces.

3. Check the chaincase bearings (**Figure 31**). Turn the bearing inner race by hand and check for roughness or excessive noise. The bearing should turn smoothly. Visually check the bearing for cracks or other abnormal conditions. If necessary, replace bearing(s) as follows:

   a. Remove the circlip (**Figure 32**).
   b. Drive the bearing out of the chaincase with a suitable size socket placed on the outer bearing race.
   c. Remove the O-ring from behind the upper bearing (1985 and later models).
   d. Clean the chaincase bearing area with solvent and thoroughly dry.
   e. Install a new O-ring in the upper bearing bore (1985 and later models).
   f. Install a new bearing by driving it squarely into the chaincase with a socket placed on the outer bearing race (**Figure 33**).
   g. Secure the bearing with the circlip (**Figure 32**). Make sure the circlip seats in the cover groove completely.

4. Check the chain adjuster bolt and nut (**Figure 34**) for deformed or worn threads. Check the threads in the chaincase for damage. Replace worn or damaged parts as required. Clean the chaincase threads with a tap.

Jackshaft Inspection

Refer to **Figure 16** for this procedure.

1. Clean the jackshaft in solvent and thoroughly dry.

2. Check the jackshaft for bending.

3. Turn the bearing and check for excessive nose or roughness. Replace the bearing if worn or damaged.

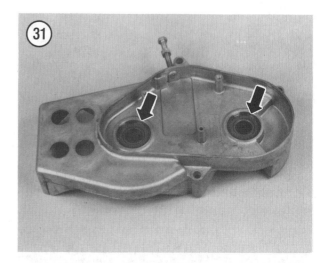

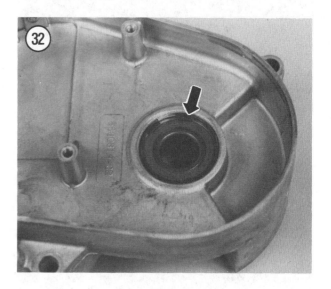

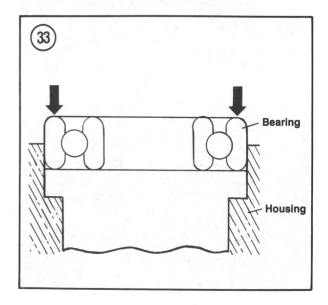

4. Check the keyway in the jackshaft for cracks or other damage.

5. Check both ends of the jackshaft for cracks, deep scoring or excessive wear.

6. If there is any doubt as to the condition of the jackshaft, repair or replace it.

Brake Disc Inspection

Refer to **Figure 16** for this procedure.

1. Visually check the brake disc for cracks, deep scoring, heat discoloration or checking.

2. Place the brake disc on a surface plate and check the flatness at the edge of the disc with a feeler gauge. If disc is severely warped, replace it.

Installation

Refer to **Figure 1** and **Figure 16** for this procedure.

1. Place the jackshaft onto the bulkhead as shown in **Figure 30**.

2. Install the first circlip (B, **Figure 29**) onto the jackshaft, if removed.

3. Insert the Woodruff key (A, **Figure 29**) into the jackshaft keyway.

4. Slide the brake disc onto the jackshaft so that it faces as shown in **Figure 35**.

5. Slide the spacer onto the jackshaft and rest it against the brake disc (A, **Figure 28**).

6. Secure the brake disc with the circlip (**Figure 27**). Make sure the circlip seats in the jackshaft groove completely.

7. Remove the wood or plastic spacer from between the brake pads, if used. Then install the brake caliper over the brake disc (**Figure 25**). Check routing of brake hose to make sure it is not kinked or damaged.

8. Install the chaincase as follows:

   a. Place the chaincase (**Figure 36**) into position on the bulkhead while at the same time aligning the jackshaft with the upper chaincase bearing.

   b. Slide the Polaris jackshaft alignment tool (part No. 2870399) over the end of the jackshaft. See **Figure 37**. Then thread the

14

bolt (with larger washer attached) into the end of the jackshaft (**Figure 38**).

c. Tighten the alignment tool bolt (**Figure 38**) to pull the jackshaft into the chaincase bearing.

d. Install the chaincase mounting bolts and tighten securely.

e. Apply Loctite 242 (blue) to the 4 brake caliper-to-chaincase mounting bolts (**Figure 39**) and install the mounting bolts. Apply the front brake and tighten the bolts to the torque specification listed in **Table 2**. Release the front brake.

9. Check jackshaft alignment as follows:

a. Bolt a bearing flange onto the bulkhead as shown in **Figure 40**.

b. With the alignment tool bolted to the end of the jackshaft (**Figure 38**), the jackshaft should be centered within the bearing flange as shown in **Figure 40**.

c. If the jackshaft is off-center, the chaincase will have to be shimmed with the Polaris shim kit (part No. 2200126).

d. If the jackshaft is centered, remove the alignment tool assembly (**Figure 38**) and the bearing flange (**Figure 40**).

e. If shimming is required, remove the chaincase and place shims between the chaincase and bulkhead. Then repeat Steps

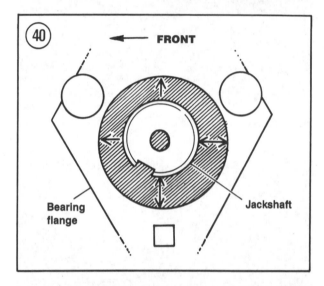

FRONT

Bearing flange

Jackshaft

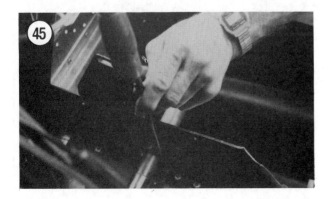

6-9 as required. When the jackshaft is centered correctly, continue with Step 10.

10. Install the jackshaft bearing assembly as follows:

   a. Slide the lock collar (**Figure 41**) onto the jackshaft so that the shoulder side faces toward the brake disc.

   b. Install the inner bearing collar and bearing (**Figure 42**) onto the jackshaft.

   c. Install the outer bearing flange (**Figure 43**) onto the jackshaft.

   d. Assemble the bearing flanges around the bearing and install the 3 bearing flange bolts and nuts. Tighten the nuts securely. See **Figure 44**.

   e. Slide the lock collar so that it engages against the bearing shoulder. Then turn the lock collar with a punch to tighten it (**Figure 45**).

   f. Tighten the lock collar set screw with an Allen wrench (**Figure 46**).

11. Reinstall the drive shaft as described in this chapter.

12. Install the secondary sheave and drive belt as described in Chapter Twelve.

13. Install the air box as described in Chapter Seven.

14. Install the chain and sprockets as described in this chapter.

14

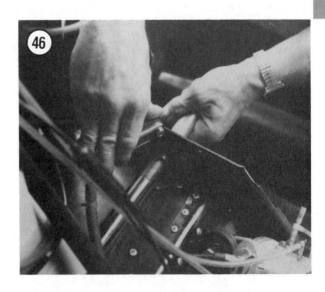

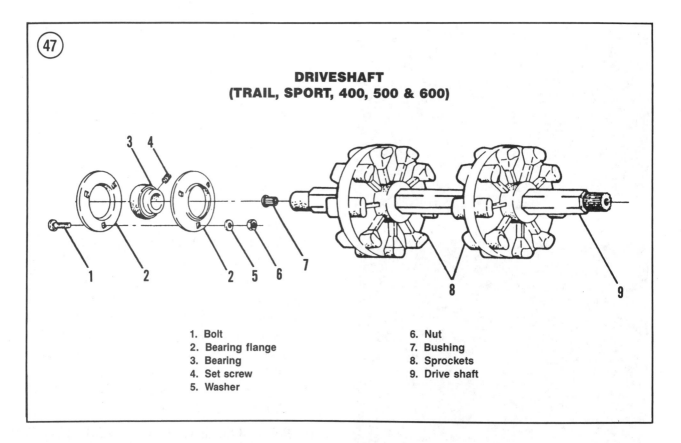

(47)

**DRIVESHAFT
(TRAIL, SPORT, 400, 500 & 600)**

1. Bolt
2. Bearing flange
3. Bearing
4. Set screw
5. Washer
6. Nut
7. Bushing
8. Sprockets
9. Drive shaft

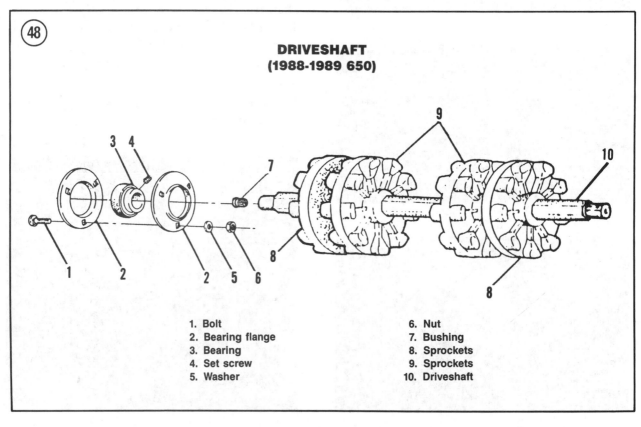

(48)

**DRIVESHAFT
(1988-1989 650)**

1. Bolt
2. Bearing flange
3. Bearing
4. Set screw
5. Washer
6. Nut
7. Bushing
8. Sprockets
9. Sprockets
10. Driveshaft

DRIVESHAFT

The driveshaft receives power from the jackshaft through the drive chain assembly on the right-hand side. Refer to **Figure 47** or **Figure 48**.

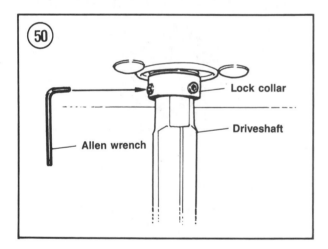

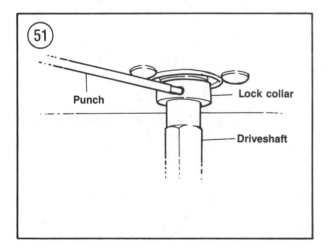

Removal

1. Remove the chain and sprockets as described in this chapter.

2. Remove the secondary sheave as described in Chapter Twelve.

3. Remove the rear suspension as described in Chapter Sixteen.

4. Disconnect the speedometer cable at the cable housing (**Figure 49**) on the left-hand side.

5. Working inside the bulkhead, loosen the driveshaft lock collar set screw (**Figure 50**). Then tap the lock collar with a punch (**Figure 51**) to loosen it. Slide the lock collar away from the bearing.

6. Remove the following parts in order from the outside of the bulkhead:

    a. Remove the bolts holding the speedometer gear housing to the bulkhead. Remove the housing (**Figure 52**).

    b. Remove the bearing (**Figure 53**).

14

c. Remove the inner bearing flange and lock collar (**Figure 54**).

7. Remove the driveshaft by pulling it from the chaincase and then from the left-hand side of the snowmobile.

Inspection

This section describes driveshaft inspection. A press is required to disassemble and reassemble the driveshaft.

1. Hold the bearing (**Figure 55**) by its outer race and spin the inner race by hand. Check for roughness or excessive noise. Also check the bearing seal for damage. Replace the bearing if necessary.

2. Check the driveshaft (**Figure 56**) for bending.

3. Check the driveshaft splines for cracks or other damage.

4. Visually inspect the sprocket wheels for excessive wear, cracks, distortion or other damage. Replace the sprocket wheels as described in this chapter.

5. If there is any doubt as to the condition of any part, replace it with a new one.

Sprocket Wheel Replacement

A press is required for this procedure. Refer to **Figure 47** or **Figure 48**.

1. Purchase all new parts to have on hand before disassembly.

2. Remove the snap rings from each sprocket wheel (**Figure 57**).

> *NOTE*
> *Before removing the sprocket wheels, note the manufacturer's logo on each sprocket wheel (**Figure 57**). These marks are used for alignment purposes. During assembly, the mark on each sprocket wheel must align with each other. Highlight marks with chalk before removal.*

> *NOTE*
> *When removing the sprocket wheels in Step 3, make sure the sprocket wheels are pressed off opposite the spline end.*

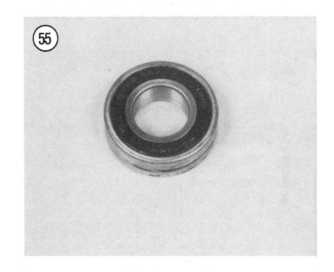

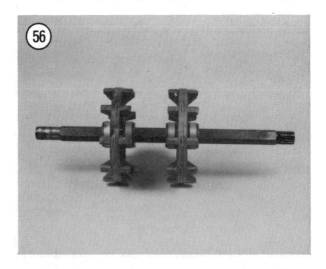

3. Using a press, press off each sprocket wheel.

4. Clean the driveshaft thoroughly. Remove all nicks with a file.

5. Installation is the reverse of these steps. Note the following when installing the sprocket wheels:

    a. When installing the wheels onto the driveshaft, make sure the manufacturer's logo on each sprocket wheel aligns with each other.

    b. Press the sprocket wheels onto the jackshaft to the dimensions shown in **Figure 58** or **Figure 59**.

    c. After the sprocket wheels are installed and positioned correctly, secure them with the snap rings (**Figure 57**).

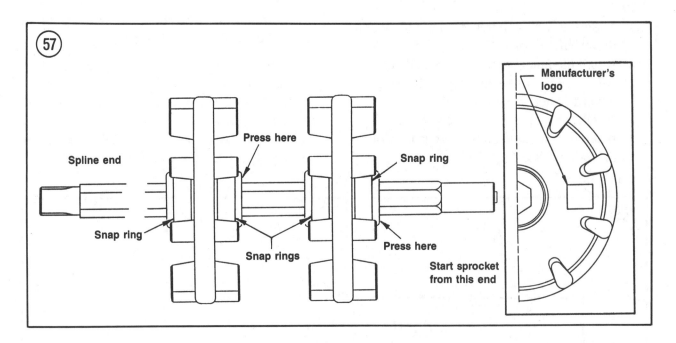

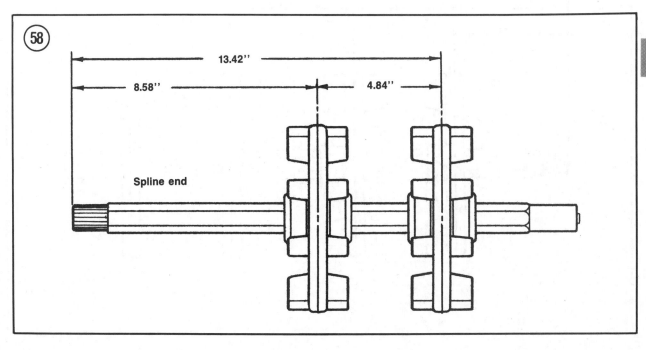

Installation

1. When installing the track, orient the track lugs to run in the direction shown in **Figure 60**.

2. Align the driveshaft so that the splined end faces toward the right-hand side, then install the driveshaft.

3. Make sure the lugs engage the track correctly.

4. Install the driveshaft bearing assembly as follows:

   a. Slide the lock collar onto the end of the driveshaft (**Figure 54**).

   b. Install the inner bearing flange and bearing (**Figure 53**). Install bearing so flange faces toward inside of snowmobile.

   c. Install the speedometer cable housing, making sure pin in housing engages slot in driveshaft. See **Figure 52**.

   d. Assemble the speedometer housing/bearing flange around the bearing and install the 3 bearing flange bolts and nuts. Tighten the nuts securely. See **Figure 49**.

   e. Slide the lock collar so that it engages against the bearing shoulder. Then turn the lock collar with a punch to tighten it (**Figure 51**).

   f. Tighten the lock collar set screw with an Allen wrench (**Figure 50**).

5. Install the rear suspension as described in Chapter Sixteen.

6. Install the chain and sprockets as described in this chapter.

7. Install the secondary sheave and drive belt as described in Chapter Twelve.

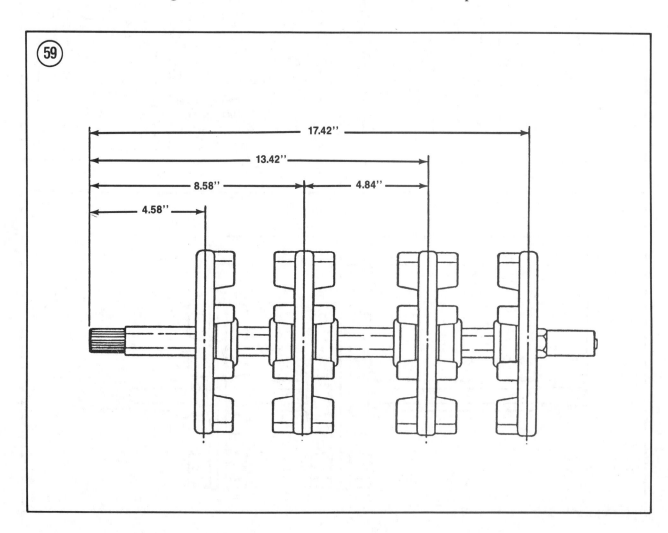

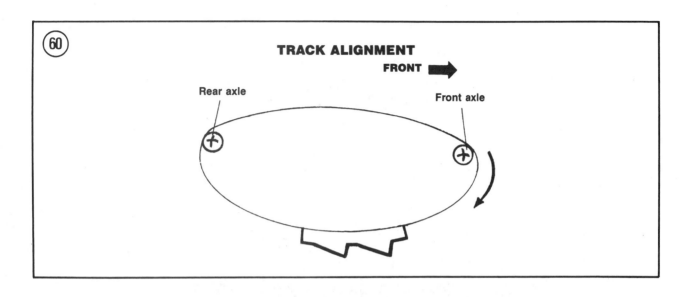

Table 1 CHAIN AND SPROCKET SPECIFICATIONS

Model	Gearing	Chain pitch
Sport	17/35	62
Trail		
1984-1985	19/35	64
1986-on	21/35	64
1988 SKS	19/35	64
400		
1985-on	19/35	64
1988 SKS	18/35	64
500		
1989 Indy and Classic	19/35	64
1989 SKS	18/35	64
600		
1983	21/35	64
1984-1985	19/35	64
1986-1987	21/35	64
650		
1988-on	21/35	64
1988-on SKS	19/35	64

Table 2 TIGHTENING TORQUES

	N·m	ft.-lb.
Brake caliper through bolt	34-41	25-30

Chapter Fifteen

Front Suspension and Steering

This chapter describes service procedures for the skis, front suspension and steering components.

Table 1 is at the end of the chapter.

SKIS

The skis are equipped with wear bars, or skags, that aid in turning the machine and protect the bottoms of the skis from wear and damage caused by road crossings and bare terrain. The bars are expendable and should be checked for wear and damage at periodic intervals and replaced when they are worn to the point they no longer protect the skis or aid in turning.

Removal/Installation

Refer to **Figure 1** or **Figure 2** for this procedure.

1. Support the front of the machine so both skis are off the ground.

NOTE
Mark the skis with a ''L'' (left-hand) or ''R'' (right-hand). The skis should be installed on their original mounting side.

2. Remove the cotter pin from the end of the ski pivot bolt. Loosen and remove the nut (A, **Figure 3**) and bolt.
3. Turn the ski sideways (**Figure 4**) and remove it from the spindle.
4. Remove the bushing (**Figure 5**) from the spindle.
5. Remove the rubber stop from the ski.
6. Inspect the skis as described in this chapter.
7. Installation is the reverse of these steps. Note the following.
8. Wipe all old grease from the pivot bolt and bushing. Reinstall the bushing (**Figure 5**) into the spindle.
9. Apply a low-temperature grease to the pivot bolt and bushing before installation.
10. Reinstall the rubber stop into the ski.

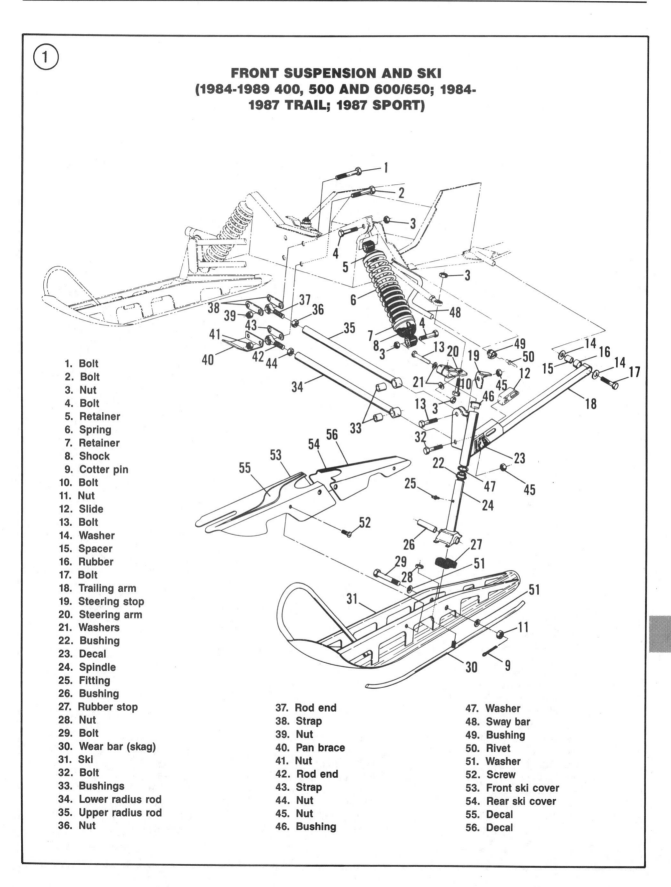

①

**FRONT SUSPENSION AND SKI
(1984-1989 400, 500 AND 600/650; 1984-
1987 TRAIL; 1987 SPORT)**

1. Bolt
2. Bolt
3. Nut
4. Bolt
5. Retainer
6. Spring
7. Retainer
8. Shock
9. Cotter pin
10. Bolt
11. Nut
12. Slide
13. Bolt
14. Washer
15. Spacer
16. Rubber
17. Bolt
18. Trailing arm
19. Steering stop
20. Steering arm
21. Washers
22. Bushing
23. Decal
24. Spindle
25. Fitting
26. Bushing
27. Rubber stop
28. Nut
29. Bolt
30. Wear bar (skag)
31. Ski
32. Bolt
33. Bushings
34. Lower radius rod
35. Upper radius rod
36. Nut

37. Rod end
38. Strap
39. Nut
40. Pan brace
41. Nut
42. Rod end
43. Strap
44. Nut
45. Nut
46. Bushing

47. Washer
48. Sway bar
49. Bushing
50. Rivet
51. Washer
52. Screw
53. Front ski cover
54. Rear ski cover
55. Decal
56. Decal

15

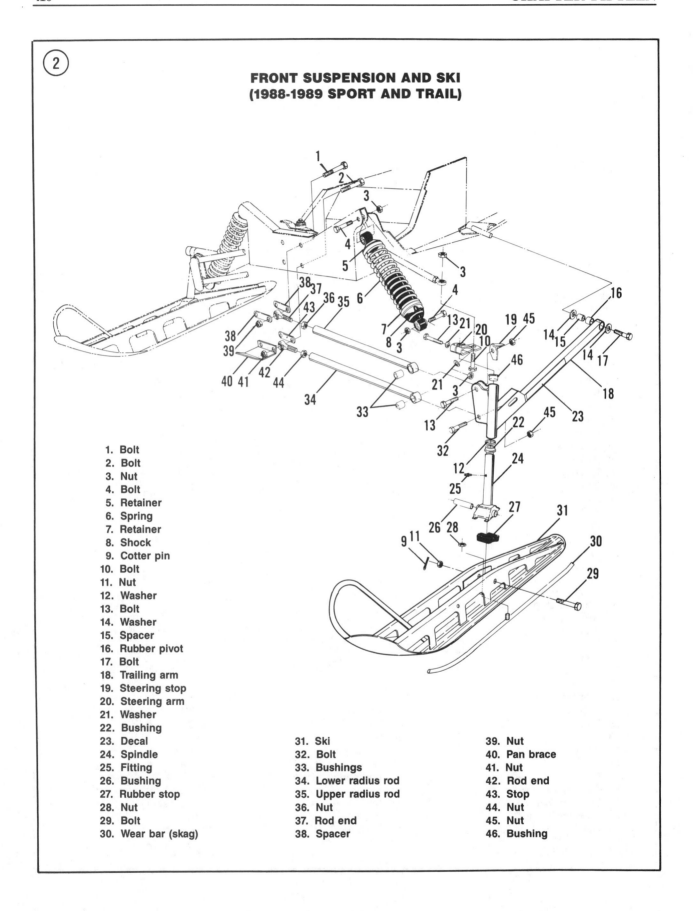

② **FRONT SUSPENSION AND SKI
(1988-1989 SPORT AND TRAIL)**

1. Bolt
2. Bolt
3. Nut
4. Bolt
5. Retainer
6. Spring
7. Retainer
8. Shock
9. Cotter pin
10. Bolt
11. Nut
12. Washer
13. Bolt
14. Washer
15. Spacer
16. Rubber pivot
17. Bolt
18. Trailing arm
19. Steering stop
20. Steering arm
21. Washer
22. Bushing
23. Decal
24. Spindle
25. Fitting
26. Bushing
27. Rubber stop
28. Nut
29. Bolt
30. Wear bar (skag)

31. Ski
32. Bolt
33. Bushings
34. Lower radius rod
35. Upper radius rod
36. Nut
37. Rod end
38. Spacer

39. Nut
40. Pan brace
41. Nut
42. Rod end
43. Stop
44. Nut
45. Nut
46. Bushing

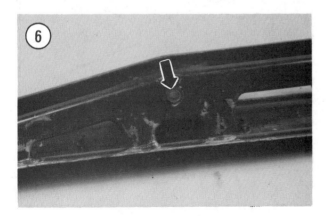

11. When installing the skis, refer to the alignment marks made before removal.

12. Bring the ski into contact with the spindle (**Figure 4**) and then turn the ski so that the ski tip faces to the front of the snowmobile. Make sure the rubber stop faces against the spindle as shown in B, **Figure 3**.

13. Align the ski pivot bolt hole with the spindle and install the pivot bolt. Install the pivot bolt from the direction shown in **Figure 1** or **Figure 2**. Install the nut and tighten to the torque specification listed in **Table 1**.

14. Check that the ski pivots up and down with slight resistance. If the ski is tight, remove the ski and check parts.

15. Install a new cotter pin through the pivot bolt. Bend the ends of the cotter pin over to lock it.

Inspection

Refer to **Figure 1** or **Figure 2** for this procedure.

1. Check the rubber stop for wear, cracking or deterioration. Replace if necessary.

2. Check the skis for metal fatigue or damage. Check the ski pivot bolt hole (**Figure 6**) for cracks or damage. Repair or replace the ski(s) as required.

3. Check the wear bars (skags) on the bottom of the skis for severe wear or damage (**Figure 7**). If necessary, replace the wear bars as follows:

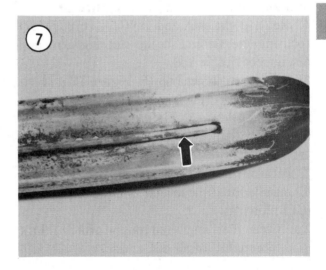

15

a. Remove the nuts (**Figure 8**) holding the wear bars to the ski.

b. Push the wear bar bolt through the ski and remove the wear bar (**Figure 7**).

c. Install new wear bar by reversing these steps. Apply Loctite 242 (blue) to the wear bar nut (**Figure 8**) and tighten securely.

FRONT SUSPENSION

The Polaris type IV Independent Front Suspension is used on all models covered in this manual. Refer to **Figure 1** or **Figure 2** when performing procedures in this section. Because of the number of parts used in the suspension, identify parts during removal as well as storing them in separate boxes.

Steering Arm and Spindle Removal/Installation

Refer to **Figure 9**. The spindle (A) rides inside the trailing arm (B) and is supported by 2 bushings. The steering arm (C) is bolted to the top of the spindle and connects the spindle to the front tie rod (D). To maintain steering alignment, the steering arm and spindle should be marked before they are separated.

1. Support the front of the machine so both skis are off the ground.

2. Remove the skis as described in this chapter.

3. Scribe an alignment mark across the steering arm and spindle as shown in **Figure 10**.

4. Remove the nut and bolt (A, **Figure 11**) securing the tie rod to the steering arm and disconnect the tie rod.

5. Remove the pinch bolt (B, **Figure 11**) and nut securing the steering arm to the spindle. Remove the steering arm (C, **Figure 11**).

6. Slide the spindle (**Figure 12**) out of the trailing arm and remove it.

7. Perform the *Inspection* in this section.

8. Installation is the reverse of these steps. Note the following:

a. Clean the spindle and trailing arm bushings thoroughly of all old grease.

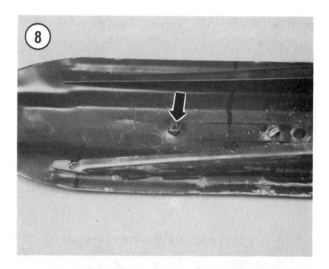

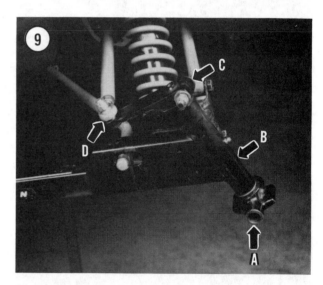

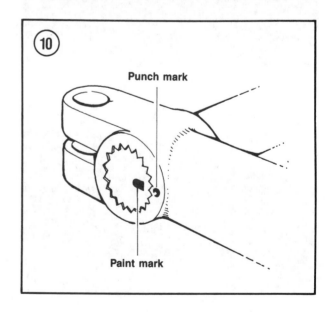

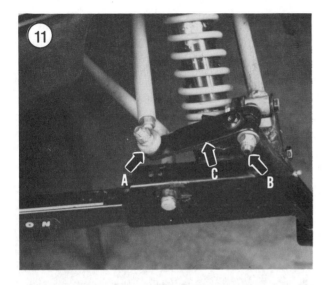

b. Apply a low-temperature grease to the spindle before installation.

c. Align the steering arm to spindle alignment marks (**Figure 10**) and install the steering arm (C, **Figure 11**).

d. Tighten the tie rod bolt to the torque specification listed in **Table 1**.

e. Tighten the steering arm bolt at the spindle to the torque specification listed in **Table 1**.

f. If new parts were installed, check steering alignment as described in this chapter.

Inspection

1. Clean all components thoroughly in solvent. Remove all dirt and other residue from all surfaces.

> *NOTE*
> *If paint has been removed from the steering arm during cleaning, touch up areas as required before installation.*

2. Visually check the steering arm (**Figure 13**) for cracks, excessive wear or other damage. Check splines for damage. Check mating splines in spindle (**Figure 14**).

3. Check the spindle (**Figure 14**) for cracks or other damage.

4. Check the spindle bushing (**Figure 14**) for cracks, deep scoring or excessive wear.

5. Replace worn or damaged parts as required.

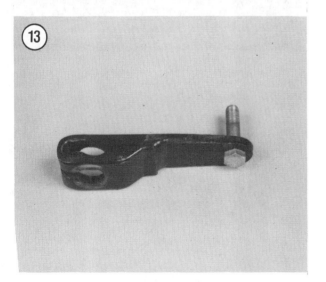

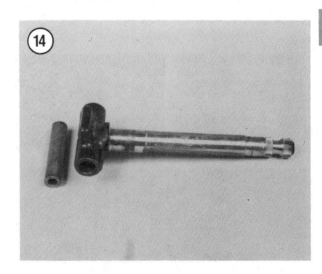

15

Trailing Arm
Removal/Installation

Refer to **Figure 1** or **Figure 2** for this procedure.

1. Support the front of the machine so both skis are off the ground.

2. Remove the steering arm and spindle as described in this chapter.

3. Disconnect the upper and lower radius arms (**Figure 15**) at the trailing arm. Remove the 2 bushings (**Figure 16**).

4. Remove the bolt, washers, rubber pivot and spacer securing the trailing arm to the frame and remove the trailing arm (A, **Figure 17**).

5. Remove the trailing arm (A, **Figure 17**) while at the same time disconnecting the torsion bar (B, **Figure 17**) from the trailing arm.

6. Perform the *Inspection* as described in the following procedure.

7. Installation is the reverse of these steps. Note the following:

   a. Position trailing arm into position, making sure to engage the torsion bar with the trailing arm. See B, **Figure 17**.

   b. Tighten all fasteners to the suspension mounting torque specification listed in **Table 1**.

Inspection

1. Clean the trailing arm in solvent and dry thoroughly.

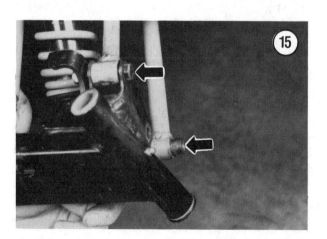

2. If paint has been removed from the trailing arm during cleaning or through use, touch up areas before installation.

3. Check the trailing arm (**Figure 18**) for cracks or other damage. Replace if necessary.

4. Check the upper and lower trailing arm bushings for cracks, deep scoring or excessive wear. Replace bushings as a set, if necessary.

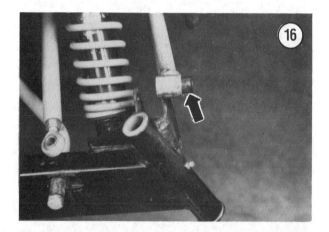

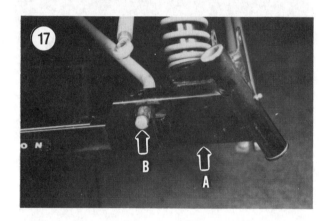

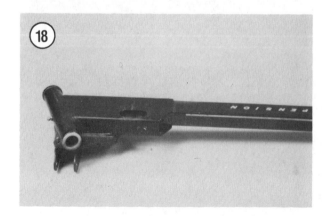

Torsion Bar
Removal/Installation

Refer to **Figure 1** or **Figure 2** for this procedure.

1. Support the front of the machine so both skis are off the ground.

2. Remove one trailing arm as described in this chapter.

3. Remove the torsion bar supports as follows:

   a. Using a small punch, knock out the rivet mandrels from the center of each rivet (**Figure 19**).

   b. After the mandrels have been removed, drill out the center of each rivet with a 1/4 in. drill bit.

   c. Remove the torsion bar supports and pull the torsion bar out of the snowmobile.

4. Installation is the reverse of these steps. Only high strength Q-type rivets should be used to secure the torsion bar supports. See your Polaris dealer for rivets.

Inspection

1. Clean the torsion bar in solvent and dry thoroughly.

2. If paint has been removed from the torsion bar during cleaning or through use, touch up areas before installation.

3. Check the torsion bar for twist as follows:

   a. Place the torsion bar on a surface plate or other flat surface.

   b. Hold one end of the bar flat against the surface plate and measure the opposite end from the bottom of the bar to the surface

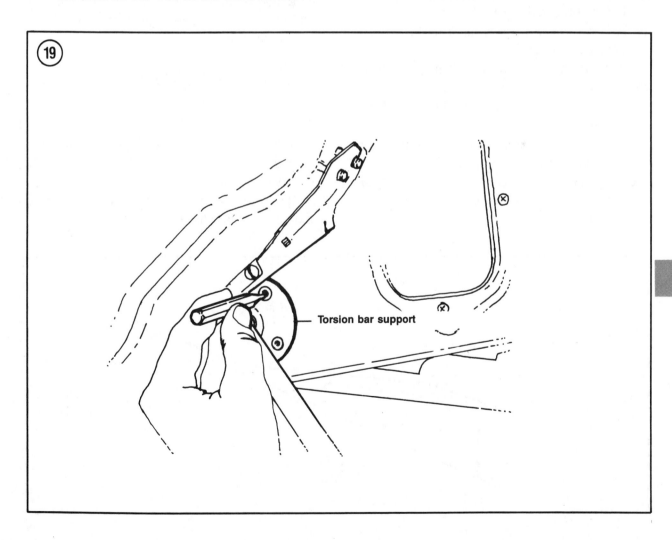

⑲

Torsion bar support

15

plate. This distance should not exceed 4.76 mm (3/8 in.). See **Figure 20**.

c. Replace the torsion bar if the twist exceeds the specification given in sub-step b.

Radius Arms
Removal/Installation

Refer to **Figure 1** or **Figure 2** for this procedure.

1. Support the front of the machine so both skis are off the ground.

2. Disconnect the radius arms at the trailing arms (**Figure 15**). Remove the 2 bushings (**Figure 16**).

3. Disconnect the radius arm rod ends at the pan brace. Do not loosen the rod end-to-radius arm nuts.

4. Remove the radius arms.

5. Perform the *Inspection* as described in the following procedure.

6. Installation is the reverse of these steps. Note the following:

   a. Clean the radius arm bushings of all old grease.

   b. Apply a low-temperature grease to the radius arm bushings before installation.

   c. Tighten all fasteners to the suspension mounting torque specification listed in **Table 1**.

   d. If the rod ends were removed from the radius rods, check steering adjustment as described in this chapter.

Inspection

1. Clean the radius arms in solvent and dry thoroughly.

2. If paint has been removed from the radius arms during cleaning or through use, touch up areas before installation.

3. Check the radius arms for wear, cracks or other damage.

4. Check the rod ends for excessive wear or damage. If damaged, loosen the nut and unscrew the ball joint. Reverse to install.

5. Replace worn or damaged parts as required.

6. If one or more rod ends were loosened or replaced, check steering adjustment as described in this chapter.

SHOCK ABSORBER

Removal/Installation

Refer to **Figure 1** or **Figure 2** for this procedure.

1. Open the shroud.

2. Remove the 2 shock absorber mounting bolts and remove the shock absorber.

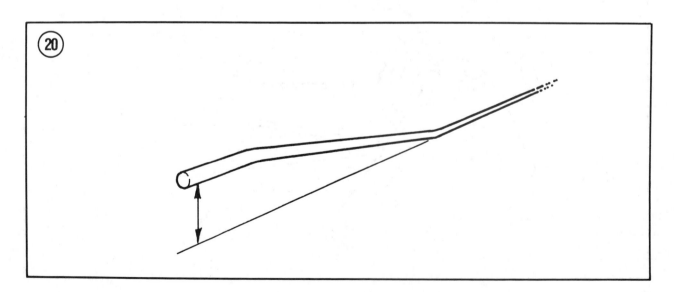

3. Perform the *Inspection* as described in this chapter.

4. Installation is the reverse of these steps. Tighten the shock absorber bolts securely.

Spring
Removal/Installation

Refer to **Figure 1** or **Figure 2** for this procedure.

1. Remove the shock absorber as described in this chapter.

2. Secure the bottom of the shock absorber in a vise with soft jaws.

3. Turn the spring adjuster to its softest position.

WARNING
Do not attempt to remove or install the shock spring without the use of a spring compressor. Attempting to remove the spring without the use of a spring compressor may cause severe personal injury. If you do not have access to a spring compressor, refer spring removal to a Polaris dealer.

4. Install a spring compressor onto the shock absorber following the manufacturer's instructions. Then compress the spring with the tool and remove the spring stopper from the top of the shock. Remove the spring.

CAUTION
Do not attempt to disassemble the shock damper housing.

5. Inspect the shock and spring as described in the following procedure.

6. Installation is the reverse of these steps. Make sure the spring stopper secures the spring completely.

Shock Inspection

1. Remove the shock spring as described in this chapter.

2. Clean all components thoroughly in solvent and allow to dry.

3. Check the damper rod as follows:
   a. Check the damper housing for leakage. If the damper is leaking, replace it.
   b. Check the damper rod for bending or damage. Check the damper housing for dents or other damage.
   c. Operate the damper rod by hand and check its operation. If the damper is operating correctly, a small amount of resistance should be felt on the compression stroke and a considerable amount of resistance felt on the return stroke.
   d. Replace the damper assembly if necessary.

4. Check the damper housing for cracks, dents or other damage. Replace the damper housing if damaged. Do not attempt to repair or straighten it.

5. Visually check the spring for cracks or damage.

STEERING ASSEMBLY

This section describes service to the handlebar, steering column and tie rods.

Steering Column/Handlebar
Removal/Installation

Refer to **Figure 21** for this procedure.

1. Remove the air box as described in Chapter Seven.

2. Remove the handlebar pad (A, **Figure 22**).

3. Remove the throttle and brake housings at the handlebar.

4. Remove the steering column nut and bolt at the steering hoop (A, **Figure 23**).

5. Remove the drag link nut and bolt at the steering column (**Figure 24**). Disconnect the drag link.

6. Remove the steering column/handlebar assembly (B, **Figure 23**).

7. Installation is the reverse of these steps. Note the following.

15

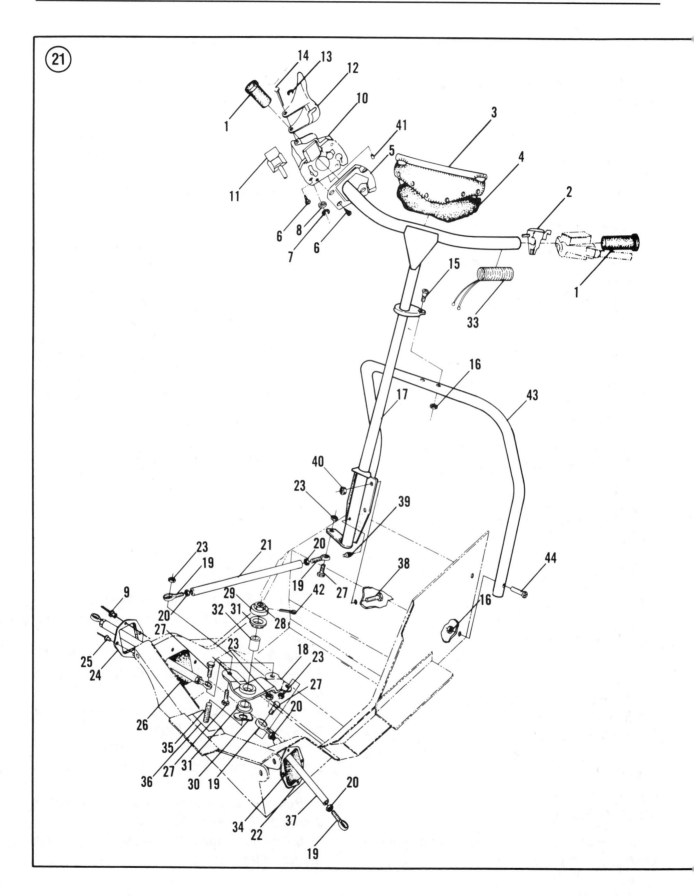

STEERING ASSEMBLY

1. Grip
2. Dimmer switch
3. Cover
4. Pad
5. Adaptor
6. Screw
7. Circlip
8. Washer
9. Rivet
10. Throttle block
11. Stop switch
12. Throttle lever
13. Circlip
14. Retaining pin
15. Bolt
16. Nut
17. Steering post
18. Steering arm
19. Rod end
20. Nut
21. Drag link rod
22. Left-hand boot plate
23. Nut
24. Right-hand boot plate
25. Rivet
26. Right-hand boot
27. Bolt
28. Nut
29. Washer
30. Wave washer
31. Bushing
32. Spacer
33. Grip heater
34. Left-hand boot
35. Bolt
36. Steering block
37. Tie rod
38. Bolt
39. Grease fitting
40. Nut
41. Allen screw
42. Cotter pin
43. Steering hoop
44. Bolt

15

8. Check all controls for proper operation. Make sure the front brake is operating properly.

9. Check steering adjustment if any parts were replaced.

Inspection

Refer to **Figure 21** for this procedure.

1. Clean all components thoroughly in solvent.

2. Check the handlebar subassembly for cracks, bending or other damage.

3. Visually check the steering column surfaces for cracks or deep scoring. Also check the welds at the top and bottom of the shaft for cracks or breakage.

Handlebar Grips

Note the following when replacing handlebar grips (B, **Figure 22**):

   a. *Grips without heating element*: These grips can be removed and installed easily by blowing compressed air into the opposite handlebar end while covering the hole in opposite grip. To install grips with air, align the grip with the handlebar and direct air from the opposite side through the handlebar. If you do not have access to compressed air, insert a thin-tipped screwdriver underneath the grip and squirt some electrical contact cleaner between the grip and handlebar or twist grip. Quickly remove the screwdriver and twist the grip to break its hold on the handlebar; slide the grip off. When installing new grips, squirt contact cleaner into the grip as before and quickly twist it onto the handlebar or twist grip. Allow plenty of time for the contact cleaner to evaporate and the grip to take hold before riding snowmobile.

   b. *Grips with heating element*: To remove a grip, locate the heating element wires on the grip. Then cut the grip on the opposite side away from the wires. Slowly peel the grip back and locate the gap between the heating element. Cut along this gap to

completely remove grip. Install new grips with a hammer. When installing new grip, route heating element wires so that they do not interfere with brake or throttle operation.

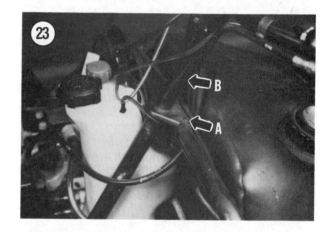

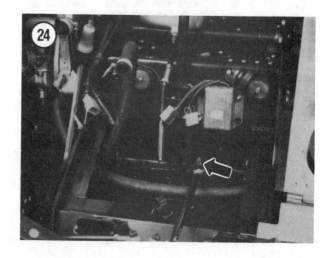

c. *All models*: Make sure grips are *tight* before riding snowmobile.

WARNING
Do not use any type of grease, soap or other lubricant to install grips.

WARNING
Loose grips can cause you to crash. Always replace damaged or loose grips before riding.

Tie Rod
Removal/Installation

The tie rods control steering movement from the center steering arm to the spindle. Refer to **Figure 21** for this procedure.

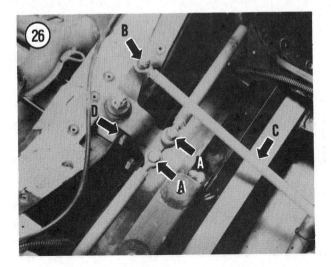

NOTE
If both tie rods are going to be removed, identify them so that they can be reinstalled in their original position.

1. Support the front of the machine so both skis are off the ground.
2. Disconnect the tie rod from the steering arm at the spindle (**Figure 25**).
3. Disconnect the tie rod at the center steering arm (A, **Figure 26**).
4. Remove the tie rod.
5. Inspect the tie rod as described in the following procedure.
6. Installation is the reverse of these steps. Note the following.
7. After installing the tie rods, check ski alignment as described in this chapter. When ski alignment is correct, tighten tie rod bolts to the torque specification in **Table 1**.

Inspection

1. Clean the tie rod in solvent and dry thoroughly.
2. If paint has been removed from the tie rod during cleaning or through use, touch up areas before installation.
3. Check the tie rod for wear, cracks or other damage.
4. Check the ball joints for excessive wear or damage. If damaged, loosen the nut and unscrew the ball joint. Reverse to install.
5. Replace worn or damaged parts as required.
6. If one or more ball joints were loosened or replaced, check ski alignment as described in this chapter.

Drag Link
Removal/Installation

1. Remove the engine as described in Chapter Five or Chapter Six.
2. Support the front of the machine so both skis are off the ground.
3. Disconnect the drag link at the steering column (**Figure 24**).

15

4. Disconnect the drag link at the center steering arm (B, **Figure 26**).

5. Remove the drag link (C, **Figure 26**).

6. Inspect the drag link as described in the following procedure.

7. Installation is the reverse of these steps. Note the following.

8. After installing the drag link, check ski alignment as described in this chapter. When ski alignment is correct, tighten drag link bolts securely.

Inspection

1. Clean the drag link in solvent and dry thoroughly.

2. If paint has been removed from the drag link during cleaning or through use, touch up areas before installation.

3. Check the drag link for wear, cracks or other damage.

4. Check the ball joints for excessive wear or damage. If damaged, loosen the nut and unscrew the ball joint. Reverse to install.

5. Replace worn or damaged parts as required.

6. If one or more ball joints were loosened or replaced, check steering adjustment as described in this chapter.

Steering Arm
Removal/Installation

Refer to **Figure 21** for this procedure.

1. Support the front of the machine so both skis are off the ground.

2. Disconnect both tie rods from the steering arm (A, **Figure 26**).

3. Disconnect the drag link at the steering arm (B, **Figure 26**).

4. Remove the cotter pin and nut securing the steering arm (D, **Figure 26**) to the steering block bolt. Remove the following parts in order:

    a. Washer.

    b. Bushing.

    c. Spacer.

    d. Steering arm.

    e. Bushing.

    f. Wave washer.

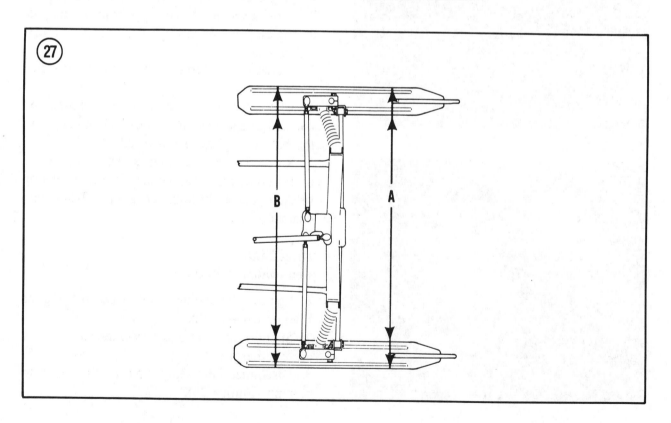

5. Inspect the steering arm as described in the following procedure.

6. Installation is the reverse of these steps. Note the following.

7. Wipe off all of the old grease on the steering block bolt and steering arm bushings. Apply a low-temperature grease before installation.

8. Assemble the steering arm assembly in the order shown in **Figure 21**. Make sure the steering arm faces in the direction shown in D, **Figure 26**. Install the steering arm nut and tighten to the torque specification listed in **Table 1**. Then check steering arm movement by turning arm by hand. Steering arm should pivot smoothly with no signs of roughness or other damage. Install a new cotter pin through the nut and bolt and bend the ends over to lock it.

Inspection

1. Clean the steering arm in solvent and dry thoroughly.

2. If paint has been removed from the steering arm during cleaning or through use, touch up areas before installation.

3. Check the steering arm for wear, cracks or other damage.

4. Check the steering block bolt for thread damage or cracks along the bolt. Replace bolt if necessary.

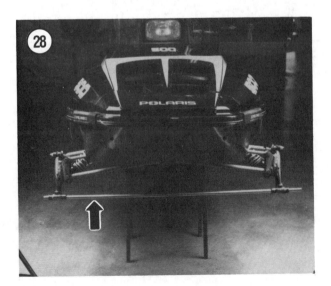

5. Check the steering arm bushings for cracks, deep scoring or excessive wear; replace bushings as a set, if necessary.

STEERING ADJUSTMENT

Before checking the steering adjustment, check the steering assembly for loose or missing fasteners or damaged parts.

Preliminary Inspection

1. Park the snowmobile on a level surface.

2. Center handlebars so that they face straight ahead.

3. With a tape measure, measure the distances indicated by A and B in **Figure 27**. The distance should be the same. If not, proceed to Step 4.

NOTE
*An alignment bar (**Figure 28**) is required to perform many of the procedures related to steering adjustment. However, because Polaris does not sell an alignment bar, you must purchase a 5/8 in. steel rod from a machine shop. The rod must meet the specifications shown in **Figure 29**.*

4. Raise the snowmobile so that the front of the machine is approximately 1-2 in. off the ground. Remove the skis as described in this chapter.

5. Perform the following:
   a. Center handlebars so that they face straight ahead.
   b. Try to insert the alignment bar through both spindles as shown in **Figure 28**. If the bar will go easily through both spindles, toe out and camber adjustments are correct. If the bar cannot be inserted through the second spindle, perform Step 6.

6. Perform the following:
   a. With a tape measure, measure the distance from a point on the spindle to the center of the chassis (at the steering arm). Repeat for the opposite side. See **Figure 30**.

15

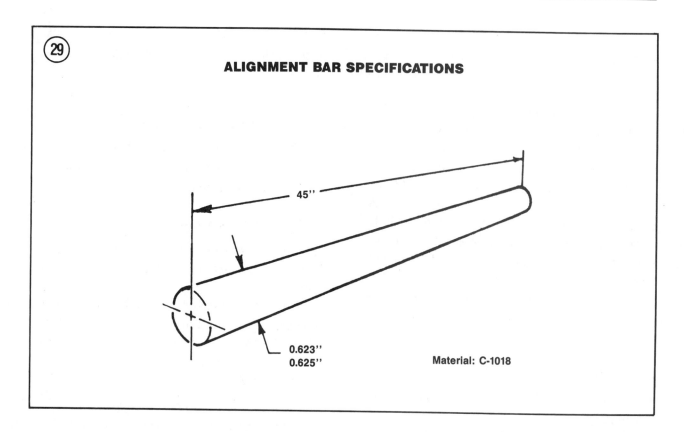

ⓐ **ALIGNMENT BAR SPECIFICATIONS**

45"

0.623"
0.625"

Material: C-1018

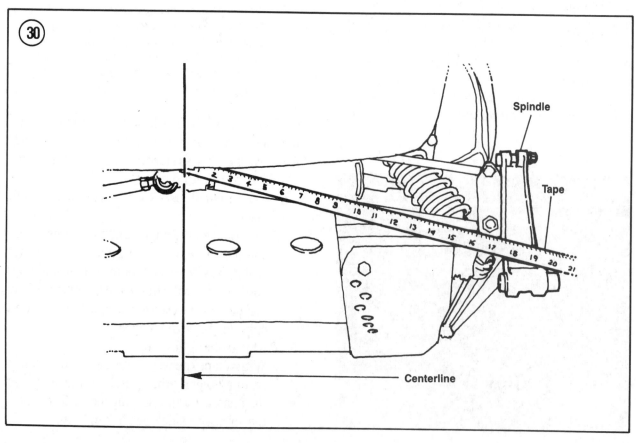

㉚

Spindle

Tape

Centerline

b. The measurements taken in sub-step a should be within ±1/8 in. If the measurements are within specification, perform Step 7. If the difference in the measurements exceed the service limit, adjustment of the radius arm(s) will be required. Loosen the radius arm ball joint locknut (**Figure 1** or **Figure 2**) and turn the radius arm(s) as required to obtain an equal distance when performing the measurement in sub-step a. When the measurements are within specification, perform Step 7.

7. Perform the following:
  a. Center handlebars so they face straight ahead.
  b. With a tape measure, measure from the center of one spindle to the other (**Figure 31**). The correct spindle distance is 36 1/2 in. ±1/8 in.
  c. If the distance is incorrect, adjust the radius arm(s) as described in Step 6.

8. Recheck suspension alignment with the alignment bar. If the alignment bar will fit through both spindles, steering adjustment is correct. If the bar will not fit through both spindles as shown in **Figure 28**, note the following:
  a. If the alignment bar is off horizontally with the spindle, adjust the toe out as described in this chapter.
  b. If the alignment bar is off vertically with the spindle, adjust the camber as described in this chapter.

9. If the steering alignment is correct, reinstall the skis as described in this chapter.

Toe Out Adjustment

1. Perform the preliminary adjustment as described in this chapter. If adjustment shows that toe out is incorrect, perform the following.

2. Loosen the locknuts on the end of both tie rods (**Figure 32**). Then turn the tie rods in or out until the adjustment bar will fit through both spindles easily (**Figure 28**). If you cannot get the alignment rod to fit through both spindles, the camber adjustment is incorrect. Adjust camber as described in this chapter.

3. If the adjustment rod fits through both spindles, suspension adjustment is now correct. Tighten the tie rod locknuts securely.

15

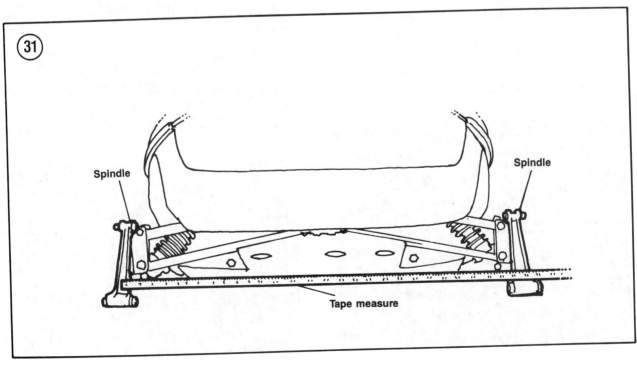

(31)

Spindle   Spindle

Tape measure

4. Install the skis as described in this chapter.

Camber Adjustment

1. Perform the preliminary adjustment as described in this chapter. If adjustment shows that camber adjustment is incorrect, perform the following.

2. Correct camber adjustment is obtained when both spindles are adjusted to a vertical or 0° position (**Figure 33**). Insert the adjustment bar through one spindle and then remove it and install it through the opposite spindle. Determine which spindle camber adjustment is off the most. Then, after determining which spindle is out of alignment, perform the following.

3. Disconnect the lower radius arm from the spindle most out of alignment. See **Figure 34**, typical. Adjust camber by first loosening the radius arm locknut and then change the length of the radius arm by turning it by hand clockwise or counterclockwise. Continue adjustment of radius arm until the alignment bar fits through both spindles. When the alignment bar will fit through both spindles, leave the bar in place and perform Step 4.

4. Check handlebar alignment. If handlebars do not face straight ahead, loosen the drag link locknut and turn the drag link (C, **Figure 26**) to align handlebars. When handlebars face straight ahead, tighten drag link locknut. Remove the alignment bar and perform Step 5.

5. Turn the handlebars all the way to the right. Then loosen the right-hand upper radius arm bolt (**Figure 35**) and adjust the steering stop so that it contacts the steering arm. Tighten the radius arm bolt.

6. Turn the handlebars all the way to the left. Perform Step 5 for the left-hand steering stop. Tighten the radius arm bolt.

7. Ski alignment is now correct. Install the skis as described in this chapter.

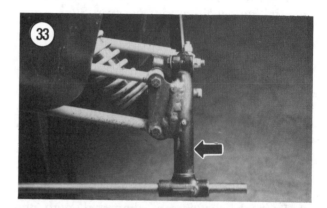

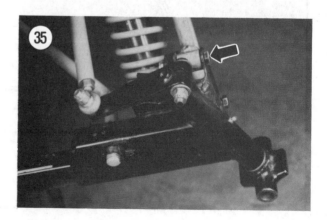

Table 1 FRONT SUSPENSION AND STEERING TIGHTENING TORQUES

	N·m	ft.-lb.
Handlebar adjuster block	15-18	11-13
Skag bolt	19-22	14-16
Spring saddle	48-54	35-40
Steering arm	49-54	36-40
Tie rod		
Attaching bolt	34-41	25-30
Locknut	14-16	10-12
Ski spindle pivot bolt	49	36

15

Chapter Sixteen

Track and Rear Suspension

All models are equipped with a slide-rail suspension. Suspension adjustment is found in Chapter Four.

REAR SUSPENSION

Removal

The rear suspension is secured to the tunnel with 4 bolts and lockwashers.

1. Loosen the locknuts on the track adjusting bolts (**Figure 1**) and back the bolts out to relieve track tension.

> *NOTE*
> *If you do not have a jack to raise the snowmobile when removing the rear suspension, place cardboard next to the snowmobile and turn it on its side. To prevent fuel leakage, drain the fuel tank before turning it on its side.*

2. Place a jack at the back of the snowmobile and raise the rear of the snowmobile slightly to remove tension from the track.

3. Remove the bolts and lockwashers holding the rear suspension to the tunnel. See **Figure 2** (front) and **Figure 3** (rear). Repeat for both sides.

4. Turn the snowmobile on its side and remove the rear suspension (**Figure 4**) from the track.

5. Lower the snowmobile to the ground.

Inspection

1. Remove the front (**Figure 5**) and rear (**Figure 6**) shafts.

2. Inspect the suspension mounting bolts for thread damage or breakage. Replace damaged bolts as required.

3. Clean all bolts thoroughly in solvent.

4. Visually inspect the shafts for any sign of wear, cracks, breakage or other abnormal conditions.

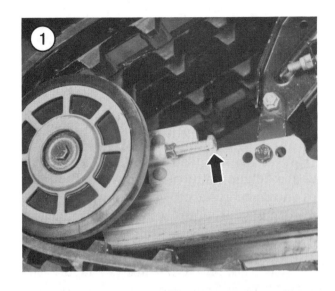

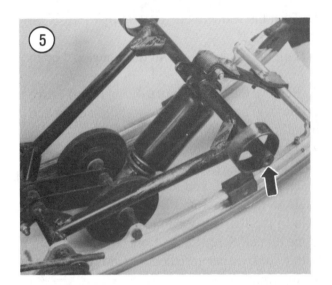

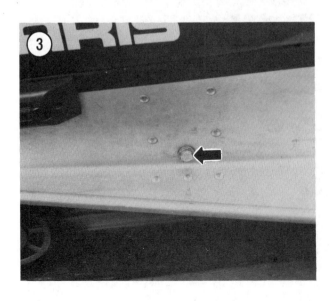

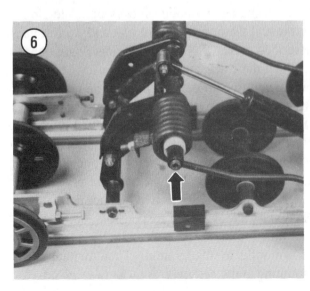

16

5. Check the shafts for bending.

6. If there is any doubt as to the condition of a shaft, replace it.

7. Lightly grease the shafts with a low-temperature grease before installation.

Installation

NOTE
Before turning the snowmobile on its side in Step 1, place a large piece of cardboard next to the snowmobile to prevent body damage.

1. Turn the snowmobile over so that it rests on its side.

2. Pull the track away from the tunnel and open it.

3. Remove the front (**Figure 4**) and rear (**Figure 5**) shafts to prevent them from sliding out when installing the rear suspension.

4. Install the rear suspension into the track and then install the front and rear shafts.

5. Hold the track and rear suspension at a 45° angle to the tunnel and line up the front shaft hole in the rear suspension with the front hole in the tunnel. Screw in the front bolt (**Figure 2**), with a lockwasher installed, but do not tighten at this time.

6. Turn the snowmobile onto its opposite side and install the opposite front bolt and washer as described in Step 5. Tighten bolt finger tight only.

7. Push the track and rear suspension into the tunnel and line up the rear hole in the rear shaft with the hole in the tunnel. Install the bolt (**Figure 3**), with a lockwasher installed, but do not tighten at this time.

8. Turn the snowmobile onto its opposite side and install the opposite rear bolt and washer as described in Step 7. Tighten bolt finger tight only.

9. When all 4 rear suspension bolts and lockwashers have been installed, lower the snowmobile to the ground. Then tighten the mounting bolts securely.

10. Adjust track tension and alignment as described in Chapter Three.

RAIL TIPS
(1986-ON)

Inspection/Replacement

The rail tips are mounted at the front of the rail and held in position with a rivet. The rail tips should be replaced when worn or if they show signs of abnormal wear or damage.

1. Remove the rear suspension as described in this chapter.

2. Visually inspect the rail tips (A, **Figure 7**) for cracks or other damage.

3. If a rail tip is damaged, remove the rivet holding the tip in place and replace the rail tip. Secure the rail tip with a new rivet burr and rivet.

4. Reinstall the rear suspension as described in this chapter.

HI-FAX

Refer to **Figure 8** (1984), **Figure 9** (1985), **Figure 10** (1986) or **Figure 11** (1987-1989).

Inspection/Replacement

The Hi-Fax strips are mounted at the bottom of the rail and held in position with a nut and

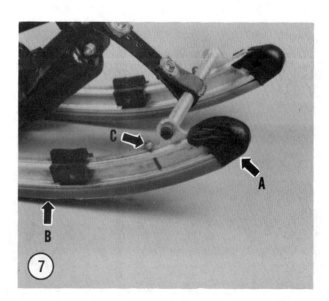

bolt. The Hi-Fax should be replaced when the thickness is 9.5 mm (3/8 in.) or less or if they show signs of abnormal wear or damage. See **Figure 12**.

1. Remove the rear suspension as described in this chapter.

2. Visually inspect the Hi-Fax (B, **Figure 7**) for cracks or other damage. If a crack is detected, replace the Hi-Fax as described in this procedure.

NOTE
The Hi-Fax must be replaced as a set.

3. Remove the Hi-Fax as follows:
   a. *1986-1989*: Remove the rail tips as described in this chapter.
   b. Turn the rear suspension over and rest it upside down on the workbench.
   c. Remove the nuts and bolts (C, **Figure 7**) at the front of each Hi-Fax.
   d. Working at the front of the Hi-Fax, drive the strip to the rear of the rail with a block of wood and a hammer as shown in **Figure 13**.

CAUTION
Do not use a steel punch to remove the Hi-Fax or the rail may become damaged.

4. Inspect the rail as follows:
   a. Clean the rail with solvent and dry thoroughly.
   b. Place a straightedge along side the runner and check for bends. If a rail is bent, it must be straightened. If the bend is severe or if the rail is cracked or damaged, replace it.
   c. Check the rail for gouges or cracks along the rail path. Smooth rough surfaces with a file or sandpaper.

5. Install the Hi-Fax as follows:
   a. Lightly grease the rail mating surface to ease installation.
   b. Working from the back of the rail, align the Hi-Fax with the rail so that the hole in the Hi-Fax faces to the front. Drive the Hi-Fax

onto the rail with the same wood block and hammer used during removal.
   c. Continue to drive the Hi-Fax onto the rail until the bolt hole in the Hi-Fax aligns with the hole in the rail. Install the bolt and nut and tighten securely.

REAR SUSPENSION ASSEMBLY

Refer to **Figure 8** (1984), **Figure 9** (1985), **Figure 10** (1986) or **Figure 11** (1987-1989).

Front Shock Absorber Removal/Installation

Because of the differences among the rear suspension assemblies, refer to **Figures 8-11** for your model when performing this procedure.

1. Remove the rear suspension as described in this chapter.

2. Record the limiter strap (A, **Figure 14**) hole position, then disconnect the limiter strap at the front torque arm.

3. Remove the lower shock absorber mounting bolt and nut at the front torque arm.

4. Remove the upper shock mounting bolt and nut (B, **Figure 14**).

5. Remove the shock absorber (A, **Figure 15**).

6. Inspect the shock absorber as described in this chapter.

7. Installation is the reverse of these steps. Note the following.

8. Apply a light coat of low-temperature grease to the shock bolts before assembly.

9. Place the shock absorber into the rear suspension and install its mounting bolts and nuts. Tighten the shock absorber bolts securely.

10. Reconnect the limiter strap (A, **Figure 14**) at the front torque arm. Reinstall the bolt through the hole recorded prior to disassembly.

Rear Shock Absorber Removal/Installation

1. Remove the rear suspension as described in this chapter.

16

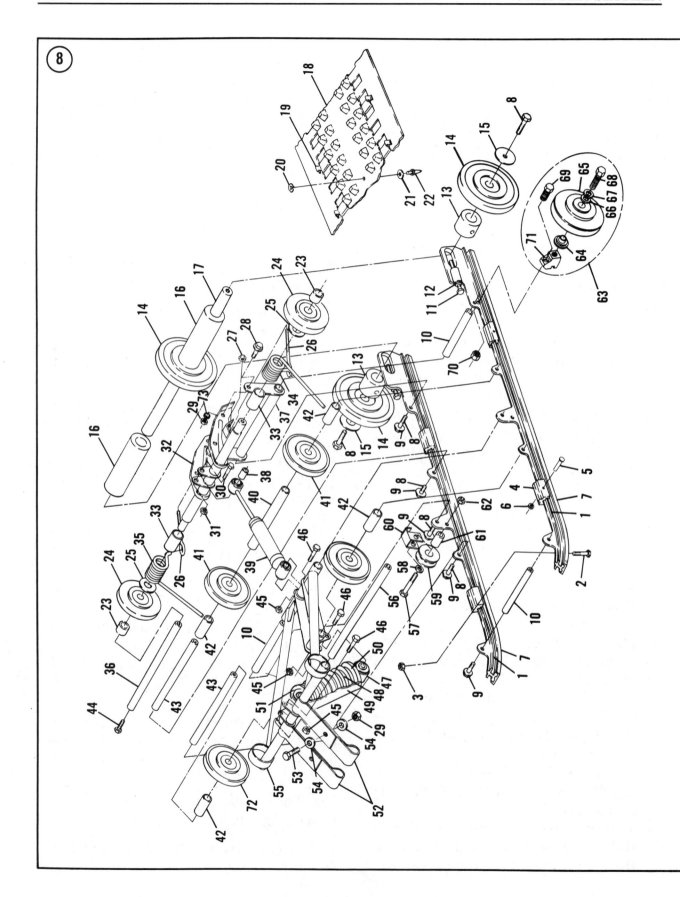

REAR SUSPENSION
(1984)

1. Rail
2. Bolt
3. Nut
4. Bumper
5. Rivet
6. Push-on nut
7. Hi-Fax
8. Bolt
9. Washer
10. Shaft
11. Bolt
12. Nut
13. Outer spacer
14. Wheel
15. Washer
16. Inner spacer
17. Shaft
18. Track
19. Clip
20. Tee nut
21. Washer
22. Stud
23. Spacer
24. Wheel
25. Washer
26. Eyebolt
27. Nut
28. Bolt
29. Nut
30. Bolt
31. Nut
32. Rear torque arm
33. Spring tube
34. Spring
35. Spring
36. Shaft
37. Pivot arm
38. Spacer
39. Shock
40. Inner spacer
41. Wheel
42. Outer spacer
43. Shaft
44. Bolt
45. Nut
46. Bolt
47. Front shock
48. Spring
49. Spring shield
50. Retainer
51. Retainer
52. Limiter straps
53. Bolt
54. Washer
55. Front torque arm
56. Shaft
57. Bolt
58. Washer
59. Spring roller
60. Spring retainer
61. Spacer
62. Nut
63. Wheel assembly (parts 64-71)
64. Wheel insert
65. Wheel
66. Washer
67. Lockwasher
68. Bolt
69. Bolt
70. Nut
71. Bracket
72. Wheel
73. Spacer

16

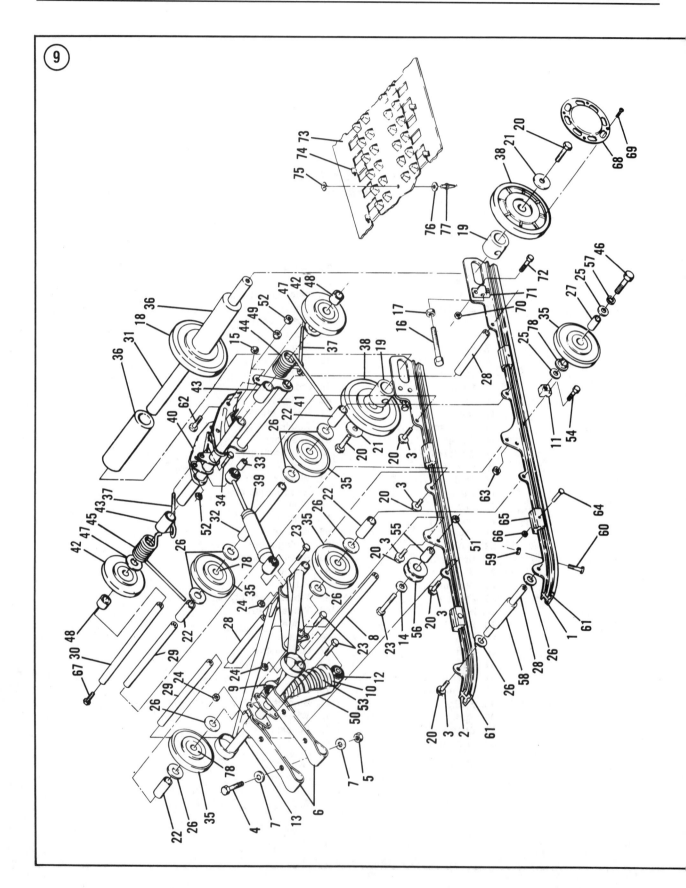

REAR SUSPENSION
(1985)

1. Left-hand rail	40. Rear torque arm
2. Right-hand rail	41. Pivot arm
3. Washer	42. Wheel
4. Bolt	43. Tube
5. Nut	44. Spring
6. Limiter straps	45. Spring
7. Washer	46. Bolt
8. Shaft	47. Washer
9. Retainer	48. Spacer
10. Retainer	49. Shoulder nut
11. Bracket	50. Spring shield
12. Front shock	51. Nut
13. Front torque arm	52. Nut
14. Washer	53. Front shock
15. Nut	54. Bolt
16. Bolt	55. Spacer
17. Nut	56. Spring roller
18. Center idler wheel	57. Washer
19. Outer spacer	58. Spacer
20. Bolt	59. Nut
21. Washer	60. Bolt
22. Outer spacer	61. Hi-Fax
23. Bolt	62. Bolt
24. Nuts	63. Nut
25. Washer	64. Rivet
26. Bushing	65. Bumper
27. Spacer	66. Push-on nut
28. Shaft	67. Bolt (600 SE)
29. Shaft	68. Hub cap (600 SE)
30. Shaft	69. Screw (600 SE)
31. Shaft	70. Nut
32. Spacer	71. Adjustment block
33. Spacer	72. Bolt
34. Bolt	73. Track
35. Wheel	74. Clip
36. Inner spacer	75. Tee nut (600)
37. Eyebolt	76. Washer (600)
38. Outer wheel	77. Stud (600)
39. Shock	78. Bearing insert

16

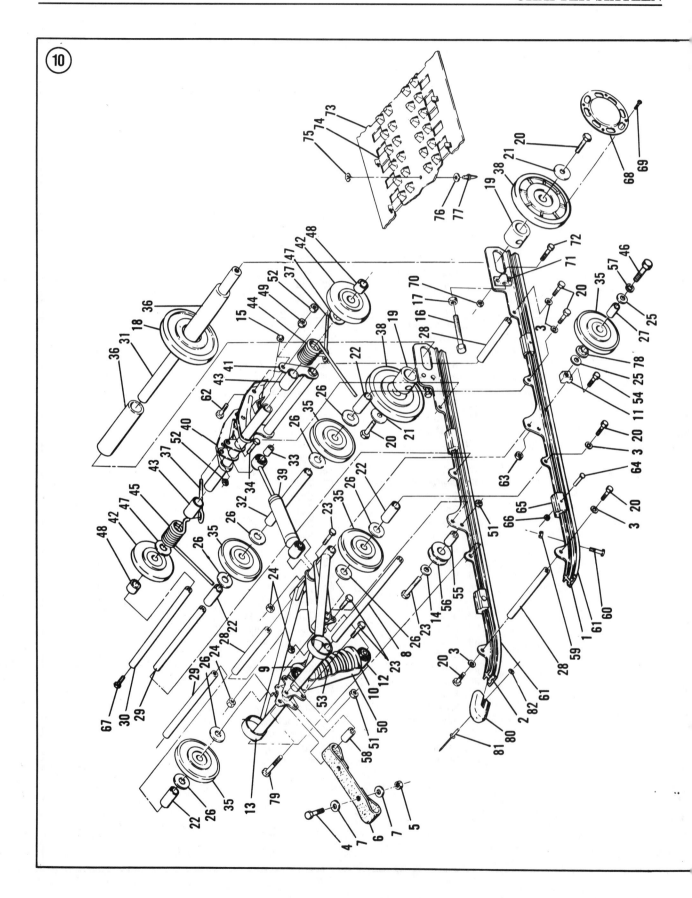

REAR SUSPENSION
(1986)

1.	Left-hand rail	42.	Wheel
2.	Right-hand rail	43.	Tube
3.	Washer	44.	Spring
4.	Bolt	45.	Spring
5.	Nut	46.	Bolt
6.	Limiter strap	47.	Washer
7.	Washer	48.	Spacer
8.	Shaft	49.	Shoulder nut
9.	Retainer	50.	Spring shield
10.	Retainer	51.	Nut
11.	Bracket	52.	Nut
12.	Front shock	53.	Front shock
13.	Front torque arm	54.	Bolt
14.	Washer	55.	Spacer
15.	Nut	56.	Spring roller
16.	Bolt	57.	Washer
17.	Nut	58.	Spacer
18.	Center idler wheel	59.	Nut
19.	Outer spacer	60.	Bolt
20.	Bolt	61.	Hi-Fax
21.	Washer	62.	Bolt
22.	Outer spacer	63.	Nut
23.	Bolt	64.	Rivet
24.	Nuts	65.	Bumper
25.	Washer	66.	Push-on nut
26.	Bushing	67.	Bolt (600 SE)
27.	Spacer	68.	Hub cap (600 SE)
28.	Shaft	69.	Screw (600 SE)
29.	Shaft	70.	Nut
30.	Shaft	71.	Adjustment block
31.	Shaft	72.	Bolt
32.	Spacer	73.	Track
33.	Spacer	74.	Clip
34.	Bolt	75.	Tee nut (600)
35.	Wheel	76.	Washer (600)
36.	Inner spacer	77.	Stud (600)
37.	Eyebolt	78.	Bearing insert
38.	Outer wheel	79.	Bolt
39.	Shock	80.	Rail tip cover
40.	Rear torque arm	81.	Rivet
41.	Pivot arm	82.	Rivet retaining washer

16

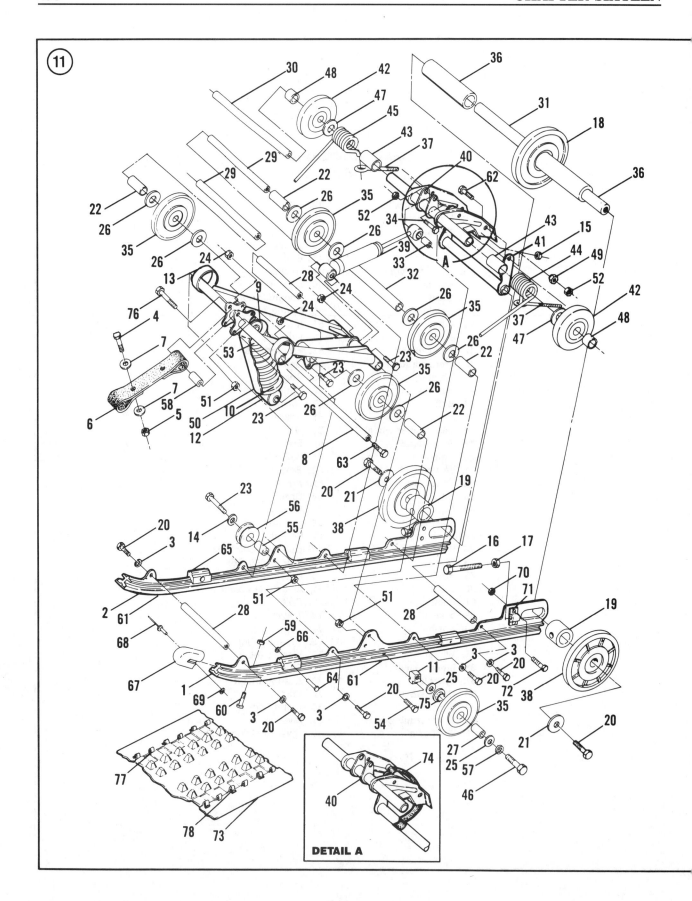

DETAIL A

REAR SUSPENSION
(1987-1989)

1. Left-hand rail
2. Right-hand rail
3. Washer
4. Bolt
5. Nut
6. Limiter strap
7. Washer
8. Shaft
9. Retainer
10. Retainer
11. Bracket
12. Front shock
13. Front torque arm
14. Washer
15. Nut
16. Bolt
17. Nut
18. Center idler wheel
19. Outer spacer
20. Bolt
21. Washer
22. Outer spacer
23. Bolt
24. Nut
25. Washer
26. Bushing
27. Spacer
28. Shaft
29. Shaft
30. Shaft
31. Shaft
32. Spacer
33. Spacer
34. Bolt
35. Wheel
36. Inner spacer
37. Eyebolt
38. Outer wheel
39. Shock
40. Rear torque arm
41. Pivot arm
42. Wheel
43. Tube
44. Spring
45. Spring
46. Bolt
47. Washer
48. Spacer
49. Shoulder nut
50. Spring shield
51. Nut
52. Nut
53. Front shock
54. Bolt
55. Spacer
56. Spring roller
57. Washer
58. Spacer
59. Nut
60. Bolt
61. Hi-Fax
62. Bolt
63. Bolt
64. Rivet
65. Bumper
66. Push-on nut
67. Rail tip cover
68. Rivet
69. Rivet burr
70. Nut
71. Adjustment block
72. Bolt
73. Track
74. Limiter strap (1988-1989 Trail only)
75. Bearing insert
76. Bolt
77. Metal clip
78. Guide clip

16

2. Remove the shock bolt at the front torque arm (B, **Figure 15**).

3. Remove the shock bolt at the rear torque arm (A, **Figure 16**).

4. Remove the shock absorber (B, **Figure 16**).

5. Inspect the shock absorber as described in this chapter.

6. Installation is the reverse of these steps. Note the following.

7. Apply a light coat of low-temperature grease to the shock bolts before assembly.

8. Place the shock absorber into the rear suspension and install its mounting bolts and nuts. Tighten the shock absorber bolts securely.

Front Shock Spring
Removal/Installation

1. Remove the shock absorber as described in this chapter.

2. Secure the bottom of the shock absorber in a vise with soft jaws.

3. Turn the spring adjuster to its softest position.

> *WARNING*
> *Do not attempt to remove or install the shock spring without the use of a spring compressor. Attempting to remove the spring without the use of a spring compressor may cause severe personal*

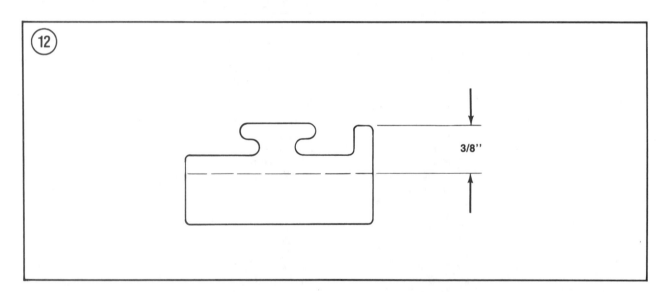

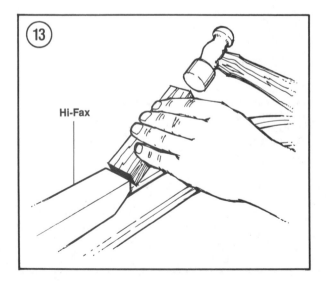

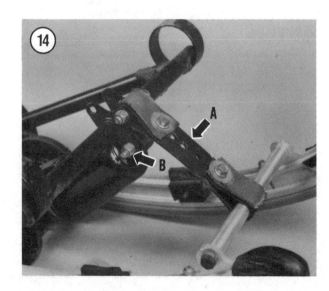

injury. If you do not have access to a spring compressor, refer spring removal to a Polaris dealer.

4. Install a spring compressor onto the shock absorber following the manufacturer's instructions. Then compress the spring with the tool and remove the spring stopper from the top of the shock. Remove the spring.

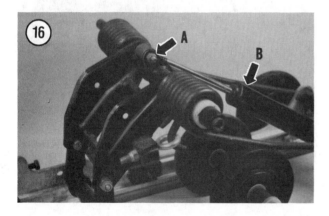

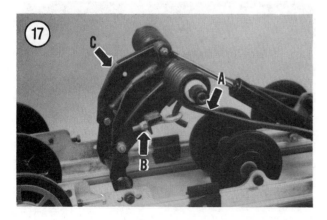

CAUTION
Do not attempt to disassemble the shock damper housing.

5. Inspect the shock and spring as described in the following procedure.
6. Installation is the reverse of these steps. Make sure the spring stopper secures the spring completely.

Shock Inspection

1. Clean all components thoroughly in solvent and allow to dry.
2. Check the damper rod as follows:
   a. Check the damper housing for leakage. If the damper is leaking, replace it.
   b. Check the damper rod for bending or damage. Check the damper housing for dents or other damage.
   c. Operate the damper rod by hand and check its operation. If the damper is operating correctly, a small amount of resistance should be felt on the compression stroke and a considerable amount of resistance felt on the return stroke.
   d. Replace the damper assembly if necessary.
3. Check the damper housing for cracks, dents or other damage. Replace the damper housing if damaged. Do not attempt to repair or straighten it.

Rear Spring
Removal/Installation

Because of the differences among the rear suspension assemblies, refer to **Figures 8-11** for your model when performing this procedure.
1. Remove the rear suspension as described in this chapter.

NOTE
The left- and right-hand springs (A, Figure 17) are different. Label springs before removal so that they will be reinstalled in their original position.

16

2. Loosen and remove the eyebolt locknut (B, **Figure 17**) to release spring tension from the eyebolt. Then disconnect the eyebolt from the spring and remove the spring (A, **Figure 17**).

3. Installation is the reverse of these steps. Note the following.

4. Adjust spring tension as described in Chapter Four.

Front Torque Arm
Removal/Installation

Because of the differences among the rear suspension assemblies, refer to **Figures 8-11** for your model when performing this procedure.

1. Remove the front shock absorber as described in this chapter.

2. Disconnect the rear shock absorber (B, **Figure 15**) at the front torque arm.

3. Remove the rear springs (A, **Figure 17**) as described in this chapter.

4. Remove the wheel assembly (C, **Figure 15**) from the front torque arm.

5. Remove the front torque arm pivot bolt assembly and remove the front torque arm.

6. Inspect the front torque arm as described in this chapter.

7. Installation is the reverse of these steps. Note the following.

8. Apply a low-temperature grease to all pivot shafts and bolts.

9. Tighten all nuts and bolts securely.

Rear Torque Arm
Removal/Installation

Because of the differences among the rear suspension assemblies, refer to **Figures 8-11** for your model when performing this procedure.

1. Remove the rear springs (A, **Figure 17**) as described in this chapter.

2. Record the rear torque arm pivot bolt hole position (**Figure 18**) before removing the bolt.

3. Remove the rear torque arm pivot bolt and remove the rear torque arm (C, **Figure 17**).

4. Inspect the rear torque arm as described in this chapter.

5. Installation is the reverse of these steps. Note the following.

6. Apply a low-temperature grease to the rear torque arm pivot shaft.

7. Mount the rear torque arm pivot shaft into the hole position (**Figure 18**) recorded prior to disassembly.

8. Tighten all nuts and bolts securely.

SUSPENSION AND GUIDE WHEELS

Refer to the following exploded views for your model when servicing the suspension and guide wheels:

a. **Figure 8**: 1984 models.
b. **Figure 9**: 1985 models.
c. **Figure 10**: 1986 models.
d. **Figure 11**: 1987-1989 models.

Inspection

1. Remove the rear suspension as described in this chapter.

2. Spin the suspension and guide wheels. See **Figure 19** and **Figure 20**, typical. The wheels should spin smoothly with no sign of roughness, binding or noise. These abnormal conditions indicate worn or damaged bearings. Replace worn or damaged bearings as described in this chapter.

3. Check the wheel hubs for cracks.

4. Check the outer wheel surface for cracks, deep scoring or excessive wear.

5. If there is any doubt as to the condition of any part, replace it with a new one. Note the following:

a. Remove circlips from wheels where used (**Figure 21**).

b. Remove pressed on wheels by first removing the wheel from the bearing with a bearing puller. Then remove the bearing with a puller.

c. Align the new bearing (**Figure 22**) with the wheel. Drive bearing into wheel with a socket placed on the *outer* bearing race (**Figure 23**).

d. Install circlip (**Figure 21**) into wheel grooves.

e. Install wheel by driving the bearing onto the shaft with a socket placed on the *inner* bearing race. Do not install by driving on the outer bearing race or bearing and/or wheel damage may occur.

6. When installing non-pressed on wheels, note the following:

a. Coat all sliding surfaces with a low-temperature grease.

b. When installing circlips, make sure the circlip seats in the shaft groove or wheel groove completely.

c. Spin each wheel and check its operation. If a wheel is tight or binding, remove the wheel and check the bearing.

16

Wheel Bearing Replacement

1. Remove the wheel from its shaft.

2. Remove the circlip from the groove in the wheel.

3. Using a socket or bearing driver, remove the wheel bearing.

4. Clean the wheel in solvent and thoroughly dry.

5. Check the circlip groove in the wheel hub for cracks. Check the hub area for breakage or other damage. Replace the wheel if the hub is damaged.

6. Install the new bearing as follows:

   a. Align the new bearing with the hub.

   b. Drive the bearing into the hub with a socket or bearing driver placed on the *outer* bearing race.

   c. Drive the bearing squarely into the hub until it bottoms out.

7. Secure the bearing with the circlip. Make sure the bearing seats in the hub groove completely.

TRACK

Removal/Installation

1. Remove the rear suspension as described in this chapter.

2. Remove the driveshaft as described in Chapter Fourteen.

3. Remove the track.

4. Installation is the reverse of these steps. Note the following.

5. When installing the track, orient the track lugs to run in the direction shown in **Figure 24**.

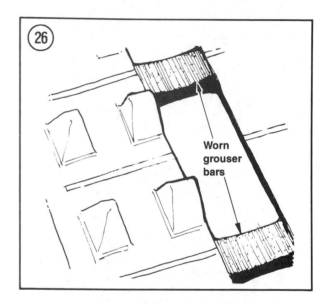

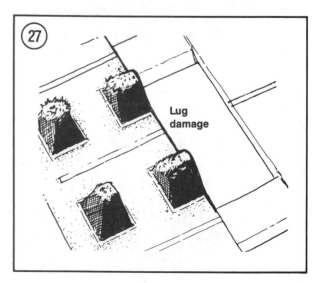

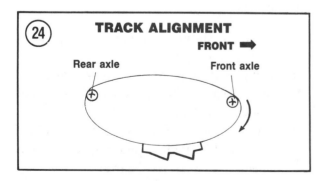

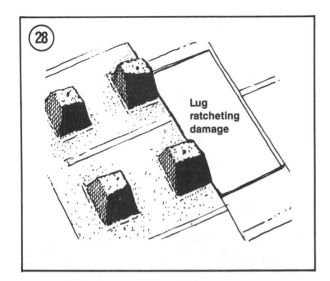

Lug ratcheting damage

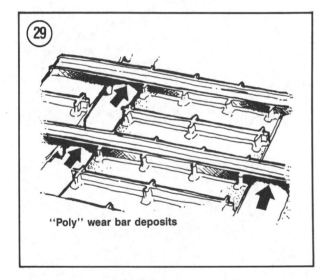

"Poly" wear bar deposits

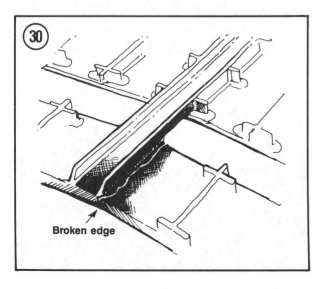

Broken edge

Inspection

1. Check for missing or damaged track cleats. Replace cleats as described in this chapter.

2. Visually inspect the track for the following conditions:

   a. Obstruction damage: Cuts, slashes and gouges in the track surface are caused by hitting obstructions. These could include broken glass, sharp rocks or buried steel. See **Figure 25**.

   b. Worn grouser bars: Excessively worn grouser bars are caused by snowmobile operation over rough and non-snow covered terrain such as gravel roads and highway roadsides. See **Figure 26**.

   c. Lug damage: The lug damage shown in **Figure 27** is caused by lack of snow lubrication.

   d. Ratcheting damage: Insufficient track tension is a major cause of ratcheting damage to the top of the lugs (**Figure 28**). Ratcheting can also be caused by too great a load and constant "jack-rabbit" starts.

   e. Over-tension damage: Excessive track tension can cause too much friction on the wear bars. This friction causes the wear bars to melt and adhere to the track grouser bars. See **Figure 29**. An indication of this condition is a "sticky" track that has a tendency to "lock up."

   f. Loose track damage: A track adjusted too loosely can cause the outer edge to flex excessively. This results in the type of damage shown in **Figure 30**. Excessive weight can also contribute to the damage.

   g. Impact damage: Impact damage as shown in **Figure 31** causes the track rubber to open and expose the cord. This frequently happens in more than one place. Impact damage is usually caused by riding on rough or frozen ground or ice. Also, insufficient track tension can allow the track to pound against the track stabilizers inside the tunnel.

16

h. Edge damage: Edge damage as shown in **Figure 32** is usually caused by tipping the snowmobile on its side to clear the track and allowing the track edge to contact an abrasive surface.

Cleat Replacement

Follow the same cleat pattern when installing new cleats.

A hand grinder, safety glasses and a universal track clip installer will be required to remove and install new cleats. See **Figure 33**.

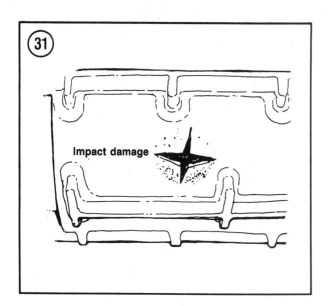

Impact damage

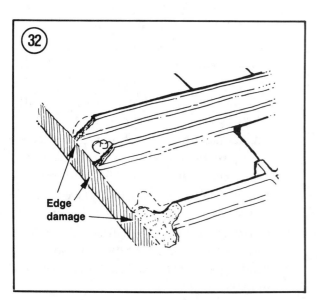

Edge damage

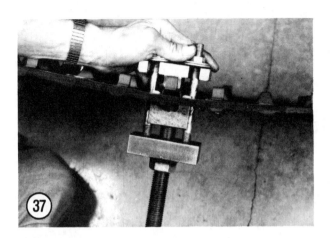

1. Using a hand grinder (**Figure 34**), grind a slit in the corner of the cleat. See **Figure 35**.
2. Pry the cleat off of the track (**Figure 36**).
3. Align the new cleat onto the track. Then install the cleat with the cleat installation tool (**Figure 37**). Check the cleat to make sure it is tight.

16

Chapter Seventeen

Off-Season Storage

One of the most critical aspects of snowmobile maintenance is off-season storage. Proper storage will prevent engine and suspension damage and fuel system contamination. Improper storage will cause various degrees of deterioration and damage.

Preparation for Storage

Careful preparation will minimize deterioration and make it easier to restore the snowmobile to service later. When performing the following procedure, make a list of replacement or damaged parts so that they can be ordered and installed before next season.

1. Remove the seat and clean the area underneath the seat thoroughly. Wipe the seat off with a damp cloth and wipe a preservative over the seat to keep it from drying out. If you are concerned about the seat during storage, store it away from the snowmobile in a safe place.

2. Flush the cooling system as described in Chapter Three. Before refilling the cooling system, check all of the hoses for cracks or deterioration. Replace hoses as described in Chapter Ten. Make sure all hose clamps are tight. Replace questionable hose clamps as required.

3. Change the chaincase oil as described in Chapter Three.

CAUTION
Do not allow water to enter the engine when performing Step 4.

4. Clean the snowmobile from front to back. Remove all dirt and other debris from the pan and tunnel. Clean out debris caught in the track.

5. Check the frame, skis and other metal parts for cracks or other damage. Apply paint to all bare metal surfaces.

6. Check all fasteners for looseness and tighten as required. Replace loose or damaged rivets.

NOTE
Refer to the appropriate chapter for the specified tightening torque as required.

7. Lubricate all pivot points with a low-temperature grease as described in Chapter Three.

8. Unplug all electrical connectors and clean both connector halves with electrical contact cleaner. Check the electrical contact pins for damage or looseness. Repair connectors as required. After the contact cleaner evaporates, apply a dielectric grease to one connector half and reconnect the connectors.

CAUTION
Dielectric grease is formulated for electrical use. Do not use a regular type of grease on electrical connectors.

9. To protect the engine from rust buildup during storage, the engine must be fogged in. Perform the following:
a. Jack up the snowmobile so that the track clears the ground.
b. Open and secure the shroud. Remove the air box as described in Chapter Seven.
c. Start the engine and allow it to warm to normal operating temperature.

WARNING
The exhaust gases are poisonous. Do not run the engine in a closed area. Make sure there is plenty of ventilation.

NOTE
A cloud of smoke will develop in sub-step d. This is normal.

d. Spray an engine preservative (storage oil) into *both* carburetors (**Figure 1**) until the engine smokes heavily or stalls.
e. Turn all switches OFF.
f. Reinstall the air box.

CAUTION
To prevent expensive engine damage, clean the area around the spark plug holes before removing the spark plugs.

g. Remove the spark plugs (**Figure 2**). Reconnect the spark plugs at their plug caps to ground them.
h. Pour approximately 30 cc (1 fl. oz.) of Polaris 40:1 injector oil into each spark plug

17

hole. Pull the recoil starter handle (approximately 10 times) to properly distribute oil onto cylinder walls.

i. Wipe a film of oil on the spark plug threads and reinstall the spark plugs. Reconnect the high tension leads at the plugs.

CAUTION
During the storage period, do not run engine.

10. *Electric start models*: Remove the battery and coat the cable terminals with petroleum jelly. See Chapter Eight. Check the electrolyte level and refill with distilled water if it is low. Store the battery in an area where it will not freeze and recharge it once a month.

11. Plug the end of the muffler with a rag to prevent moisture from entering. Then tag the machine with a note to remind you to remove the rag before restarting the engine next season.

12. Remove the drive belt (**Figure 3**) and store it on a flat surface. Refer to Chapter Twelve.

13. Apply a light coat of oil or Polaris Clutch and Cable Lubricant to sheave faces, ramps and shafts. See Chapter Twelve.

14. Clean the jackshaft thoroughly.

15. Reinstall the primary and secondary sheaves. Tighten the primary sheave bolt to the torque specification listed in Chapter Twelve. Tighten the secondary sheave bolt securely.

16. Close and secure the belt shield.

WARNING
Some fuel may spill in the following procedure. Work in a well-ventilated area at least 50 feet from any sparks or flames, including gas appliance pilot lights. Do not smoke in the area. Keep a B:C rated fire extinguisher handy.

17. Using a suitable siphon tool, siphon fuel out of the fuel tank and into a gasoline storage tank.

18. When the fuel tank is empty, remove the drain plug (**Figure 4**) on each carburetor and drain the carburetors. Wipe up spilled gasoline immediately.

19. Protect all glossy surfaces on the chassis, hood and dash with an automotive type wax.

20. Then raise the track off of the ground with wood blocks. Make sure the snowmobile is secure. Release track tension while snowmobile is in storage.

21. Cover the snowmobile with a heavy cover that will provide adequate protection from dust and damage. Do not cover the snowmobile with plastic as moisture can collect and cause rusting. If the snowmobile has to be stored outside, block the entire vehicle off the ground.

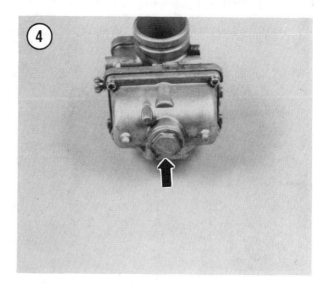

Removal From Storage

Preparing the snowmobile for use after storage should be relatively easy if proper storage procedures were followed.

1. Remove the plug from the end of the muffler.

2. Adjust track tension as described in Chapter Three.

3. Inspect the drive belt for cracks or other abnormal conditions. Then reinstall the drive belt as described in Chapter Twelve.

4. Check the chaincase oil level. Refill as described in Chapter Three. If the oil was not changed before storage, change the oil as described in Chapter Three.

5. Check and adjust drive belt tension.

6. Check the coolant level and refill if necessary.

7. Check all of the control cables for proper operation. Adjust as described in Chapter Three.

8. Fill the oil injection tank as described in Chapter Three. If the tank was dry or if a hose was disconnected, bleed the oil pump as described in Chapter Nine.

9. *Electric start models*: Check battery electrolyte level and fill with distilled water as necessary. Make sure the battery has a full charge (recharge if necessary). Clean the battery terminals and install the battery. Connect the positive, then the negative cable. Apply petroleum jelly to the battery terminals.

10. Perform an engine tune-up as described in Chapter Three.

11. Check and fill the fuel system.

12. Make a thorough check of the snowmobile for loose or missing nuts, bolts or screws.

13. Start the engine and check for fuel or exhaust leaks. Make sure the lights and all switches work properly. Turn the engine off.

WARNING
The exhaust gases are poisonous. Do not run the engine in a closed area. Make sure there is plenty of ventilation.

14. After the engine has been initially run for a period of time, install new spark plugs as described in Chapter Three.

17

Index

Wiring Diagrams

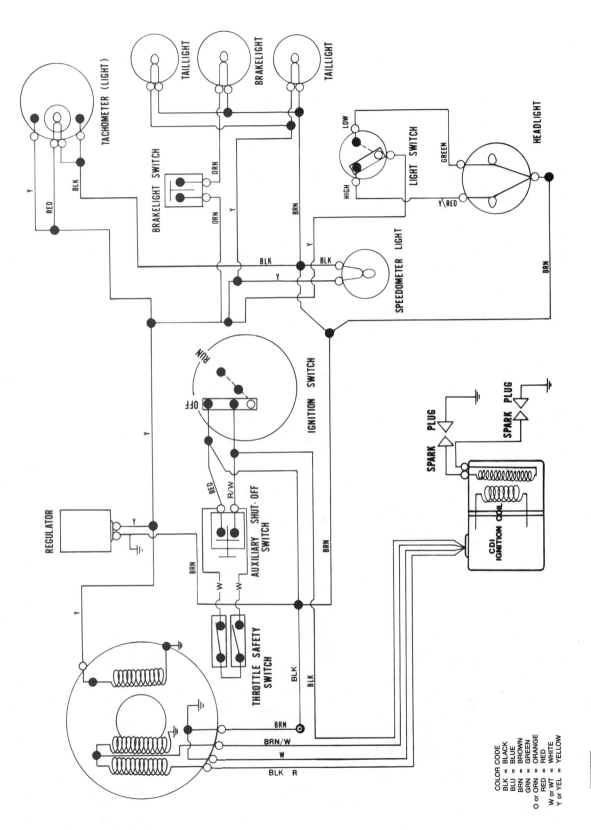

1984 INDY TRAIL

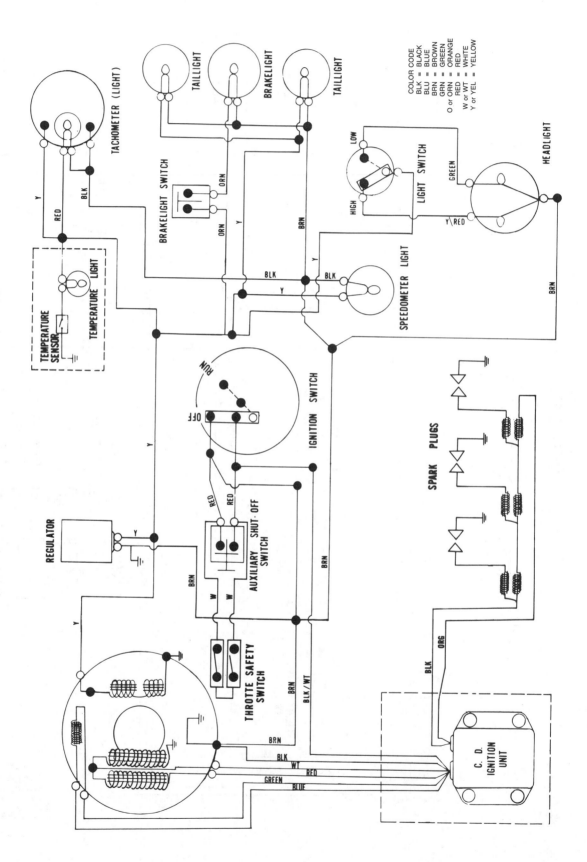

1984 INDY 600

COLOR CODE
BLK = BLACK
BLU = BLUE
BRN = BROWN
GRN = GREEN
O or ORN = ORANGE
RED = RED
W or WT = WHITE
Y or YEL = YELLOW

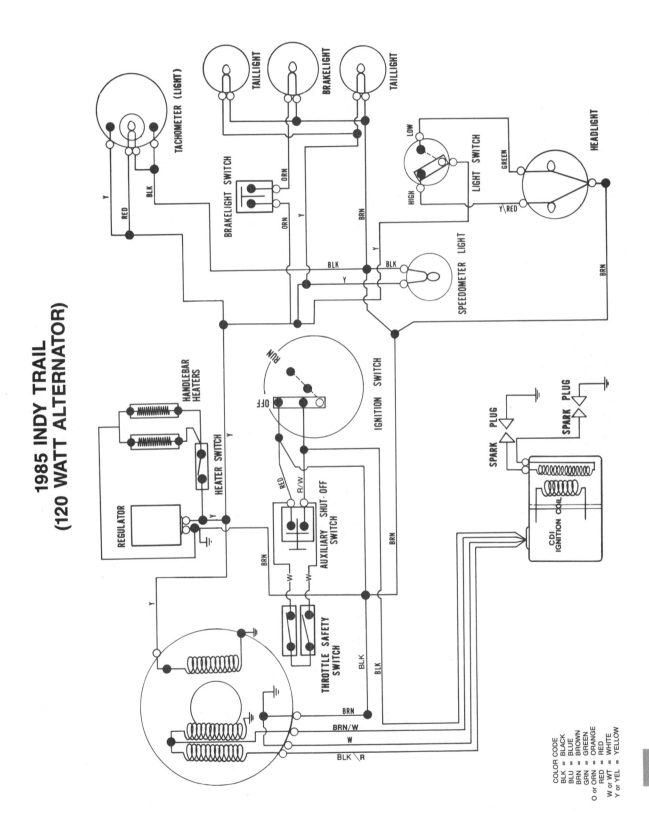

1985 INDY TRAIL
(120 WATT ALTERNATOR)

TACHOMETER (LIGHT)

TAILLIGHT

BRAKELIGHT

TAILLIGHT

LIGHT SWITCH

HEADLIGHT

BRAKELIGHT SWITCH

SPEEDOMETER LIGHT

IGNITION SWITCH

HANDLEBAR HEATERS

HEATER SWITCH

REGULATOR

AUXILIARY SHUT-OFF SWITCH

SPARK PLUG

SPARK PLUG

CDI IGNITION COIL

THROTTLE SAFETY SWITCH

COLOR CODE
BLK = BLACK
BLU = BLUE
BRN = BROWN
GRN = GREEN
O or ORN = ORANGE
RED = RED
W or WT = WHITE
Y or YEL = YELLOW

19

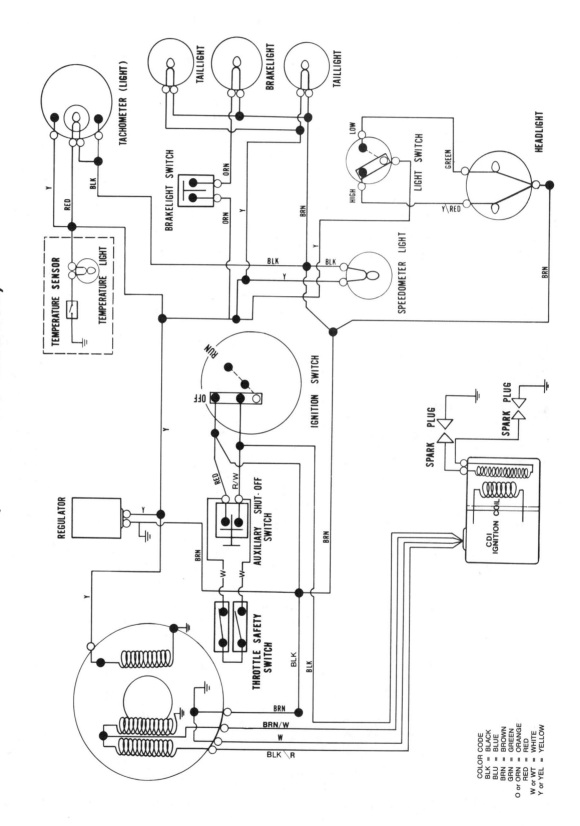

**1985 INDY 400
(120 WATT ALTERNATOR)**

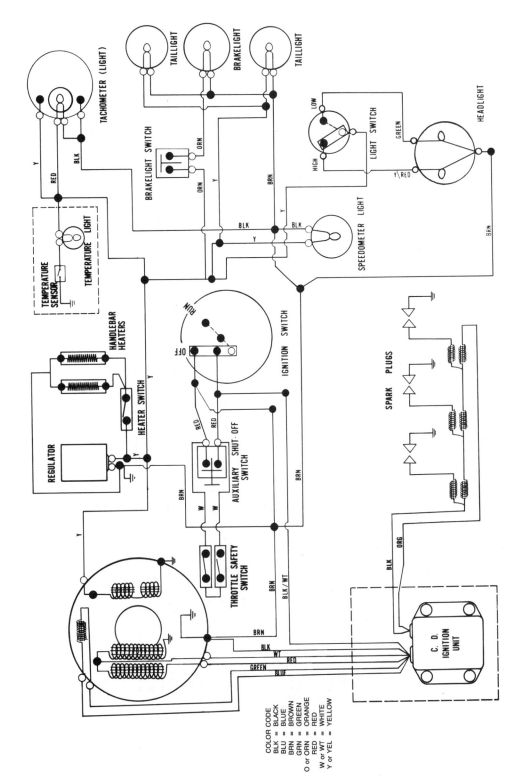

1985 INDY 600 & INDY 600 SE (120 WATT ALTERNATOR)

Handlebar Heaters standard on Special Edition
Optional on Indy 600

COLOR CODE
BLK = BLACK
BLU = BLUE
BRN = BROWN
GRN = GREEN
O or ORN = ORANGE
RED = RED
W or WT = WHITE
Y or YEL = YELLOW

19

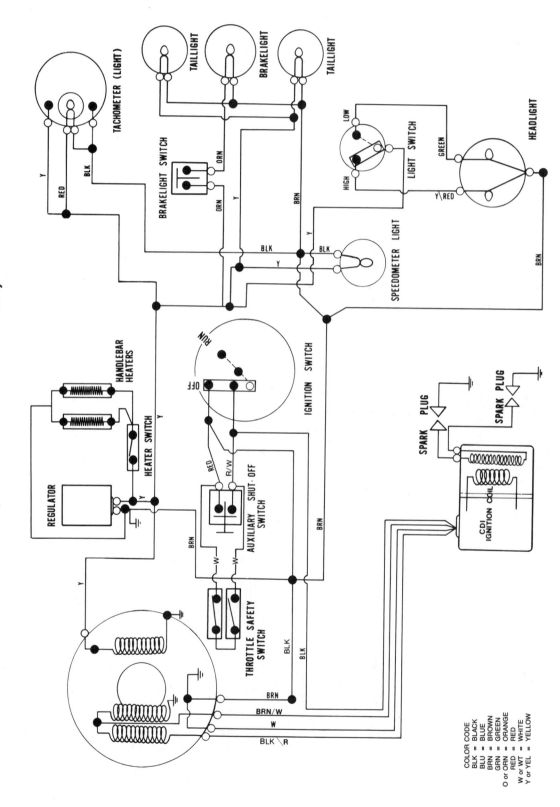

1986 INDY TRAIL
(120 WATT ALTERNATOR)

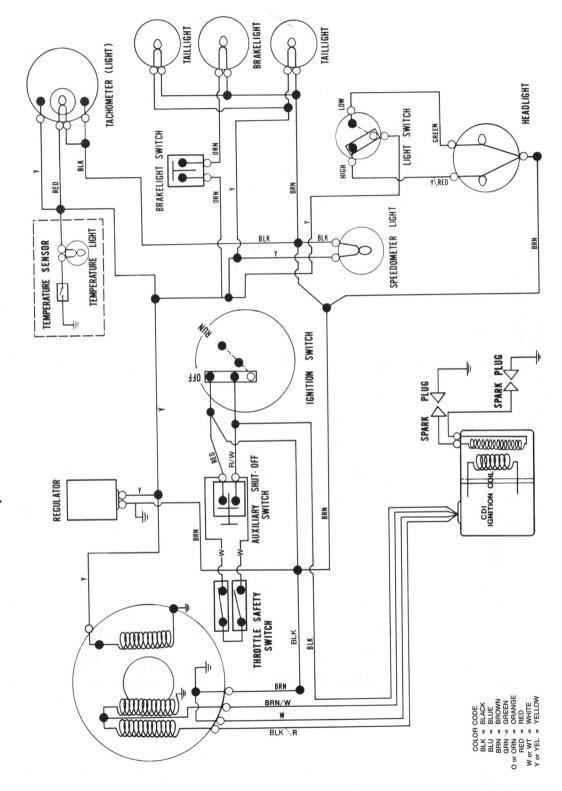

**1986 INDY 400
(120 WATT ALTERNATOR)**

TACHOMETER (LIGHT)

TAILLIGHT

BRAKELIGHT

TAILLIGHT

HEADLIGHT

LOW

HIGH

LIGHT SWITCH

GREEN

Y\RED

BRAKELIGHT SWITCH

ORN

ORN

Y

BRN

BLK

Y

BLK

SPEEDOMETER LIGHT

BRN

Y

RED

BLK

TEMPERATURE SENSOR

TEMPERATURE LIGHT

RUN

OFF

IGNITION SWITCH

RED

R/W

AUXILIARY SHUT-OFF SWITCH

SPARK PLUG

SPARK PLUG

CDI IGNITION COIL

REGULATOR

Y

Y

BRN

W

W

THROTTLE SAFETY SWITCH

BRN

BLK

BLK

Y

BRN

BRN/W

W

BLK\R

COLOR CODE
BLK = BLACK
BLU = BLUE
BRN = BROWN
GRN = GREEN
O or ORN = ORANGE
RED = RED
W or WT = WHITE
Y or YEL = YELLOW

19

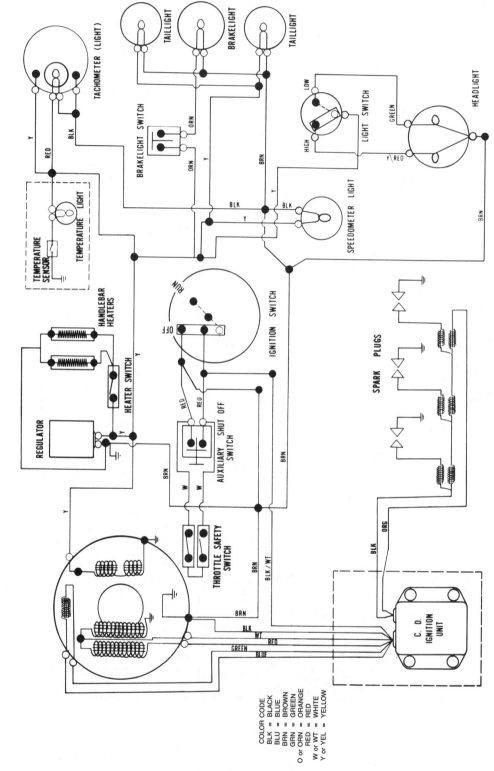

1986 INDY 600 & INDY 600 LE (120 WATT ALTERNATOR)

Handlebar Heaters standard on Limited Edition
Optional on Indy 600

COLOR CODE
BLK = BLACK
BLU = BLUE
BRN = BROWN
GRN = GREEN
O or ORN = ORANGE
RED = RED
W or WT = WHITE
Y or YEL = YELLOW

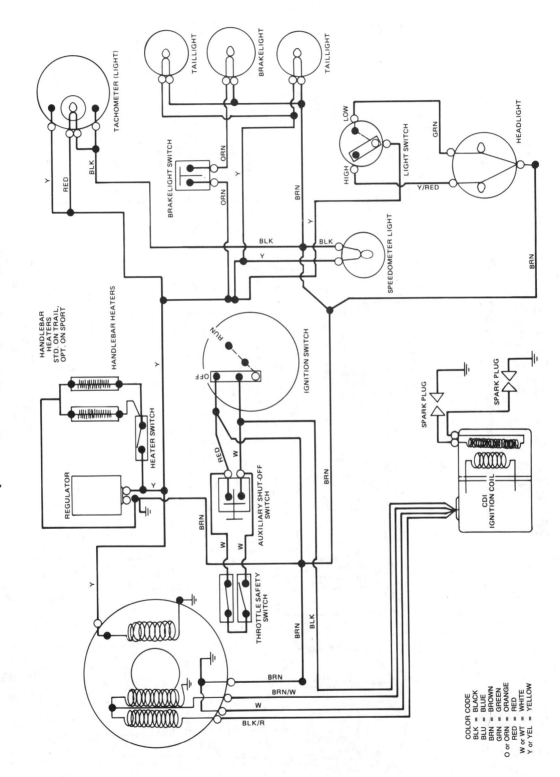

1987 INDY SPORT & INDY TRAIL (133 SKS) (120 WATT ALTERNATOR)

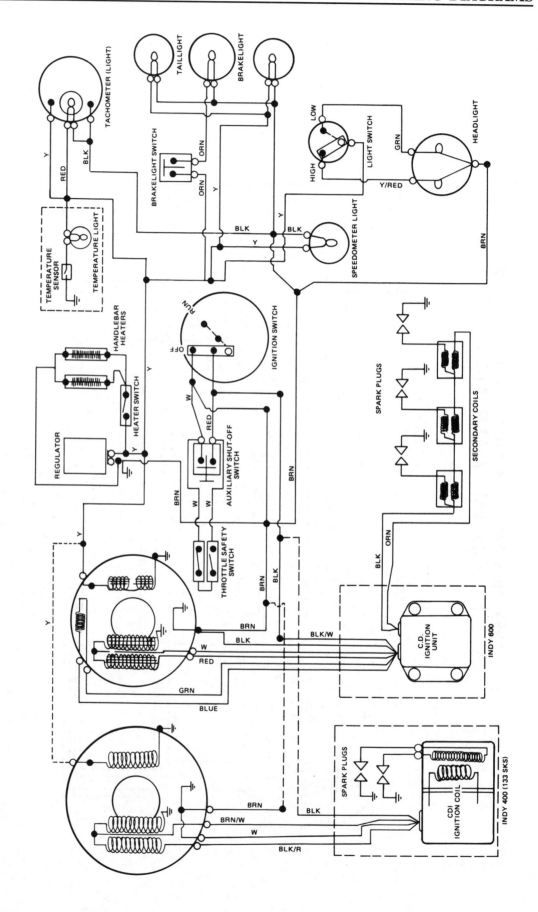

1987 INDY 400 (133 SKS) & INDY 600 (120 WATT ALTERNATOR)

Handlebar heaters optional on both models

COLOR CODE
BLK = BLACK
BLU = BLUE
BRN = BROWN
GRN = GREEN
O or ORN = ORANGE
RED = RED
W or WT = WHITE
Y or YEL = YELLOW

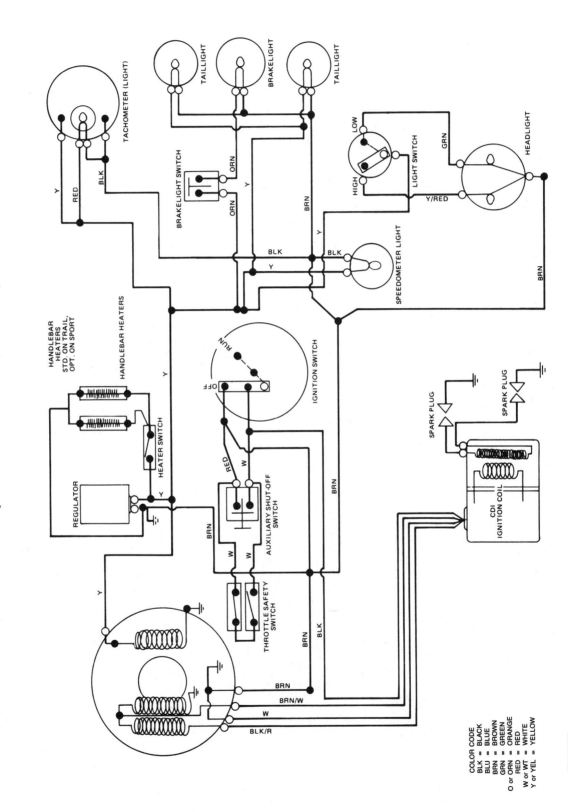

1988 INDY SPORT & INDY TRAIL (ALL EXCEPT ES) (120 WATT ALTERNATOR)

COLOR CODE
BLK = BLACK
BLU = BLUE
BRN = BROWN
GRN = GREEN
O or ORN = ORANGE
RED = RED
W or WT = WHITE
Y or YEL = YELLOW

19

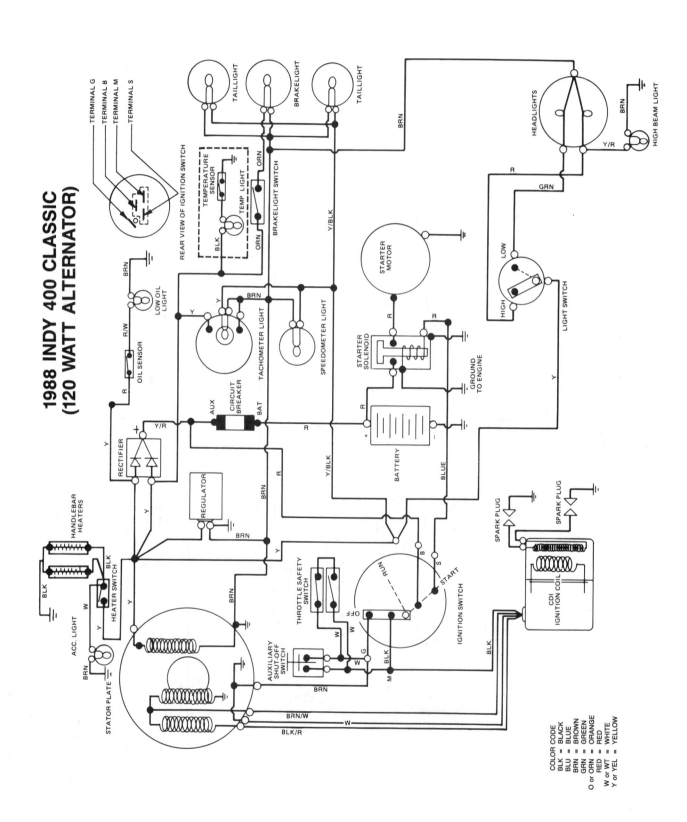

1988 INDY 400 CLASSIC
(120 WATT ALTERNATOR)

COLOR CODE
BLK = BLACK
BLU = BLUE
BRN = BROWN
GRN = GREEN
O or ORN = ORANGE
RED = RED
W or WT = WHITE
Y or YEL = YELLOW

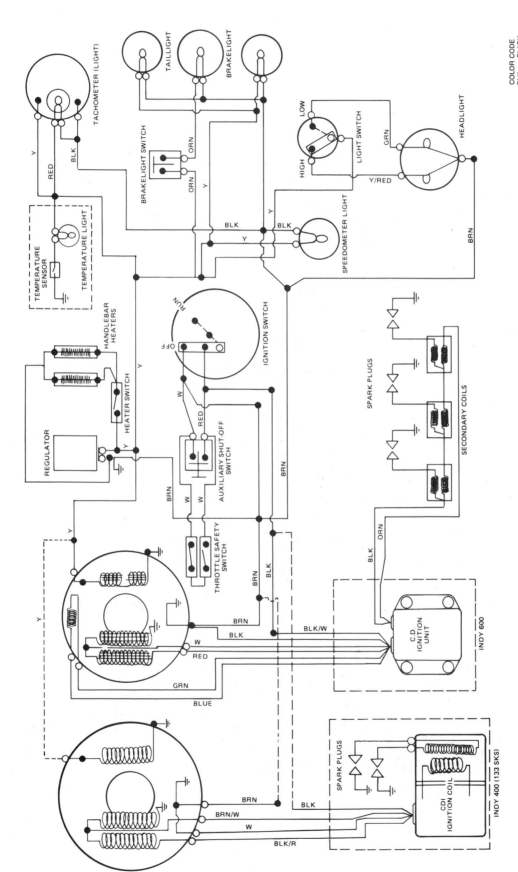

1988 INDY 400 (SKS) & INDY 650 (SKS) (120 WATT ALTERNATOR)

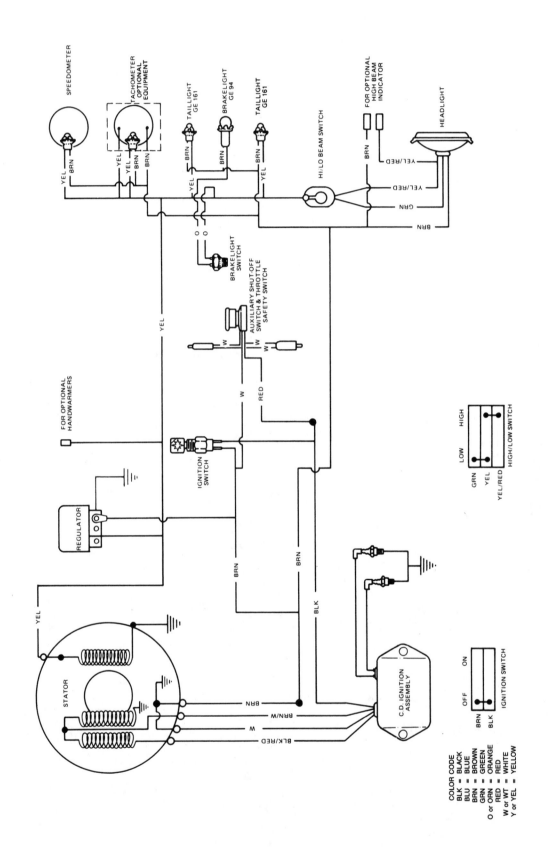

1989 INDY SPORT & INDY SPORT GT

1989 INDY 400, INDY 500 & INDY 500 SP

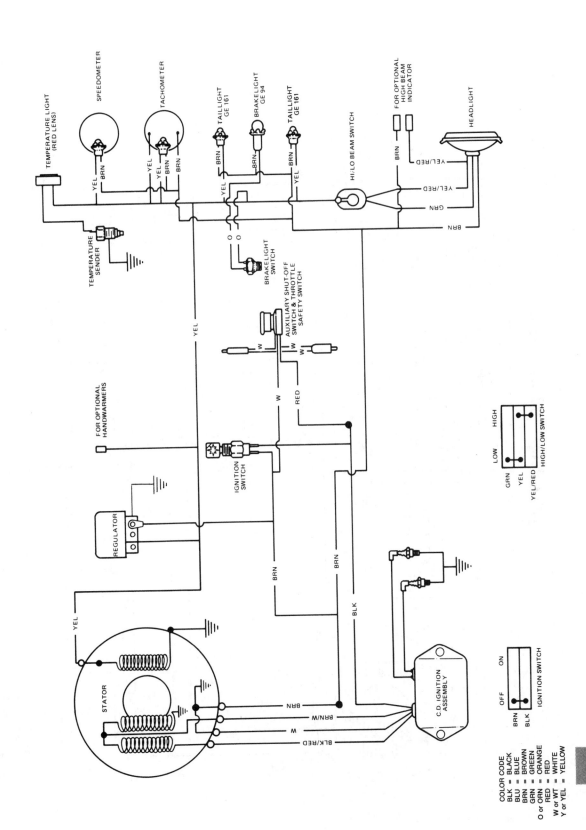

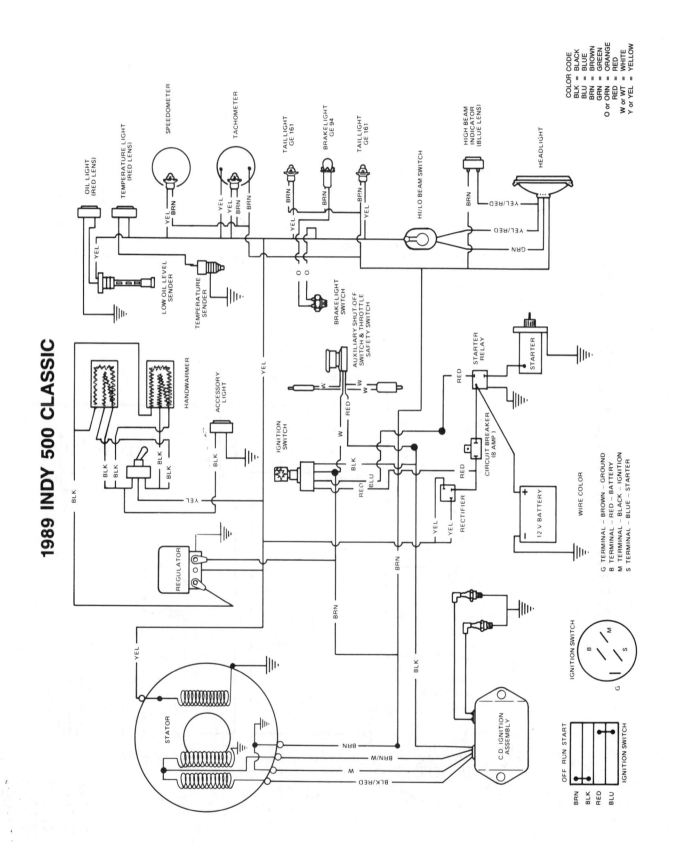

1989 INDY 500 CLASSIC

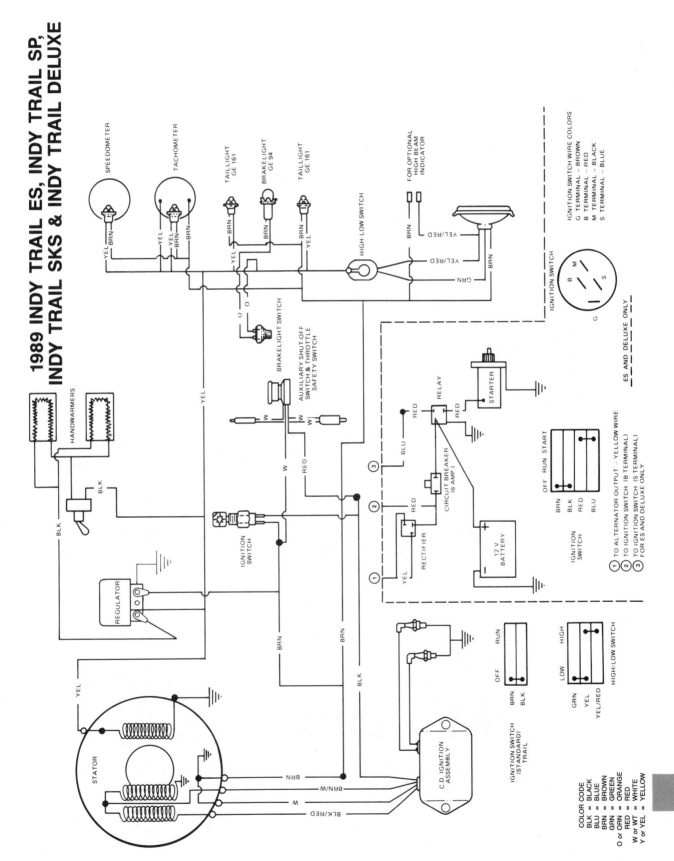

1989 INDY TRAIL ES, INDY TRAIL SP, INDY TRAIL SKS & INDY TRAIL DELUXE

SPEEDOMETER

TACHOMETER

TAILLIGHT GE 161

BRAKELIGHT GE 94

TAILLIGHT GE 161

HIGH LOW SWITCH

FOR OPTIONAL HIGH BEAM INDICATOR

IGNITION SWITCH WIRE COLORS
G TERMINAL – BROWN
B TERMINAL – RED
M TERMINAL – BLACK
S TERMINAL – BLUE

IGNITION SWITCH

ES AND DELUXE ONLY

BRAKELIGHT SWITCH

AUXILIARY SHUT-OFF SWITCH & THROTTLE SAFETY SWITCH

HANDWARMERS

IGNITION SWITCH

REGULATOR

STATOR

C.D. IGNITION ASSEMBLY

RELAY

STARTER

CIRCUIT BREAKER (8 AMP.)

RECTIFIER

12 V BATTERY

	OFF	RUN	START
BRN			
BLK			
RED			
BLU			

IGNITION SWITCH

① TO ALTERNATOR OUTPUT – YELLOW WIRE
② TO IGNITION SWITCH (B TERMINAL)
③ TO IGNITION SWITCH (S TERMINAL) FOR ES AND DELUXE ONLY

	OFF	RUN
BRN		
BLK		

IGNITION SWITCH (STANDARD) TRAIL

	LOW	HIGH
GRN		
YEL		
YEL/RED		

HIGH/LOW SWITCH

COLOR CODE
BLK = BLACK
BLU = BLUE
BRN = BROWN
GRN = GREEN
O or ORN = ORANGE
RED = RED
W or WT = WHITE
Y or YEL = YELLOW

19

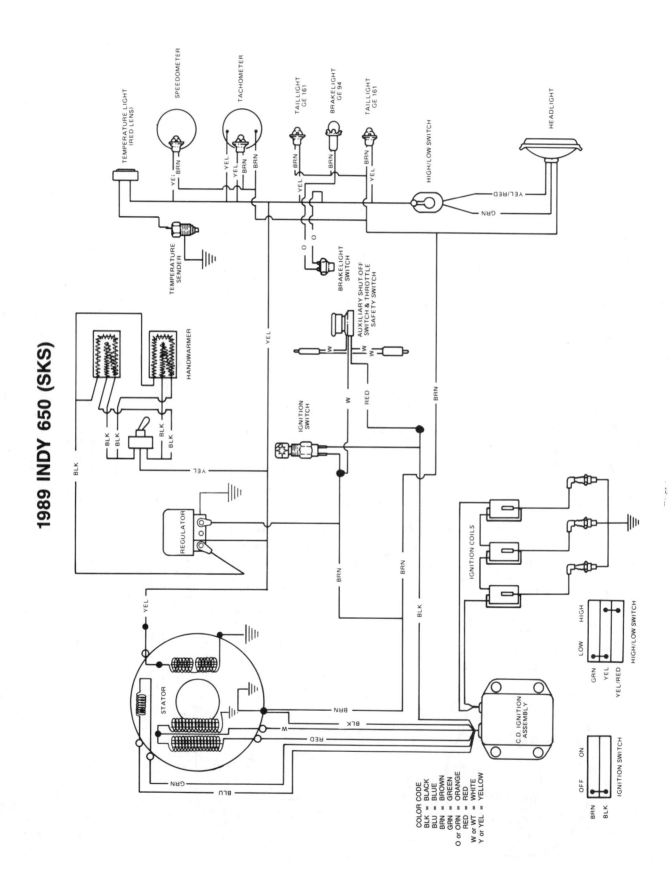

1989 INDY 650 (SKS)

COLOR CODE
BLK = BLACK
BLU = BLUE
BRN = BROWN
GRN = GREEN
O or ORN = ORANGE
RED = RED
W or WT = WHITE
Y or YEL = YELLOW

NOTES

NOTES